Physical Chemistry Experiments

(Chinese and English Edition)

物理化学实验

（中英文对照）

孙 越 / 主编

北京大学出版社
PEKING UNIVERSITY PRESS

图书在版编目(CIP)数据

物理化学实验：汉英对照 / 孙越主编. --北京：北京大学出版社，2024. 10. --ISBN 978-7-301-35676-0

Ⅰ. O64-33

中国国家版本馆 CIP 数据核字第 2024EU7053 号

书　　名	物理化学实验(中英文对照) WULI HUAXUE SHIYAN(ZHONG-YING WEN DUIZHAO)
著作责任者	孙　越　主编
责 任 编 辑	王斯宇
标 准 书 号	ISBN 978-7-301-35676-0
出 版 发 行	北京大学出版社
地　　址	北京市海淀区成府路 205 号　100871
网　　址	http://www.pup.cn　　新浪微博：@北京大学出版社
电 子 邮 箱	编辑部 lk2@pup.cn　　总编室 zpup@pup.cn
电　　话	邮购部 010-62752015　发行部 010-62750672 编辑部 010-62764976
印 刷 者	北京圣夫亚美印刷有限公司
经 销 者	新华书店 787 毫米×1092 毫米　16 开本　20.75 印张　532 千字 2024 年 10 月第 1 版　2024 年 10 月第 1 次印刷
定　　价	75.00 元

辽宁师范大学
规划教材建设项目

PREFACE
前　言

我国“十四五”规划和2035年远景目标纲要提出，要建设高质量教育体系，要提高高等教育质量，要建设高质量本科教育。化学是一门以实验为基础的自然科学，化学实验教学对于学生理解和巩固所学知识，培养观察能力、思维能力、创新能力等具有重要意义。物理化学实验是物理化学学科教学的重要组成部分，是培养学生物理化学基本素质、动手能力和科研能力的重要教学环节。

本教材在内容上包括绪论、基础实验、综合实验、研究设计型实验、思政教学建议和附录六个部分。绪论部分介绍了物理化学实验的基本要求、安全知识、数据处理等内容。实验部分编排了涉及热力学、动力学、电化学等知识的30个实验。思政教学建议挖掘了这些实验的思政元素及思政教学思路。附录则给出了实验时会用到的一些文献数据。

本教材在吸收国内同类教材经典内容的同时，还具以下特色：

1. 为了推动物理化学实验双语教学课程建设，使学生在掌握物理化学实验技能的同时提高专业英语水平，本教材在编写过程中融入双语教学内容，对所有实验进行相应的英文表述，为学生今后直接使用英语从事科学研究、阅读英文文献及撰写科技论文打下良好基础。

2. 为响应习近平总书记提出的“其他各门课都要守好一段渠、种好责任田，使各类课程与思想政治理论课同向而行，形成协同效应”的重要指示，本教材在编写过程中增加了课程思政的重要元素，并给出了各实验项目开展课程思政的相关建议。

3. 为适应化学学科发展，本教材编排了近20%的计算化学实验，以培养学生“计算化学已成为学科发展支柱”的重要认识，进一步提高学生的想象思维能力和综合实验素质。

4. 更加注重绿色化学的教学理念，将传统实验中有毒有害的试剂及实验方法进行了改进，以减少环境污染，提高安全性。例如，将“最大泡压法测定溶液表面张力实验”中常用的正丁醇溶液改成乙醇溶液等。

5. 引入了科技工作者广泛使用的Origin软件和虚拟实验等内容，突出了现代信息技术在物理化学实验教学中的应用。

6. 重视实验安全。在实验安全理论知识的基础上，引入数字化实验安全资源，通过扫描教材中二维码，可打开实验安全测试等电子资源。

本教材由孙越主编、姜笑楠副主编，参加编写的还有张德君、郝强、王磊、于沫涵。其中绪论由孙越副教授、张德君博士、姜笑楠博士、王磊博士共同执笔；实验 11、12、13、14、15、18、19、20、21、22、23、25、26、27、28、30，课程思政建议，附录的中文由孙越副教授执笔，实验内容的英文由孙越副教授和张德君博士共同执笔；实验 1、2、3、4、5、6、7、8、9、10、29 的中文由姜笑楠博士执笔，英文部分由姜笑楠博士和郝强博士共同执笔；实验 17、24 的中英文由王磊博士执笔；实验 16 的中英文由于沫涵博士执笔；全书由孙越统稿。本教材的编写还得到了玉占君、冯春梁两位老师的帮忙，在此表示感谢。本教材的编写过程中，也参考了国内外的相关教材和文献，在此，我们向有关作者谨表谢意。

本教材内容新颖，叙述力求简洁，不仅可以作为高等院校化学、应用化学、化工、材料、生物、药学等相关专业的物理化学实验教材，也可作为相关专业的实验人员和科技人员参考资料。本书的正式出版得益于北京大学出版社及辽宁师范大学的大力支持，在此表示衷心感谢。感谢大连慕乐科技有限公司在实验安全方面的技术支持。

限于我们的水平，书中的疏漏之处在所难免，真诚希望兄弟院校的同行和读者不吝赐教。

编　者

2023 年 10 月

CONTENTS

目 录

第一章 绪 论 ………… 001

1.1 物理化学实验的目的和要求 ………… 001

1.2 物理化学实验室安全知识 ………… 002

1.3 误差分析和数据处理 ………… 008

1.4 常用实验数据处理软件 Origin 的使用 ………… 017

第二章 基础实验 ………… 027

实验 1 燃烧焓的测定 ………… 027

Experiment 1 Determination of the heat of combustion ………… 032

实验 2 氨基甲酸铵分解反应平衡常数的测定 ………… 039

Experiment 2 Determination of equilibrium constants for the decomposition reaction of ammonium carbamate ………… 042

实验 3 凝固点降低法测定葡萄糖的摩尔质量 ………… 045

Experiment 3 Determination of molar mass of glucose by freezing point depression ………… 049

实验 4 双液系的气-液平衡相图 ………… 053

Experiment 4 Binary gas-liquid phase diagram ………… 059

实验 5 KCl-HCl-H_2O 三组分系统相图的绘制 ………… 065

Experiment 5 Construction of the phase diagram for the KCl-HCl-H_2O ternary system ………… 069

实验 6 蔗糖水解反应速率常数的测定 ………… 073

Experiment 6 Determination of the rate constant for sucrose hydrolysis reaction ………… 076

实验 7 乙酸乙酯皂化反应速率常数的测定 ………… 080

Experiment 7 Determination of the rate constant for the saponification reaction of ethyl acetate …… 084
实验 8 原电池电动势的测定和相关热力学函数的计算 …… 091
Experiment 8 Determination of electromotive force of galvanic cell and calculation of thermodynamic functions …… 096
实验 9 电势-pH 曲线的测定与应用 …… 103
Experiment 9 Determination and application of potential-pH curve …… 109
实验 10 循环伏安法研究铁氰化钾的电化学行为 …… 116
Experiment 10 Study of the electrochemical behavior of potassium ferrocyanide using cyclic voltammetry …… 120
实验 11 最大泡压法测定溶液的表面张力 …… 126
Experiment 11 Determination of surface tension of solution by maximum bubble pressure method …… 131
实验 12 溶液吸附法测量固体物质的比表面 …… 138
Experiment 12 The specific surface area determination of solid materials by solution adsorption methods …… 141
实验 13 溶胶的制备和性质研究 …… 144
Experiment 13 Preparation and properties of sol …… 149
实验 14 黏度法测定水溶性高聚物的平均摩尔质量 …… 154
Experiment 14 Determination of average molar mass for water-soluble macromolecules by viscosity method …… 159
实验 15 配合物的磁化率测定 …… 165
Experiment 15 Determination of the magnetic susceptibility for the complex …… 171
实验 16 甲醛分子的结构和性质的计算化学研究 …… 180
Experiment 16 Computational chemistry study of the structure and properties of formaldehyde …… 190
实验 17 甲烷分子的结构和性质的计算化学研究 …… 202
Experiment 17 Computational chemistry study of the structure and properties of methane …… 209
实验 18 富勒烯在不同工作介质中的电化学行为研究(虚拟仿真实验) …… 216
Experiment 18 Study on the electrochemical behavior of fullerene in different working media (virtual simulation experiment) …… 219

第三章 综合实验 …… 223

实验 19 气相色谱法测定非电解质溶液的热力学函数 …… 223

Experiment 19 Determination of thermodynamic function of non-electrolyte solution by gas chromatography …… 228

实验 20 生物酶催化反应动力学常数的测定 …… 236

Experiment 20 Determination of kinetic constants of reaction catalyzed by bioenzyme …… 239

实验 21 电导法测定水溶性表面活性剂的临界胶束浓度 …… 244

Experiment 21 Determination of the critical micelle concentration of water-soluble surfactants by conductivity …… 247

实验 22 溶液法测定极性分子的偶极矩 …… 251

Experiment 22 Determination of dipole moments of polar molecules by solution method …… 256

实验 23 常压顺-丁烯二酸催化氢化 …… 263

Experiment 23 Catalytic hydrogenation of maleic acid at normal pressure …… 267

实验 24 氨分子与水分子的二聚体稳定结构及其氢键强度预测 …… 271

Experiment 24 Prediction of stable structure and hydrogen bond strength for dimer of ammonia and water molecules …… 275

第四章 研究设计型实验 …… 279

实验 25 氧氟沙星在固体电极表面的电化学行为 …… 279

Experiment 25 Electrochemical behaviour of ofloxacin on the surface of solid electrode …… 281

实验 26 稀土金属直接加氢制备纳米稀土金属氢化物 …… 283

Experiment 26 Preparation of nano rare earth metal hydride by direct hydrogenation of rare earth metal …… 284

实验 27 低温等离子体直接分解 NO …… 286

Experiment 27 Direct decomposition of NO by low-temperature plasma …… 288

实验 28 氢质子交换膜燃料电池的组装与性能测试 …… 290

Experiment 28 Fabrication and properties determination of hydrogen proton exchange membrane fuel cells …… 292

实验 29 甲烷生成焓和燃烧焓的理论计算 …… 294

Experiment 29 Theoretical calculation of enthalpy of formation and combustion enthalpy of methane ······ 297
实验 30 分子中原子电荷的理论计算 ······ 300
Experiment 30 Theoretical calculation of atomic charge in molecule ······ 302

第五章 课程思政建议 ······ 307

附 录 ······ 315

第一章
绪　论

1.1 物理化学实验的目的和要求

1.1.1 物理化学实验课程的学习目的

物理化学实验是在无机化学实验、有机化学实验和分析化学实验基础上，独立开设的一门化学实验课，具有综合性、研究性较强及定量化程度较高的特点。它利用物理学的方法和手段，综合了化学中所需的几大基本研究方法和工具，探究物质的物理性质及其与化学变化和相变化之间的关系。通过对选定的化学热力学、相图、化学动力学、电化学、表面与胶体化学及结构化学等相关实验内容的学习，学生可获取物理化学研究的基本知识和技能，掌握物理化学过程的基本研究方法和实验原理、反应条件控制与检测技术、实验数据的处理方法、各种理化参数的测量技术等。该门实验课程对学生了解物理化学的研究思路，了解近代大型仪器的性能及在物理化学中的应用，培养学生的动手能力、观察能力、思维能力、想象能力、表达能力、查阅文献能力和科研创新能力等具有重要意义。

1.1.2 物理化学实验课程的学习要求

实验过程中，学生应以提高自己实际工作能力为目的，勤于动手，开动脑筋，钻研问题，做好每个实验。相关要求一般如下：

（1）实验课前，要认真阅读实验教材及有关参考资料，归纳总结，写出实验预习报告。预习报告应包括实验目的，基本原理，实验步骤，注意事项等关键内容。

（2）进入实验室，必须穿实验服，佩戴防护镜，戴手套。进入实验室后，要认真核对仪器、试剂，对不熟悉的仪器设备，应仔细阅读说明书，请教老师。实验时应按教材和仪器说明进行操作，如有更改意见，须与指导教师讨论，经指导教师同意后方可进行。

（3）实验过程中，要节约用电、水和试剂等所有相关材料。保持实验室安静，不得随意走动。应仔细观察实验现象,严格控制实验条件,准确记录实验原始数据。废液、废物要放入指定容器,不能随意丢弃。

(4) 实验原始数据应随时、如实记录在专门的记录本上,记录本应编有页码,写明实验题目、实验温度、压强等实验条件。记录的实验数据要详细、准确,且注意整洁、清楚,不得用铅笔、圆珠笔、红色笔填写,应使用蓝色(黑色)碳素笔或签字笔。记录本不得随意涂改,当记录中出现错误时,不得使用修正液或修正带,应画"—"杠修改(使原字迹可辨认),并在旁边标上正确值,同时说明错误的原因,如笔误、计算错误、容器破损等,并在旁边签名,标明日期。

(5) 实验结束后,应认真清理实验台面,清洗并核对仪器,若有损坏,应自觉报告登记,按赔偿制度进行赔偿。要保持实验室的整洁,经指导教师同意后,方可离开实验室。

1.2 物理化学实验室安全知识

在化学实验室中,常常潜藏着爆炸、着火、中毒、灼伤、割伤、触电等危险,安全是首先要注意的问题,因此与实验室工作无关的人员未经允许不得擅自进入实验室。进入实验室后要先熟悉逃生安全通道和消防物资(灭火毯和二氧化碳灭火器)的存放位置,当然,在未发生危险时,不得随意移动消防物资;未经允许,实验室以外的设备不得带入实验室使用;实验室内要保持安静,不允许大声喧哗、嬉戏和打闹;不得随意碰触实验室内标注的危险源,如墙壁上的配电箱、液氮容器、气体钢瓶和其他未经允许使用的仪器设备等;最后离开实验室的人员要进行水、电、门窗等相关安全检查;实验室发生突发状况时,要听从实验室负责人员的调动和指挥。除此之外,具体还应注意下面内容。

1.2.1 安全用电常识

电在实验室中经常使用,看不见摸不着,但却最容易成为安全隐患,若用电不规范,很容易导致触电,严重的将会付出惨痛的代价。人体通过 50 Hz 的交流电,1 mA 就有感觉,10 mA 以上肌肉强烈收缩,25 mA 以上呼吸困难,100 mA 以上则会使心脏的心室产生纤维性颤动,以致无法救活。直流电对人体也有相似的危害,因此我们在用电的时候需要注意:

(1) 电器仪表在使用前,应先了解要求使用的电源是交流电还是直流电,要求的电压的大小(380 V、220 V、110 V 或 6 V)及直流电器仪表的正、负极。

(2) 不能用试电笔去试高压电。使用高压电源应有专门的防护措施。

(3) 电线的安全通电量应大于用电功率,使用的保险丝要与实验室允许的用电量相符。

(4) 大功率仪器(包括空调等)应使用专用插座,电器长期不用时,应切断电源。

(5) 配电箱前不应有物品遮挡,应当便于操作,周围不应放置烘箱、电炉、易燃易爆气瓶、易燃易爆化学试剂、废液桶等。

(6) 实验之前要检查线路连接是否正确,经指导教师检查同意后方可接通电源。实验结束时,先切断电源再拆线路。修理或安装电器时,要先切断电源。

(7) 实验室内的插排或实验台上的插座不得擅自连接其他电子产品。

(8) 实验过程中不要用潮湿的手接触电器,电线、电器不要被水淋湿或浸在导电液体中。

(9) 电源裸露部分应有绝缘装置(例如电线接头处应裹上绝缘胶布),所有电器的金属外壳都应接地保护。

(10) 若室内有氢气、煤气等易燃易爆气体,应避免产生电火花。继电器工作和开关电闸时,易产生电火花,要特别小心。

(11) 在电器仪表使用过程中,如发现有不正常声响、局部升温、电线烧焦的气味或电器接触点(如电插头)接触不良时,应立即切断电源,并报告指导教师进行检查。

(12) 如遇电线起火,立即切断电源,用沙子或二氧化碳、四氯化碳、水基灭火器灭火,禁止用水或泡沫灭火器等导电液体灭火。

(13) 如有人触电,应迅速切断电源,然后进行抢救。

1.2.2 使用化学药品的安全防护

防毒

化学药品一般多具有不同程度的毒性,因此,要尽量杜绝和减少直接接触化学药品,以免其中有毒成分通过皮肤、呼吸道和消化道进入人体内。

(1) 禁止在实验室内喝水、吃东西,饮食用具不要带进实验室,以防毒物污染,离开实验室要洗净双手。

(2) 药品柜和试剂溶液均应避免阳光直射及靠近暖气等热源。要求避光的试剂应装于棕色瓶中或用黑纸或黑布包好存于暗柜中。

(3) 发现试剂瓶上标签掉落或将要模糊时应立即贴好标签。无标签或标签无法辨认的试剂都要当成危险物品重新鉴别后小心处理,不可随便乱扔,以免引起严重后果。

(4) 化学药品及试剂要定位放置,用后复位,节约使用。但用后多余的化学试剂不得倒回原瓶。

(5) 按规定量取用药品,取用完毕后及时密封储存,避免种类混淆或玷污。

(6) 易制毒、易制爆化学品是管制化学品,要实施双人双锁管理,存量合规,并有防盗与防爆措施。

(7) 高汞盐[$HgCl_2$,$Hg(NO_3)_2$ 等]、可溶性盐($BaCO_3$,$BaCl_2$)、重金属盐(镉盐,铅盐)以及氰化物、三氧化二砷等剧毒物,应限量存储,并妥善保管。

(8) 实验前应了解所用药品的性能(尤其是毒性)和防护措施。

(9) 操作有毒气体(如 H_2S、Cl_2、Br_2、NO_2)及浓盐酸、氢氟酸等,应在通风橱或通风柜中进行。

(10) 防止天然气管、天然气灯漏气,使用完天然气后,一定要把闸门关好。

(11) 苯、四氯化碳、乙醚、硝基苯等的蒸气会引起中毒,虽然它们都有特殊气味,但经常吸入会使人嗅觉减弱,必须高度警惕。

(12) 若不慎吸入刺激性气体,应迅速脱离现场至新鲜空气处,保持呼吸通畅,必要时就医。

(13) 有些药品(如苯、有机溶剂、汞)能经皮肤渗透入体内,应避免直接与皮肤接触。

(14) 盐酸、硫酸的配制使用过程中,有刺激性气体产生,需配合风机、风道等设备设施,将酸雾排出室外。少量泄漏时,可用沙土、干燥石灰或苏打灰混合。因硫酸与水接触会产生大量的热,严禁用水清洗硫酸,以免造成液体飞溅伤人。

(15) 若不慎将酸或碱溅在皮肤或衣服上,可用大量流动清水冲洗。如溅到眼睛里,应立即提起眼睑,用大量流动清水冲洗后就医,以免损伤视力。如浓硫酸滴在皮肤上,须用干布轻轻擦净,再用清水冲洗。

防爆

可燃气体与空气混合,当两者比例处于爆炸极限时,只要有一个适当的热源(如电火花)诱发,就会引起爆炸。表 1 列出了某些气体与空气相混合的爆炸极限。

表 1 与空气相混合的某些气体的爆炸极限(20 ℃,1 个大气压下)

气体	爆炸上限(体积分数/%)	爆炸下限(体积分数/%)	气体	爆炸上限(体积分数/%)	爆炸下限(体积分数/%)
氢	74.2	4.0	苯	6.8	1.4
乙烯	28.6	2.8	丙酮	12.8	2.6
乙炔	80.0	2.5	一氧化碳	74.2	12.5
乙醇	19.0	3.3	水煤气	72.0	7.0
乙醚	36.5	1.9	氨	27.0	15.5

为了防止爆炸事故发生,要做到以下几点:

(1) 实验中,应尽量避免能与空气形成爆鸣混合气的气体泄漏在室内空气中,室内通风要良好。

(2) 操作大量可燃性气体时,严禁同时使用明火,还要防止发生电火花及其他撞击火花。

(3) 易燃有机溶剂(特别是低沸点、易燃溶剂)在室温时即具有较大的蒸气压,空气中混杂易燃有机溶剂的蒸气到达某一极限时,极易发生燃烧爆炸。而且,有机溶剂蒸气都比空气的密度大,会沿着桌面或地面飘移至较远处,或沉积在低洼处。因此,切勿将易燃溶剂倒入废物缸中。

(4) 严禁将强氧化剂和强还原剂放在一起。某些药品如叠氮铝、乙炔银、乙炔铜、高氯酸盐、过氧化物等受震和受热都易引起爆炸,使用要特别小心。

(5) 有些有机化合物如醚或共轭烯烃,久置后会生成易爆炸的过氧化物,须特别处理后才能使用。

(6) 常压操作时,应确保全套装置与大气相通,切勿造成密闭体系。减压蒸馏时,要用圆底烧瓶或吸滤瓶作承受器,不要用锥形瓶,否则可能会发生炸裂。加压操作时(如高压釜、封管等)要有可靠的防护措施,并应常常留意釜内压力有无超过安全负荷,选用封管的玻璃厚度是否适当、管壁是否匀称。

防火

许多有机溶剂如乙醚、丙酮、乙醇、苯等非常容易燃烧,大量使用时室内不能有明火、电火花或静电放电。有些物质如磷、金属钠、钾、电石及金属氢化物等,在空气中易氧化自燃。还有一些金属粉末如铁、锌、铝等,比表面大,也易在空气中氧化自燃。这些物质要隔绝空气保存,使用时要特别小心。

一旦发生火情,应冷静判断情况,采取措施,如采取隔绝氧的供应、降低燃烧物质的温度、将可燃物质与火焰隔离等办法。常用来灭火的有水、沙(砂)以及 CO_2 灭火器、CCl_4 灭火器、泡沫灭火器、干粉灭火器等,可根据着火原因、场所情况正确选用。

水是最常用的灭火物质,可以降低燃烧物质的温度,并且形成“水蒸气幕”,能在相当长时间内阻止空气接近燃烧物质。但是,应注意起火地点的具体情况:

(1) 有金属钠、钾、镁、铝粉、电石、过氧化钠等,采用干砂等灭火。

(2) 对易燃液体(密度比水小,如汽油、苯、丙酮等)的着火,采用泡沫灭火剂更有效,因为泡沫比易燃液体轻,覆盖在上面可隔绝空气。

(3) 在有灼烧的金属或熔融物的地方着火,应采用干砂或固体干粉灭火器来灭火。

(4) 电气设备或带电系统着火,用二氧化碳或四氯化碳灭火器较合适。

上述四种情况均不能用水,因为有的可以生成氢气等,使火势加大甚至引起爆炸,有的会发生触电;同时也不能用四氯化碳灭碱金属的着火。另外,四氯化碳有毒,在室内救火时最好不用。灭火时不能慌乱,防止在灭火过程中再打碎装有可燃物的容器。

防灼伤

强酸、强碱、强氧化剂、溴、磷、钠、钾、苯酚、冰醋酸等都会腐蚀皮肤，特别要防止溅入眼内。液氧、液氮等低温也会严重冻伤皮肤，使用时要小心，万一灼伤应及时治疗。

1.2.3 高压气体钢瓶的使用及注意事项

气体钢瓶是由无缝碳素钢或合金钢制成，适用于装介质压力在 15 MPa 以下的气体。标准气瓶类型见表 2。

表 2 标准气瓶类型

气瓶类型	气体种类	工作压力/MPa	试验压力/MPa	
			水压试验	气压试验
甲	O_2、H_2、N_2、CH_4 压缩空气和惰性气体	15.0	22.5	15.0
乙	纯净水煤气及 CO_2 等	12.5	19.0	12.5
丙	NH_3、氯、光气和异丁烯等	3.0	6.0	3.0
丁	SO_2 等	0.6	1.2	0.6

气体钢瓶的使用常识

(1) 气瓶应在通风良好的场所使用，如果在通风条件差或狭窄的场地里使用气瓶，应采取相应的安全措施，以防止出现氧气不足或危险气体浓度加大的现象。采取的安全措施主要包括强制通风、氧气监测和气体检测等。

(2) 使用中的气瓶需要由专业人员每三年检查一次，装腐蚀性气体的钢瓶每两年检查一次，不合格的气瓶不可继续使用。

(3) 使用气瓶前，使用者应对气瓶进行安全状况检查，检查重点包括：瓶体是否完好；减压器、流量表、软管、防回火装置是否有泄漏、磨损及接头松懈等现象。

(4) 气瓶应防止曝晒、雨淋、水浸，环境温度超过 40 ℃时，应采取遮阳等措施降温。

(5) 气瓶应存放在远离热源的地方，可燃性气瓶一律不准进入实验室内，并应与氧气钢瓶分开存放。

(6) 氧气钢瓶严禁与油类物质或易燃有机物接触（特别是气瓶出口和压力表上）。可燃气瓶（如 H_2、C_2H_2 气瓶）气门螺丝为反丝，不可燃或助燃气瓶（如 N_2、O_2 气瓶）为正丝。各种压力表一般不可混用。

(7) 氢气钢瓶应放在远离实验室的专用气房内，用紫铜管引入实验室，并安装防止回火的装置。

(8) 气瓶使用时要注意固定，防止倾倒，严禁卧倒使用。对已卧倒的气瓶，不准直接开气使用,使用前必须先立牢静止 15 min 后,再接减压器使用。

(9) 禁止将气瓶与电气设备及电路接触,与气瓶接触的管道和设备要有接地装置。在气、电焊混合作业的场地,要防止氧气瓶带电,如地面是铁板,要垫木板或胶垫加以绝缘。乙炔气瓶不得放在橡胶等绝缘体上。

(10) 开启或关闭气瓶阀门时,应用手或专用扳手,不准使用其他工具,以防损坏阀件。装有手轮的阀门不能使用扳手。如果阀门损坏,应将气瓶隔离并及时维修。

(11) 开启或关闭瓶阀应缓慢,特别是盛装可燃气体的气瓶,以防止产生摩擦热或静电火花。

(12) 打开钢瓶总阀门时,高压表显示瓶内贮气总压力。开启总阀门时,不要将头或身体正对总阀门,防止万一阀门或压力表冲出伤人。

(13) 停止使用时,先关闭总阀门,待减压阀中残留气体泄光后,再关闭减压阀。

(14) 冬天,气瓶的减压器和管系发生冻结时,可用 10 ℃以下温水解冻,严禁用火烘烤或使用铁器一类的东西猛击气瓶,更不能猛拧减压表的调节螺丝,以防止气体突然大量冲出,造成事故。

(15) 气瓶投入使用后,不得对瓶体进行挖补、焊接修理。严禁将气瓶用作支架等其他用途。

(16) 气瓶使用完毕,要妥善保管。气瓶上应有状态标签(“空瓶”、“使用中”、“满瓶”标签)。

(17) 钢瓶内气体不能全部用尽,要留下一些气体,以防止外界空气进入气体钢瓶,一般应保持 0.5 MPa 以上的残留压力。

(18) 使用时注意各气瓶上漆的颜色及标字,见表 3,避免混淆。

表 3 高压气体钢瓶常用的颜色和标志

气瓶类别	气瓶颜色	字样颜色	字样
氮气瓶	黑	黄	氮
氧气瓶	浅蓝	黑	氧
氢气瓶	暗绿	红	氢
压缩空气瓶	黑	白	压缩空气
二氧化碳气瓶	黑	黄	二氧化碳
氦气瓶	棕	白	氦
氨气瓶	黄	黑	氨
氯气瓶	草绿	白	氯
乙炔气瓶	白	红	乙炔

(续表)

气瓶类别	气瓶颜色	字样颜色	字样
氟氯烷气瓶	铝白	黑	氟氯烷
石油气瓶	灰	红	石油气体
纯氩气瓶	灰	绿	纯氩

气体钢瓶的搬运

(1) 搬运气瓶时,要旋紧瓶帽,以直立向上的位置来移动,注意轻装轻卸,禁止从瓶帽处提升气瓶。

(2) 近距离(5 m 内)移动气瓶,应用手扶瓶肩转动瓶底,并且要使用手套。移动距离较远时,应使用专用小车搬运,特殊情况下可采用适当的安全方式搬运。

(3) 禁止用身体搬运高度超过 1.5 m 的气瓶到手推车或专用吊篮等里面,可采用手扶瓶肩转动瓶底的滚动方式。

(4) 卸车时应在气瓶落地点铺上软垫或橡胶皮垫,逐个卸车,严禁溜放。

(5) 装卸氧气瓶时,工作服、手套和装卸工具、机具上不得粘有油脂。

(6) 当提升气瓶时,应使用专用吊篮或装物架,不得使用钢丝绳或链条吊索。严禁使用电磁起重机和链绳。

为了更安全地开展相关实验,请实验前扫下方二维码,完成相关安全测试。

1.3 误差分析和数据处理

在物理化学实验过程中,由于实验方法、实验仪器和实验条件的选择,以及实验者观察的局限性等因素,实验测得的数据都会存在误差。因此,无论是在实验之前对测量所能达到准确度的预估,还是在实验后对数据进行的合理处理,都必须具有正确的误差概念。通过对误差的分析,可寻找适当的实验方法,选用最适合的仪器及量程,给出测量的有利条件,评价实验结果的可靠性。

1.3.1 误差的概念与分类

物理量的测量可分为直接测量和间接测量两种。能够直接测得被测量结果的测量为直接测量,如温度、压力、长度的测量及物质质量的称量等等。利用多个直接测量数据并通过公式计算才能得到所需结果的测量则称为间接测量,如电动势的测定、偶极矩的测定、表面张力的测定等等。物理化学实验中的测量大多属于间接测量。

1.3.1.1 准确度和误差

真值(true value, T) 即真实值,是指在一定条件下,被测量组分客观存在的真实值。真值在不同场合有不同的含义。真值可分为理论真值和规定真值。理论真值也称绝对真值,如平面三角形内角之和等于 180°。规定真值是国际上公认的某些基准量值,如米的定义为“米等于光在真空中 1/299792458 秒时间间隔内所经路径的长度”。这个米基准就当作计量长度的规定真值。

对于被测的物理量,真值通常是个未知量。由于误差的客观存在,真值一般无法测得。测量次数无限多时,根据正负误差出现的概率相等的误差分布规律,在不存在系统误差的情况下,它们的平均值接近真值。故在实验科学中真值的定义为无限多次观测值的平均值,如式(1)所示。

$$T = \lim_{n \to \infty}\left[\frac{1}{n}\sum_{i=1}^{n} x_i\right] = \lim_{n \to \infty} \bar{x} \tag{1}$$

但实际测定的次数总是有限的,由有限次测量求出的平均值,只能近似地视为真值,可称此平均值为最佳值(或可靠值)。

常用的平均值有算术平均值和几何平均值两种,其中算术平均值最常用。

设 $x_1, x_2, \cdots, x_n$ 为各次的测量值,n 代表测量次数,则其算术平均值 $\bar{x}$ 和几何平均值 $\bar{x}_{\text{几何}}$ 可由式(2)和(3)表示。

$$\bar{x} = \frac{x_1 + x_2 + \cdots + x_n}{n} = \frac{1}{n}\sum_{i=1}^{n} x_i \tag{2}$$

$$\bar{x}_{\text{几何}} = \sqrt[n]{x_1 \cdot x_2 \cdot \cdots \cdot x_n} = \sqrt[n]{\prod_{i=1}^{n} x_i} \tag{3}$$

准确度表示测量值与真值的接近程度。准确度反映了测量中所有系统误差和随机误差的综合影响。误差越小,表明测量结果的准确度越高。

误差可以用绝对误差和相对误差来表示。绝对误差是测量值与真值之差,可用式(4)表示。其中,x_i 为第 i 次的测量值。由于测量次数通常不止一次,因此常采用测量值的算术平均值表示绝对误差,如式(5)所示。

$$E_a = x_i - T \tag{4}$$

$$E_a = \frac{1}{n}\sum_{i=1}^{n} x_i - T = \bar{x} - T \tag{5}$$

相对误差是绝对误差与真值之比，如式(6)所示。

$$E_r = \frac{E_a}{T} \times 100\% \tag{6}$$

1.3.1.2 精密度和偏差

精密度是表示测量值之间的接近程度。测量结果的准确度或准确性可以用误差评价，测量结果的精密度或重现性则可以用偏差衡量。偏差越小，说明测量结果的精密度越高。偏差可以用绝对偏差、平均偏差、相对平均偏差、标准偏差和相对标准偏差来表示。

绝对偏差是各次的测量值与其平均值之差，如式(7)。偏差可以为正数、负数和零。平均偏差是各个绝对偏差的绝对值的算术平均值，如式(8)。相对平均偏差是平均偏差与平均值的百分比，如式(9)。平均偏差和相对平均偏差均为大于零的数。

$$D_i = x_i - \bar{x} \tag{7}$$

$$\bar{D}_r = \frac{|D_1| + |D_2| + \cdots + |D_n|}{n} = \frac{1}{n}\sum_{i=1}^{n} D_i \tag{8}$$

$$\bar{D}_r = \frac{\bar{D}}{\bar{x}} \times 100\% \tag{9}$$

用数理统计方法处理实验数据时，常用标准偏差[式(10)]和相对标准偏差[式(11)]来衡量精密度。

$$s = \sqrt{\frac{1}{n-1}\sum_{i=1}^{n}(x_i - \bar{x})^2} = \sqrt{\frac{1}{n-1}\sum_{i=1}^{n} D_i^2} \tag{10}$$

$$s_r = \frac{s}{\bar{x}} \times 100\% \tag{11}$$

用标准偏差表示精密度能更好地说明测量结果的分散程度，因为单次测量结果的偏差平方之后，较大的偏差能更显著地反映出来。

1.3.1.3 准确度和精密度的关系

精密度与准确度之间存在一定关系。测量值的精密度高，表明测量条件较稳定。但是由于存在可能引起误差的因素，精密度高并不一定表示准确度也高。

由图 1 可以看出，A 的随机误差大，精密度、准确度都很差；B 的随机误差小，精密度很好，但准确度不好；C 的随机误差很小，精密度和准确度都很好。由此可以说明，一个精密度很好的测量结果，其准确度不一定很好；但要得到准确度很高的测量结果，一定要有高精密度的测量结果来保证。

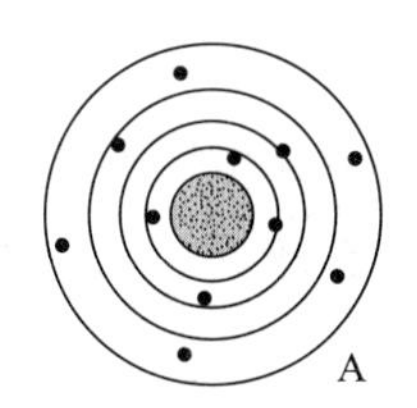

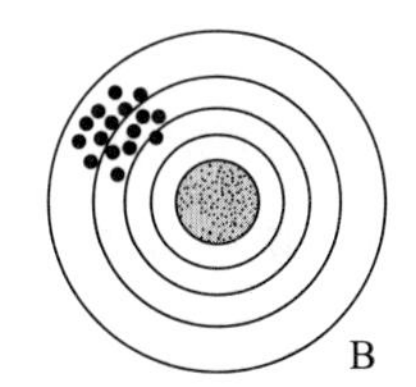

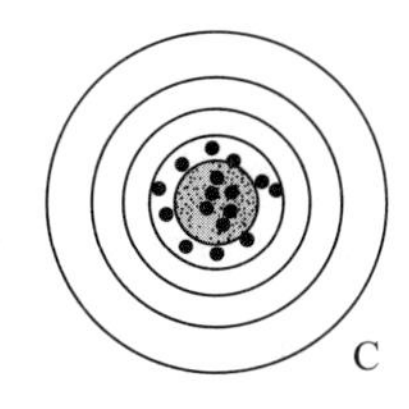

图 1　A、B、C 三组测量结果的准确度和精密度

1.3.1.4　误差的分类

根据误差产生的原因、作用规律和对实验结果产生的影响,可将误差分为系统误差、随机误差和过失误差三种类型。

(1) 系统误差(或恒定误差)

系统误差是指在相同条件下,对某一物理量测量多次时,误差值恒定不变,或在条件改变时,按某一确定的规律变化的误差。产生系统误差的原因通常有仪器误差、测量方法的问题以及个人习惯性误差。例如,仪器构造不够完善,示数部分的刻度不够准确,仪器零点偏移,试剂的纯度不符合要求,实验方法不完善或采用近似公式,记录某一信号的时间总是滞后,对颜色的感觉不灵敏或读数的姿势不正确等等。

通常可以采取几种不同的实验技术、采用不同的实验方法、改变实验条件、校准仪器、提高试剂纯度等手段,确定系统误差的大小,然后使之消除或者减小。

(2) 随机误差(或偶然误差)

在相同条件下对同一个物理量进行多次测量,其每次测量结果的误差值是不确定的,时大时小,时正时负,没有确定的方向,这类误差称为随机误差。这类误差产生的原因不明,因而无法控制和补偿。

若对某一被测量进行足够多次的等精度测量,就会发现随机误差服从统计规律,这种规律呈正态分布(图 2)。

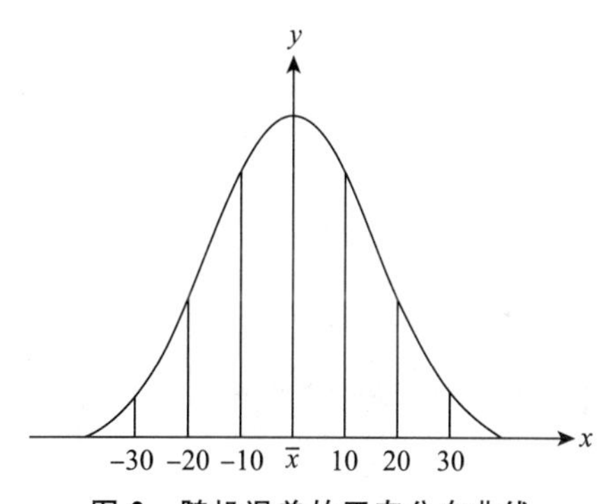

图 2　随机误差的正态分布曲线

由图 2 可以看出,以 $\bar{x}$ 为中心的正态分布曲线具有以下特征:

① 对称性:正态分布曲线以 y 轴对称,绝对值相等的正误差和负误差出现的概率几

乎相等。

② 单峰性:绝对值小的误差出现的概率多,绝对值大的误差出现的概率少。

③ 有界性:在一定测量条件下的有限次测量中,误差不会超过一定界限。通过统计分析可计算出,误差在$\pm\sigma$内出现的概率是68.3%,误差在$\pm 2\sigma$内出现的概率是95.5%,误差在$\pm 3\sigma$内出现的概率是99.7%,误差绝对值大于3σ出现的概率仅为0.3%。因此,多次重复测量中误差绝对值大于3σ的数据可以舍弃。随着测量次数的增加,随机误差的算术平均值趋近于零,所以多次测量结果的算术平均值将更接近于真值。

(3) 过失误差(或粗差)

过失误差是一种与实际事实明显不符的误差,过失误差明显地歪曲实验结果。误差值可能很大,且无一定的规律。它主要是由于实验人员粗心大意、操作不当等因素造成的,如读错数据、记错或计算错误、操作失误等。在测量或实验时,只要认真负责,是可以避免这类误差的。存在过失误差的观测值在实验数据整理时应该将其剔除。

1.3.2 误差分析

在物理化学实验数据测定中,绝大多数是要对几个物理量进行测量,代入某个函数关系式,然后进行运算才能得到结果。在间接测量中,每个直接测量值的准确度都会影响最后结果的准确性。

通过误差分析,我们可以明确直接测量的误差对结果将会产生多大影响,从而找出误差的主要来源,以便选择适当的实验方法,合理配置仪器,寻求测量的最佳条件。

1.3.2.1 仪器的精密度

仪器的精密度是影响实验结果的重要因素之一。因此,要根据误差分析结果选择适当精密度的仪器,不能盲目地使用精密仪器。如果使用几种仪器进行测试,要注意各种仪器精度的相互匹配。

数字式仪表的精密度一般由其显示的最后一位读数的一个单位来表示。如果没有精度表示,对于大多数仪器来说,最小刻度的1/5可以看作其精密度,如1/10温度计的测量误差为0.02 ℃,分析天平为0.0002 g,50 mL滴定管为0.02 mL等。

1.3.2.2 间接测量结果的误差计算

设有物理量N由直接测量值$u_1, u_2, \cdots, u_n$求得,即

$$N = f(u_1, u_2, \cdots, u_n) \tag{12}$$

直接测量值的平均误差为:$\Delta u_1, \Delta u_2, \cdots, \Delta u_n$,对式(12)进行全微分,得

$$\mathrm{d}N = \left(\frac{\partial N}{\partial u_1}\right)_{u_2, u_3, \cdots, u_n} \mathrm{d}u_1 + \left(\frac{\partial N}{\partial u_2}\right)_{u_1, u_3, \cdots, u_n} \mathrm{d}u_2 + \cdots + \left(\frac{\partial N}{\partial u_n}\right)_{u_1, u_2, \cdots, u_{n-1}} \mathrm{d}u_n \tag{13}$$

当各自变量的平均误差 Δu_i 足够小时，可用其代替 $\mathrm{d}u_i$，并考虑最不利的情况下，直接测量的误差不能抵消，从而引起误差的累积，故取其绝对值。上式变为

$$\Delta N=\left|\frac{\partial N}{\partial u_1}\right||\Delta u_1|+\left|\frac{\partial N}{\partial u_2}\right||\Delta u_2|+\cdots+\left|\frac{\partial N}{\partial u_n}\right||\Delta u_n| \tag{14}$$

如果将 $N=f(u_1,u_2,\cdots,u_n)$ 两边先取对数，再求微分，得到最大相对平均误差传递公式

$$\frac{\Delta N}{N}=\frac{1}{f(u_1,u_2,\cdots,u_n)}\left(\left|\frac{\partial N}{\partial u_1}\right||\Delta u_1|+\left|\frac{\partial N}{\partial u_2}\right||\Delta u_2|+\cdots+\left|\frac{\partial N}{\partial u_n}\right||\Delta u_n|\right) \tag{15}$$

式(14)、(15)分别是间接测量中最终实验结果的平均误差和相对平均误差的普遍公式。运用以上公式可以根据不同函数关系式进行误差传递的计算。

对于加法运算，设 $N=u_1+u_2+u_3$，其测量结果的最大相对平均误差为

$$\frac{\Delta N}{N}=\frac{|\Delta u_1|+|\Delta u_2|+|\Delta u_3|}{u_1+u_2+u_3} \tag{16}$$

对于减法运算，设 $N=u_1-u_2-u_3$，则有

$$\frac{\Delta N}{N}=\frac{|\Delta u_1|+|\Delta u_2|+|\Delta u_3|}{u_1-u_2-u_3} \tag{17}$$

对于乘、除法运算，设 $N=u_1\times u_2$ 或 $N=u_1/u_2$，则有

$$\frac{\Delta N}{N}=\frac{|\Delta u_1|}{u_1}+\frac{|\Delta u_2|}{u_2} \tag{18}$$

对于乘方、开方运算，如 $N=u^n$

$$\frac{\Delta N}{N}=n\left|\frac{\Delta u}{u}\right| \tag{19}$$

1.3.2.3 误差分析应用

例 1 由相对误差选择合适的仪器

用电热补偿法测定 KNO_3 在水中的溶解热，$\Delta H_{溶解}=\frac{MIVt}{m}$。$M$ 为 KNO_3 的相对分子质量，电流 $I=0.5$ A，电压 $V=6$ V，时间 $t=400$ s，KNO_3 的质量 $m=3$ g。如果要把相对误差控制在 3%以内，应选用什么规格的仪器？

实验结果的误差来源于 I,V,t,m 这 4 个直接测量量。可推导出误差传递公式

$$\frac{\Delta(\Delta H_{溶解})}{\Delta H_{溶解}}=\frac{\Delta I}{I}+\frac{\Delta V}{V}+\frac{\Delta t}{t}+\frac{\Delta m}{m}$$

用秒表测量时间，误差不超过 1 s，相对误差 1/400 = 0.25%；溶质质量若用台秤称量，误差将大于 3%，若用分析天平，0.0004/3，误差在 0.02%以下。所以，电流表和电压表的选择是本实验测量的关键。若要控制最大相对误差在 3%以下，I、V 的相对误差都应将控制

在 1%以下，因此应选用精度为 0.5 级的电流表和 1.0 级的电压表（准确度为最大量程值的 1%），且电流表的最大量程为 1.0 A，电压表的最大量程为 5.0 V。

$$\frac{\Delta I}{I}=\frac{1.0\times 0.005}{0.5}=1\% \qquad \frac{\Delta V}{V}=\frac{5.0\times 0.01}{4.5}=1\%$$

例 2 测量过程最有利条件的确定

在利用 Weston 电桥测定电阻时，被测电阻可由下式计算

$$R_x=R\frac{L_1}{L_2}=R\frac{L-L_2}{L_2}$$

式中，R 为已知电阻，L 是电阻丝全长，$L=L_1+L_2$，间接测量值 R_x 的误差取决于直接测量值 L_2 的误差

$$\mathrm{d}R_x=\pm\left[\frac{\partial\left(R\dfrac{L-L_2}{L_2}\right)}{\partial L_2}\right]\mathrm{d}L_2=\pm\left(\frac{RL}{L_2^2}\right)\mathrm{d}L_2$$

相对误差为

$$\frac{\mathrm{d}R_x}{R_x}=\pm\left[\left(\frac{RL}{L_2^2}\right)\frac{L_2}{R(L-L_2)}\mathrm{d}L_2\right]=\pm\left[\frac{L}{(L-L_2)L_2}\mathrm{d}L_2\right]$$

因为 L 为常数，所以当 $(L-L_2)L_2$ 为最大时，其相对误差最小。

$$\frac{\mathrm{d}}{\mathrm{d}L_2}[(L-L_2)L_2]=0$$

故 $L_2=\frac{1}{2}L$。

所以，用 Weston 电桥测定电阻，电桥上的接触点在中间时，测量误差最小。由测定电阻，可以求得电导，而电导的测定也是物理化学实验中常用的方法之一。

1.3.2.4 有效数字

有效数字一般是指实际工作中能够测量到的数字，包括确定的数字和最后 1 位不确定的数字（只能保留 1 位可疑数字）。因此，任何一次测量，都应记录到仪器刻度的最小估计读数。例如：用最小分度为 1 cm 的标尺测量两点间的距离，其结果为 9.140 m，这里是 4 位有效数字，前 3 位数的测量是准确的，而最后 1 位“0”是估读的。有效数字能反映物理量的大小及测量的精密度。因此，只有准确掌握有效数字的运算规则，才能有效反映测量结果的真实性。有效数字的运算常遵守如下规则：

（1）在运算过程中，若数值的首位数为 8 或 9，则有效数字可多算 1 位，例如 8.325 在运算过程中可看作为 5 位有效数字。

（2）在根据运算规则对有效数字进行取舍时，应用“四舍六入五成双”原则，也就是当

数字等于小于4时,应该舍去;大于5时,应该进位;等于5时,要根据与保留的有效数字的最后1位数是奇数还是偶数确定,若为奇数就进位,若为偶数就舍去。例如,将2.3654,2.3656,2.3655,2.3645,2.36451皆保留4位有效数字,其结果依次为2.365,2.366,2.366,2.364,2.365。

(3) 有效数字进行加减法运算时,各数字小数点后所取的位数与其中位数最少者的相同。进行乘除运算时,其有效数字位数与各因子中有效数字位数最少者相同。对数的有效数字的位数与其真数相同。如 lg100 = 2.000。

(4) 在所有计算式中,常数 π,e 及因子(如 $\sqrt{2}$)和一些从手册查得的常数等,可根据计算式中需保留的其他有效数字的位数来确定,需要几位就取几位。表示误差的有效数字一般只取一位,最多取两位。如 1.87±0.02,而 8954±56 应表示为 $(8.95 \pm 0.06) \times 10^3$。

(5) 在复杂计算中,应分步计算,每一步都应按有效数字运算规则,对有效数字位数进行取舍。

1.3.3 实验数据表达

表示实验数据中各变量间的关系一般有3种方法:列表法,图解法和经验公式法。

1.3.3.1 列表法

列表法是将所测得的数据列成表格,以表示各变量间的对应关系,可直观反映各变量间递增、递减或周期性变化规律,同时也是图解法和经验公式法的基础。基本要求为:

(1) 数据表应有简明完整的名称。

(2) 每一行(或列)应有名称、数量、单位和因次。

(3) 相同变量应排成一列,小数点要对齐,要注意有效数字的位数。

(4) 选择的自变量如时间、温度、浓度等,应按递增排列。

(5) 若表中数据有公共乘方因子,应将公共乘方因子写在栏头内。

1.3.3.2 图解法

图解法是利用实验数据或计算结果作图,可从图上获得最大值、最小值、转折点、周期性和变化速率等重要特性,其最大优点是一目了然。图解法的一般步骤和规则如下:

(1) 坐标纸的选择与坐标的确定

直角坐标纸最为常用,有时也用半对数坐标纸、三角坐标纸。选用直角坐标纸时,习惯上以自变量为横轴,因变量为纵轴。

(2) 坐标轴比例尺的选择

坐标轴比例尺应能表示出全部有效数字,使从图上读出的物理量的精密度与测量时的精密度一致。同时应方便易读,例如用坐标轴上 1 cm 表示 1,2,5 或 1,2,5 的 10^n(n 可以是正、负整数和零）倍，应避免用 3,4,6,7,8,9 表示。满足前两项要求的前提下，还应该考虑图的大小和充分利用图纸。

（3）画坐标轴

选定比例尺后，画上坐标轴，应注明分度值、名称、量纲，坐标的文字书写方向应与该坐标轴平行。要注意坐标单位的表示方法，如 T/K 不应写为 T，K 或 T（K），ln$p$/MPa 不应写成 ln$p$，MPa 或 ln$p$（MPa）等。

（4）作代表点

将测得的数据绘于图上，数据点要清晰，在同一图上表示不同数据时应该用不同的符号加以区别，如点、圆、矩形、叉等，但其大小应与其误差相对应。

（5）将代表点连成曲线

连曲线时要用适当的工具作图，曲线要清晰圆滑。由于每一个测量值都会存在误差，按测量数据所描的点不一定是真实值的正确位置，因此，应使各数据点均匀分布在曲线两侧。

（6）标图的名称

图作好后，要标明图的序号和名称，必要时还应标明实验条件。

1.3.3.3 经验公式法

在实验数据处理中，线性回归和曲线拟合是常用的方法，用于描述不同变量之间的关系，确定相应函数的系数，并建立经验公式或数学模型。经验公式在回归分析中被称为回归方程，用于近似表示数据，但很难完全准确地表达全部数据。

以下是建立经验公式或数学模型的一般步骤：

（1）作图

利用实验测得的变量值绘制曲线。这一步骤有助于可视化数据分布，对数据的整体趋势有一个直观的认识。

（2）曲线分析与基本形式确定

对绘制的曲线进行分析，判断其基本形式。如果数据点基本上呈直线分布，可采用一元线性回归方法确定直线方程。如果数据呈曲线分布，则需要根据曲线的特点选择相应的曲线类型，可以参考已有的数学函数形状。

（3）函数的直线化

如果确定曲线方程的类型，尽可能将其变换为直线方程，然后采用一元线性回归方法处理。这样做有助于简化数据分析过程。

（4）确定公式中的常量

对于表示测量数据的直线方程（或经过直线化后的方程），可使用各种方法，如最小二乘法，确定方程中的常量（例如，$y=a+bx$ 中的 a 和 b）。

(5)公式准确性检验

通过将测量数据中的自变量值代入公式,计算出函数值,并与实际测量值进行比较。如果差异很大,可能表明所确定的公式基本形式存在错误,需要考虑建立另一种形式的公式。

(6)多项式回归处理

如果测量曲线难以判断属于何种类型,可以考虑使用多项式回归进行处理,以更灵活地拟合数据。

总体而言,这些步骤提供了一种系统的方法来处理实验数据并建立数学模型,使得对数据之间关系的理解更为深入。随着电子计算机技术的发展,数据处理变得更加便捷和精确。

1.4 常用实验数据处理软件 Origin 的使用

图形化是显示和分析复杂实验数据的重要手段,熟练掌握计算机图表制作及相关数据的分析方法已成为科学研究人员和工程技术人员所必须具备的基本科学素养。目前常用的科技绘图及数据处理软件有 Excel、Origin、Sing、Malo 等。Origin 因其功能强大、简单易学、兼容性好,逐渐成为科技工作者制作图表及分析数据的首选工具。

1.4.1 Origin 基础知识

以 Origin Pro 2024 为例,主界面主要包括菜单栏(Menu bar)、工具栏(Tool bar)、项目管理器(Project explorer)和绘图区(Plot area)等。

菜单栏位于界面顶部,一般可以实现 Origin 软件的大部分功能。例如,"Plot"的绘图功能,操作界面下可进行二维和三维的绘图,包括但不限于气泡图、彩色映射图、统计图和面积图等,另外还提供众多样式的制图模板;"Column"可对每列数据进行对应操作,包括数据列的添加、设置、删除等。"Analysis"能够实现数学运算、图形变换、线性多项式和非线性曲线等多种拟合手段;"Statistics"可提供行列统计、T-检验、方差分析和多元分析等。

工具栏通常位于菜单栏下方,通过工具栏可进行最常用功能的快捷操作。

项目管理器类似于资源管理器,可方便切换各个窗口;Origin 项目文件默认扩展名为 *.opju,在一个项目文件中可以使用右键快捷菜单对多个子文件夹和子窗口进行创建、移动、改名和删除等操作;同一项目下,类型不同的子窗口即使处于不同的子文件夹也不允许重名,进而实现数据、图形等子窗口的统一和分类管理。

此外,绘图区通过使用工作簿(Workbook)、矩阵(Matrix)和图形(Graph)等子窗口存放不同的对象、数据和图形。工作簿与 Excel 类似,包含多个工作表(sheet),以便于数据的管理。一般可通过菜单栏“File-New”或工具栏的“New Workbook”按钮进行工作簿的创建。双击工作表顶部的列标签,可打开列属性设置对话框,在列属性对话框中可设置列的名称、宽度、数据格式,给列加上注释,以及按特定函数形式设置数据等。此外右击工作簿底部的工作标签可打开快捷菜单,工作表的添加、复制、移动、改名和删除等操作可由此实现;矩阵的功能操作包括矩阵属性、维数和数值的设置,矩阵的转置和取反,矩阵的扩展和收缩等;图形窗口用于显示和存储绘制的图形,可进行图形的定制,设置包含图层坐标轴、图形化数据注释等内容。

1.4.2 数据录入

Origin 中常用的数据录入方法有手动输入、通过剪贴板粘贴和数据导入三种。

如所需数据较少,可手动逐行逐列输入。此外还可通过数学公式计算生成所需数据,点击菜单栏“Column-Set Column Values”或双击对应列顶部的列标签进而打开“Set Values”对话框,如图 3 所示,输入对应公式就可以自动计算该列数据。

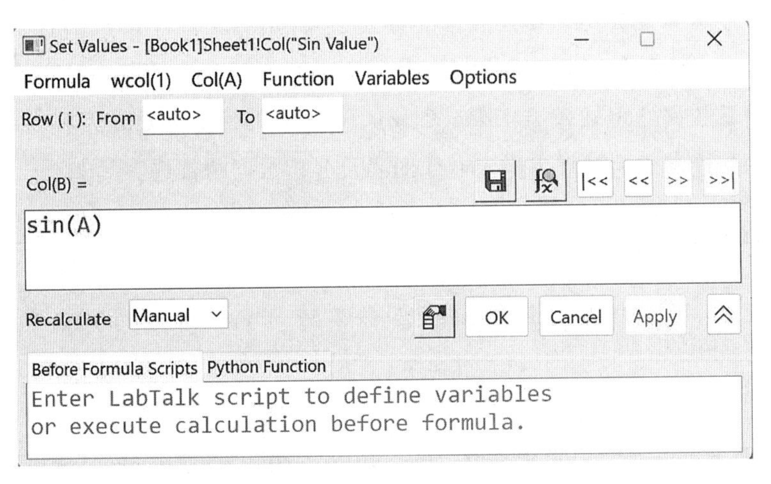

图 3 “Set Values”对话框

通过 Windows 剪贴板结合“Edit”下拉菜单中的“Cut”、“Copy”和“Paste” 命令,可以把其他应用软件(如 Excel)或其他 Origin 项目文件中的数据传送到当前工作表,也可由 Origin 工作表向其他应用软件传送数据。

数据文件导入可通过“Data-Connect to File”菜单实现,Data Connectors 支持 Text/CSV、Excel、Matlab 等众多格式的数据源导入数据。

1.4.3 绘图

首先在工作表中选中要作图的数据列或区域,然后单击二维图形工具栏上的相应作图按钮即可。例如,选中图中数据,点击“Scatter”按钮,即可得到如图 4 所示图形。

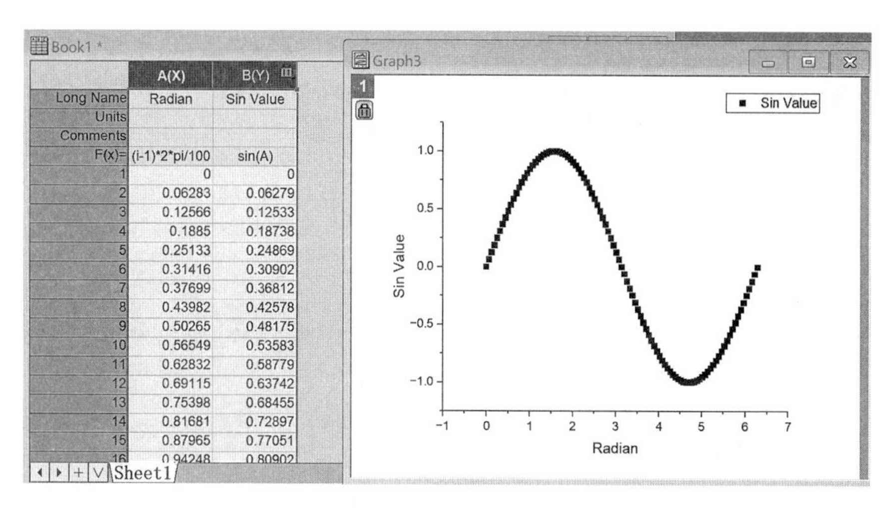

图 4 利用“Scatter”工具作图

Origin 提供可绘制的图形比较多,常见的 2D 图形包括 scatter,spline,Line+Symbol 等。同时,表中可以构建多个 Y 列,即可与同一个 X 列关联,也可与不同的 X 列各自关联。相对应的是在同一个绘图窗口中绘制坐标轴范围不同或度量单位不同的图形,需要使用多层图形来实现。Origin 提供了多种图形模板(图 5),用于快速绘制各种常用的多层图形,有双 Y 轴(Double Y)、垂直两栏(Vertical 2 Panel)、水平两栏(Horizontal 2 Panel)、4 栏(4 Panel)和 9 栏(9 Panel)等。通常选中相应的数据,再选中相应的图形模板即可绘制。

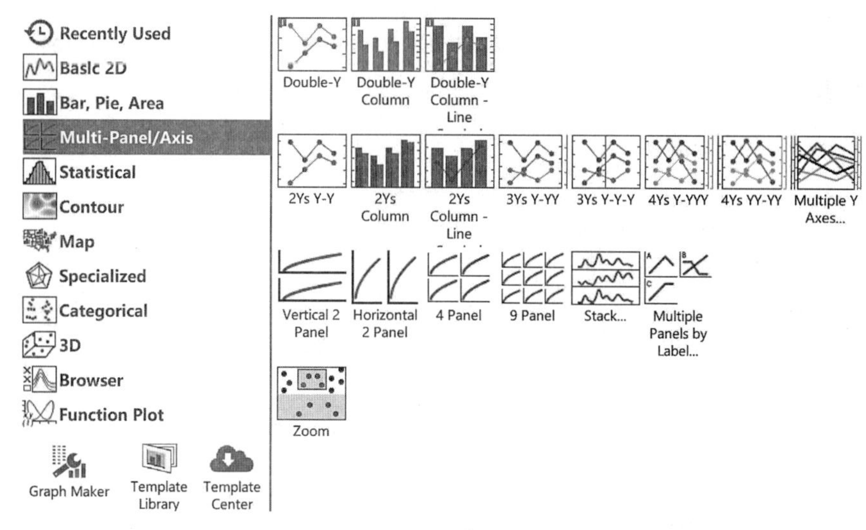

图 5 Origin 的图形模板

绘制好图形后可在所绘图形的线或符号上双击,或右键单击图线,从菜单底部选取“Plot Details”对话框进行图形定制。如图 6 所示,对话框左半部分上侧以树状结构列出当前图形窗口所包含的图层以及图层所包含的图线等信息,左半部下侧通过“Plot Type”选项卡可把曲线设置成“Line”“Scatter”“Line+Symbol”“Column/Bar”等不同格式。在“Line”

选项卡中可以设置线的连接方式“Connect”、线型“Style”、宽度“Width”、颜色“Color”以及曲线下区域填充“Fill Area Under Curve”等；在“Scatter”选项卡中可以设置符号的样式、大小“Size”、边框与填充属性等；“Line+Symbol”选项卡是对前两种图线形式的综合应用；最后在“Column/Bar”选项卡中，可设置连接数据点的垂线颜色“Color”、线型“Style”和宽度“Width”等。

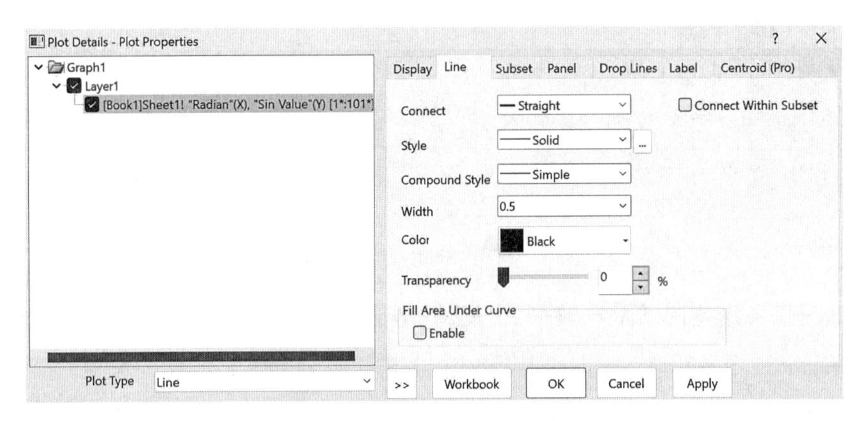

图 6　“Plot Details”对话框

在所绘制图形的坐标轴上，双击可打开如图 7 所示的坐标轴设置对话框，可以进行“Scale”“Tick Lables”“Minor Tick Labels”“Title”“Grids”“Special Ticks”和“Breaks”等选项卡的设置。其中“Tick Labels”“Minor Tick Labels”“Special Ticks”以及“Title”选项卡可以对坐标轴的刻度、标识等进行设定，“Scale”选项卡用于设定所绘图形的坐标轴范围、坐标轴类型（线性、对数）等；“Grids”选项卡用于设定网格线；“Breaks”选项卡可以将不希望显示的部分坐标区域隐藏起来。

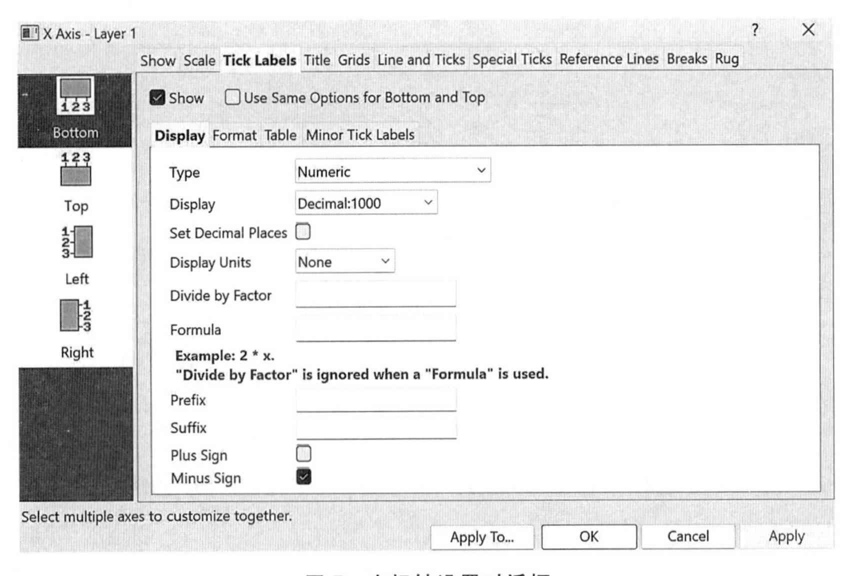

图 7　坐标轴设置对话框

1.4.4 图形输出

在 Origin 中绘制的图形通常以剪贴板传送和输出图形文件两种方式进行输出。在图形窗口激活的状态下，通过菜单栏“Edit”或右键快捷菜单命令“Copy Page”直接将 Origin 图形复制到剪贴板，然后在其他应用程序（如 Word、Powerpoint 等）中应用“开始—粘贴”命令，即可将图形粘贴到指定位置。默认情况下，粘贴命令会将 Origin 对象嵌入其他应用程序，因此，可在其他程序里双击嵌入的图形，打开 Origin 程序对相应图形进行编辑。若只需复制图形本身，可使用 Origin 菜单栏“Edit”或右键快捷菜单命令“Copy Graph as Image”，也可在正常拷贝后到应用程序的“开始”菜单中选择“选择性粘贴”命令，并指定粘贴格式为图形文件，这样粘贴的图形将不可再用 Origin 编辑。

Origin 还支持把图形输出为多种格式的矢量或位图文件，这样可将图形保存为独立的文件，便于出版印刷，输出图形文件的具体步骤如下：

在图形窗口激活的状态下，使用菜单“File-Export Graphs”打开如图 8 所示的“Export Graph：expG2img”对话框。

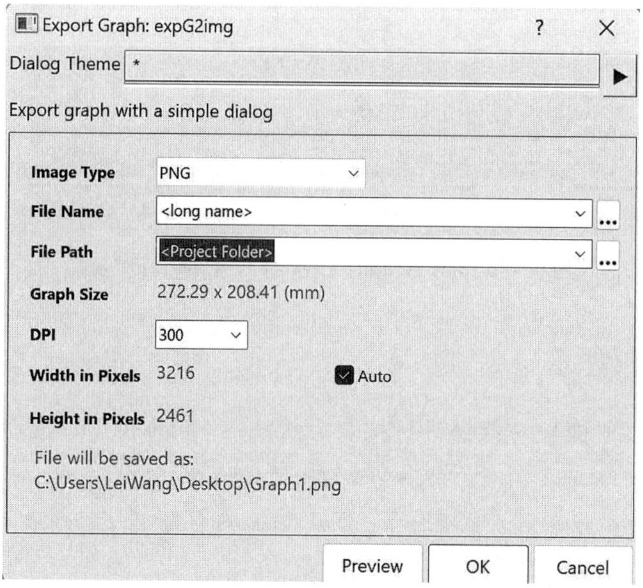

图 8 “Export Graph：expG2img”对话框

在“Image Type”下拉选项表中选择拟保存的图形格式，如（ * . png， * . emf， * . pdf 等）；在“File Name”项输入文件名；在“File Path”输入框设定文件保存位置。在“File-Export Graphs（Advanced）”菜单命令下可打开“Export Graph（Advanced）：expGraph”窗口（图 9），可进行更高级的图片导出，包括更多输出格式，“Image Size”设定输出图形的单位、尺寸等参数，及“Image Settings”设定输出图形的分辨率等。最后单击“OK”按钮即可输出图形文件,在 Word 和 Powerpoint 等软件中可使用“插入”命令嵌入 Origin 输出的图形文件。

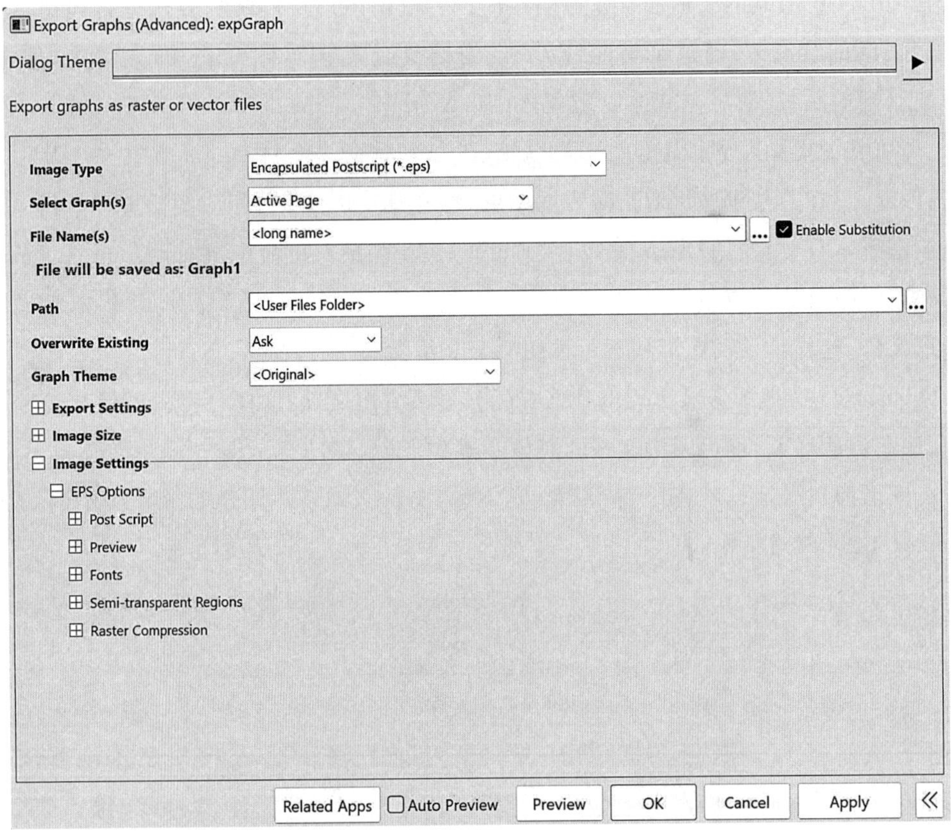

Export Graphs (Advanced): expGraph
Dialog Theme
Export graphs as raster or vector files
Image Type　Encapsulated Postscript (*.eps)
Select Graph(s)　Active Page
File Name(s)　<long name>　Enable Substitution
File will be saved as: Graph1
Path　<User Files Folder>
Overwrite Existing　Ask
Graph Theme　<Original>
Export Settings
Image Size
Image Settings
EPS Options
Post Script
Preview
Fonts
Semi-transparent Regions
Raster Compression
Related Apps　Auto Preview　Preview　OK　Cancel　Apply

图 9　“Export Graph(Advanced):expGraph”窗口

1.4.5　数据拟合

使用 Origin 可便捷地对数据进行拟合分析。在“Analysis-Fitting”菜单的子菜单中,Origin 提供了线性、多项式、指数、自定义等多种常用的数据拟合方法。

用 Analysis 菜单完成拟合步骤如下:首先选中数据,单击二维图形工具中的散点图按钮作二维散点图,之后使用菜单命令“Analysis-Fitting-Linear Fit”打开如图 10 所示的“Linear Fit”对话框,用户可在“Fit Control”选项下设定希望固定的参数如截距或斜率,若对截距和斜率均不做限定,直接单击“OK”按钮即可得到拟合结果,Origin 将在图形中自动添加拟合曲线和拟合结果报告。使用菜单命令“Analysis-Fitting-Polynomial Fit”可对数据进行多项式拟合。打开如图 11 所示的“Polynomial Fit”对话框,在 Polynomial Order 中输入 2,即指定以二次多项式进行拟合;输入 3,则三次多项式拟合。单击“OK”按钮即可得到拟合结果。

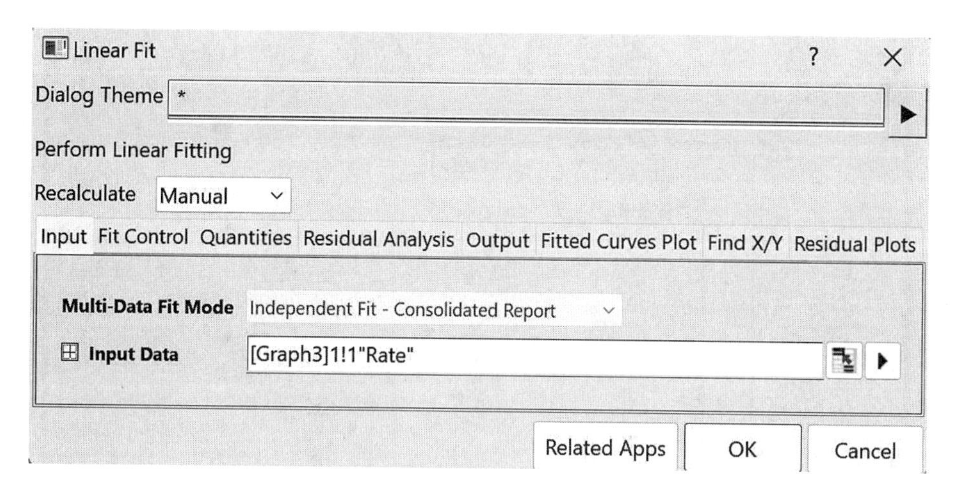

图 10 “**Linear Fit**”对话框

图 11 “**Polynomial Fit**”对话框

此外,Origin 自带软件“Garget”可实现实时变化取值范围的拟合。通过菜单栏“File-Sample Projects”打开“Learning Center”对话框后,选取“Sample Projects-Curve fitting-Linear Regression using Gadget”选取指定数据进行线性拟合,如图 12 所示,这时可左键单击拉动黄色数据框随时变化拟合数据的选取。这时界面“Gadget”菜单下“Quick Fit”可提供“Linear”“Quadratic”和“Cubic”三种最常用的拟合方式。

如需在曲线上某一点绘制切线,可点击 Origin 主界面右侧“App”窗口内的“Add Apps”按钮,打开“App Center”后搜索“Tangent”插件,下载自动安装后会出现在“App”窗口。点击打开如图 13 对话框,确定后会在绘图区的界面上出现一个浮动框,鼠标左键点击浮动框中的箭头,拖动到我们要绘制切线的位置,这时界面会返回对应点的坐标以及切线信息,如图 14。

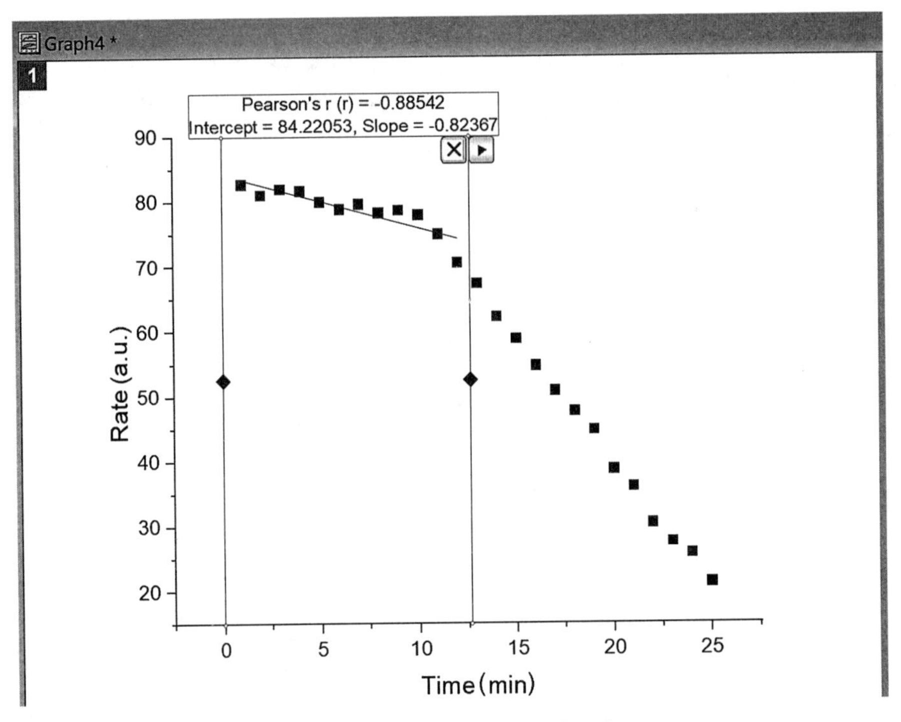

图 12　实时变化取值范围的拟合

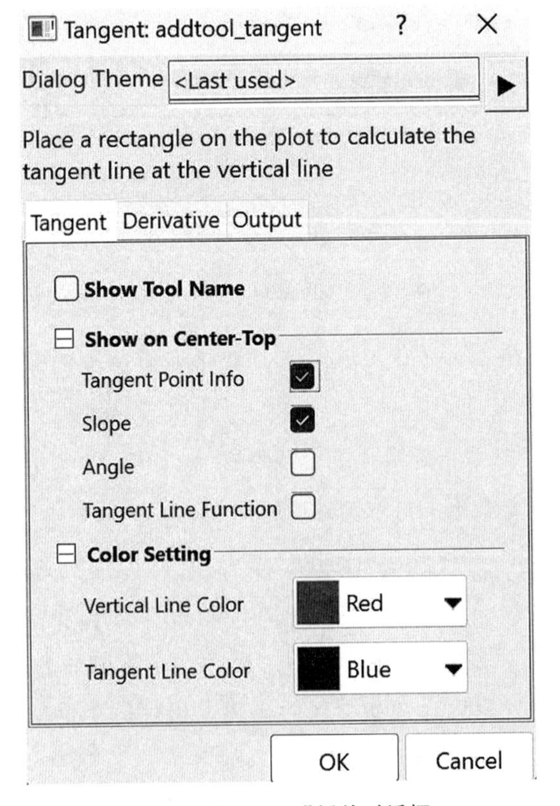

图 13　“Tangent”插件对话框

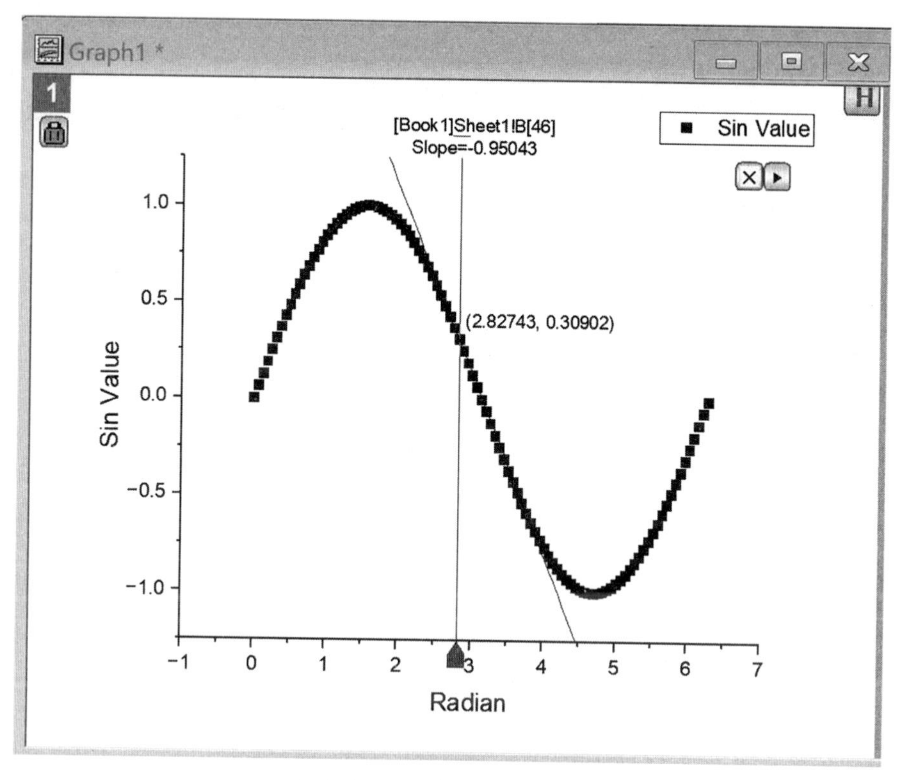

图 14 “Tangent”插件给出的切线信息

总体上，Origin 绘图功能非常强大，这里由于篇幅所限，只是简要介绍了其基本用法，很多细节无法详细介绍，其他使用方法或者细节，可以边使用边摸索，在实践中积累经验，不断提升。

第二章
基础实验

实验1　燃烧焓的测定

实验目的

1. 了解氧弹热量计构造、原理和使用方法，掌握燃烧焓的定义以及等压热效应与等容热效应的关系。

2. 用绝热氧弹热量计测量冰糖（或饼干、奶片）的燃烧焓。

3. 学会用雷诺图解法校正温度，求温度改变值。

基本原理

在标准压力下，反应温度 T 时，单位量的可燃物质 B 完全氧化为同温下的指定产物时的标准摩尔燃烧焓变，称为物质 B 的标准摩尔燃烧焓。燃烧产物指定如下：化合物中 C 燃烧变为 $CO_2(g)$，H 燃烧变为 $H_2O(l)$，Cl 变成 $HCl(aq)$，S 变为 $SO_2(g)$，N 变为 $N_2(g)$，金属变为游离状态。

系统发生化学变化之后，系统的温度回到反应前初始态的温度，系统放出或吸收的热量称为该反应的热效应。若反应是在等压条件下进行的，则该反应热是等压热效应（Q_p）；若反应是在等容条件下进行的，则该反应热是等容热效应（Q_V）。通常情况下，若没有特别注明，反应热都是指等压热效应。热效应可以按如下方法测定：物质在热量计中做绝热变化，通过测量热量计的温度改变，进而计算出从热量计中放出或需要加入多少热量才能恢复到始态温度。所得结果就是等温变化中的热效应。由此可见，我们需要知道等容热效应和等压热效应之间的关系。

由热力学第一定律可知

$$\Delta_r H = \Delta_r U + \Delta(pV) \tag{1}$$

当非体积功等于零时，$Q_V = \Delta_r U$；$Q_p = \Delta_r H$。若把参加反应的气体和反应生成的气体都看作理想气体，则它们之间存在以下关系

$$Q_p = Q_V + \Delta nRT \tag{2}$$

式中 $\Delta n = \sum n_g$（产物）$-\sum n_g$（反应物）；R 为摩尔气体常数，其值为 8.314 $J \cdot K^{-1} \cdot mol^{-1}$；$T$ 为反应时的热力学温度［$T/K=(273.15+t)/℃$，t 为环境温度，即热量计套筒的温度］。

热量计的种类很多，本实验所用的氧弹热量计是一种环境等温式的热量计。氧弹的剖面图如图 1-1 所示。

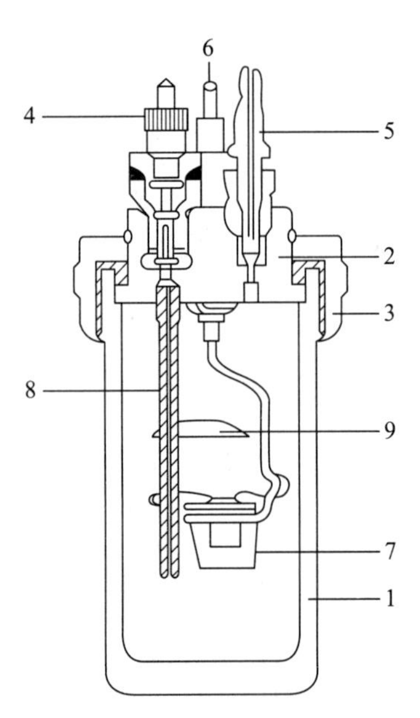

1.厚壁圆筒　2.弹盖　3.螺帽　4.进气口　5.排气口　6.电极
7.燃烧杯　8.点火电极（同时也是进气管）　9.燃烧挡板

图 1-1　氧弹剖面图

氧弹热量计的基本原理是能量守恒定律。样品完全燃烧后所释放的能量使得氧弹本身及其周围的介质和热量计有关附件温度升高，通过测量介质在燃烧前后温度的变化值，可求算样品的等容热效应。其关系式如下

$$-\frac{m(\text{样})}{M} \cdot Q_V - l \cdot Q_l = [m(\text{水}) \cdot C_{\text{水}} + C_{\text{计}}] \cdot \Delta T = b \cdot \Delta T \tag{3}$$

式中 m(样)和 M 分别为样品的质量和摩尔质量（即相对分子质量）；Q_V 为样品的等容热效应；l 是燃烧掉的铁丝长度；Q_l 是燃烧单位长度铁丝的热效应，其值为 $-2.9\ J \cdot cm^{-1}$；m(水)和 $C_{\text{水}}$ 分别是以水作为测量介质时，水的质量和比热容；$C_{\text{计}}$ 为热量计的水当量（即除水之外，系统内其他部件升高 1 ℃所需吸收的热量）；ΔT 为样品燃烧前后水温的变化值，b 为热量计常数｛$b=[m(\text{水}) \cdot C_{\text{水}}+C_{\text{计}}]$，单位 $J \cdot ℃^{-1}$｝。应当注意，只有当测定样品所用水的质量与用苯甲酸测定 b 值所用水的质量相同时，方可利用式(3)计算样品的 Q_V 值。

由于热量计与周围环境的热交换是无法完全避免的，因此需要用雷诺（Renolds）温度校正图对温度测量值进行校正。具体方法为：称取适量待测物质，估计其燃烧后可使水温

上升 1.5～2.0 ℃。要预先调节盛水桶内的水的温度低于环境温度 1.0 ℃左右。实验操作结束后,将燃烧前后所测得的一系列水温对时间作图,可得如图 1-2 所示的曲线。图中 H 点意味着燃烧开始,热传入介质;D 点为观察到的温度最高值;当环境温度为 J 点所对应的温度时,从 J 点作水平线交曲线于 I 点,过 I 点作垂直于横轴(t)的线 ab,再将 FH 线和 GD 线分别延长交直线 ab 于 A'、C'两点,其间的温度差值即为经过校正的 ΔT ($\Delta T=T_{C'}-T_{A'}$)。图中 AA'是由于环境辐射和搅拌导致的温度升高,故应予以扣除。CC'是由于热量计向环境的热漏造成的温度降低,计算时应考虑在内。故可认为,AC 两点的差值较客观地表示了样品燃烧引起的系统温度升高的数值。

在热量计的绝热性能良好的情况下,会得到如图 1-3 所示的 Renolds 温度校正图。由于热漏很小,而搅拌器功率较大,不断引进的能量使得曲线不出现温度极高点,此时由于搅拌而引起的温度升高应扣除,即 AC'之间的温度差值为经过校正的 ΔT。

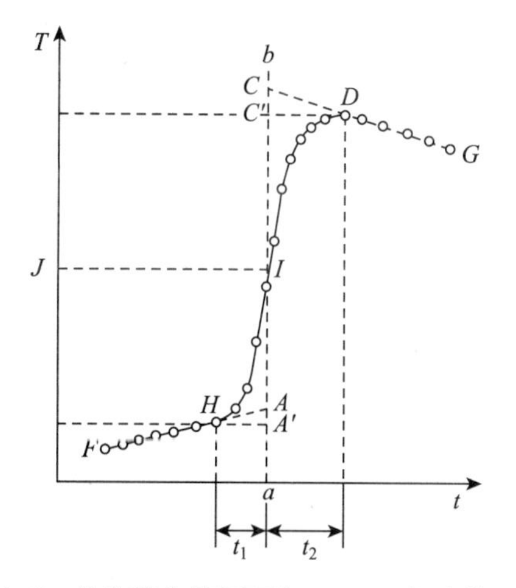

图 1-2 绝热稍差情况下的 Renolds 温度校正图

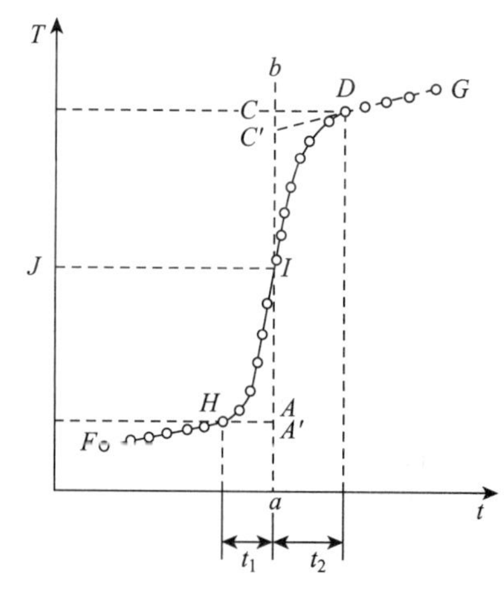

图 1-3 绝热良好情况下的 Renolds 温度校正图

本实验采用 XRY-1A 型氧弹热量计来测量温度,其面板如图 1-4 所示。控制器面板上有电源、搅拌、数据、结束、点火、复位六个电子开关按键和七位数码管,能对样品热值测定进行全过程操作和温度显示。其中左边两位数字代表测温次数,右边五位代表测量的实际温度,本仪器测温范围为 10～35 ℃。

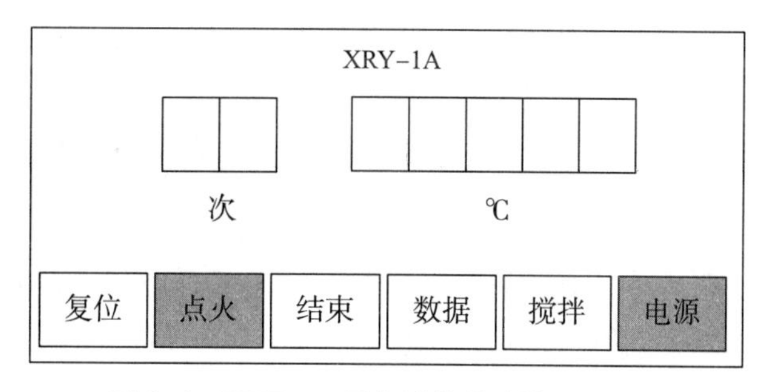

图 1-4 XRY-1A 型氧弹热量计控制器面板

开机后，只要不按“点火”键，仪器就会逐次自动显示温度数据 100 个，测温次数从 00→99 递增，每半分钟一次，并伴有蜂鸣器的鸣响，此时按动“结束”键或“复位”键能使显示测温次数复零。按动“点火”键后，氧弹内点火丝得到约 24 V 交流电压，从而烧断点火丝，点燃坩埚中的样品，同时，测量次数复零。以后每隔半分钟测温一次并贮存测温数据共 31 个，当测温次数达到 31 后，测温次数就自动复零（实验时，点火后最多读 31 个数，读完第 31 个数后按“结束”键）。按“结束”键后，仪器显示全零，然后按动“数据”键，点火后的测量温度数据重新逐一在五位数码管上显示出来。实验者可以与实验时记录的温度数据（注：电脑贮存的数据是蜂鸣器鸣响的那一秒的温度值）进行核对。当操作人员每按一次“数据”键，被贮存的温度数据和测温次数自动逐个显示出来，方便实验者核查测温记录。

仪器和试剂

XRY-1A 氧弹热量计
氧气钢瓶
氧气减压阀
压片机
塑料烧杯
直尺
剪刀
温度计（0～50 ℃）
容量瓶（1 000 mL）
精密电子天平
0.1 g 电子天平
引燃专用铁丝
苯甲酸（分析纯）
冰糖（或饼干、奶片）

实验步骤

1. 测量热量计常数 b

（1）样品压片

用 0.1 g 电子天平称取 0.9 g 左右的苯甲酸，用压片机压片。用钢尺准确量取 20 cm 的棉线、10 cm 引燃专用铁丝，并用精密电子天平称量铁丝和棉线的质量。用已称好质量的棉线将苯甲酸片缠绕捆绑牢固（最好十字花形状），将已称好质量的铁丝从棉线和苯甲酸片之间穿过，用精密电子天平准确称取捆绑后铁丝、棉线及苯甲酸片总质量，减去之前称得的铁丝和棉线质量，即得到苯甲酸样品片的精确质量。

（2）充氧气

用紫铜管将氧弹充气阀门与钢瓶减压器出口接通。先逆时针旋转钢瓶总阀门，使高压表指针指向 10 MPa 左右。再缓缓顺时针旋紧减压器，使低压表示数为 2 MPa，氧气充入氧弹内，3 min 后，观察低压表指针是否下降可判断氧弹是否漏气。若指针未下降，则表明氧弹不漏气，即可关闭减压器，将紫铜管与氧弹充气阀门连接的一端拆下。当所有同学充氧气结束后，要将钢瓶的总阀门关闭。由于总阀门与减压器之间有余气，因此要再次旋紧减压器，放掉余气，然后再次关闭（旋松）减压器，使钢瓶氧气表恢复原状。在充氧过程中，若发现异常，要查明原因并排除之。

（3）测量燃烧时的温度

① 用温度传感器测量热量计套筒温度（用于绘制 Renolds 温度校正图）。然后调节水温使其低于套筒温度（环境温度）1.0 ℃左右（用温度传感器测量）。②用容量瓶准确量取低于环境温度约 1 ℃的水 2 000 mL 放入盛水桶中，将充好氧气的氧弹也放入盛水桶中，并将盛水桶放入热量计内。③接上点火导线，连好控制箱上的所有电路导线，再准确量取低于室温约 1 ℃的水 1 000 mL 放入盛水桶中，盖上胶木盖，将测温传感器插入内桶。打开电源和搅拌开关，仪器开始显示内桶水温，每隔半分钟蜂鸣器报时一次。当内桶水温均匀上升后，每隔半分钟记下显示的温度。当记下第 10 次温度的同时按“点火”键（测量次数自动复零），每隔半分钟记一次温度数据，大约测 31 个数据（跟踪记录，实验停止后，按“结束”键，仪器显示“零”，然后按动“数据”键，点火后的数据可在仪器上重新逐个显示。实验者可以对实验记录的数据进行核对）。④检查数据无误后，停止搅拌，取出温度传感器，打开桶盖（注意：先拿出传感器，再打开桶盖）。取出氧弹，旋松氧弹放气阀放掉氧弹内气体，打开氧弹，检查燃烧是否完全（若燃烧不完全，氧弹内会有大量黑灰，需要重新做），量取剩余铁丝的长度并记录。将盛水桶中的水倒入指定的桶中。

2. 测定冰糖（或饼干、奶片）的 Q_V

用电子天平称取 1.4～1.6 g 的冰糖（或饼干、奶片），测定其燃烧热，步骤同 1。

数据记录与处理

1. 将测量的苯甲酸和冰糖（或饼干、奶片）的原始数据分别列入实验前设计的表格中。

2. 利用 Renolds 图解法求出苯甲酸的 ΔT 并代入式(3)求出 b 值。

3. 计算冰糖（或饼干、奶片）的 Q_V

由 Renolds 图解法求出冰糖（或饼干、奶片）燃烧的 ΔT，代入下式求出 Q_V。

$$-\frac{m(\text{样})}{M}\cdot Q_V - l\cdot Q_l = b\cdot\Delta T$$

4. 求冰糖（或饼干、奶片）的 Q_p。由冰糖（或饼干、奶片）的 Q_V 值计算冰糖（或饼干、奶片）的 Q_p。

思考与讨论

1. 苯甲酸样品为什么要压成片状？

2. 在量热学测定中，为什么要利用 Renolds 温度校正图对温度测量值进行校正？

3. 讨论本实验的误差来源，如何使实验的误差降到最小？

4. 本实验的设计思想有哪些方面值得借鉴？

文献值

表 1-1　几种物质燃烧热的文献值

等压燃烧焓	kcal · mol^{-1}	kJ · mol^{-1}	J · g^{-1}	测定条件
苯甲酸	−771.24	−3 226.9	−26 460	$p^\ominus$,20 ℃
蔗糖	−1 348.7	−5 643	−16 486	$p^\ominus$,25 ℃
萘	−1 231.8	−5 153.8	−40 205	$p^\ominus$,25 ℃

注：1cal=4.18 J。

参考资料

1. Shoemaker D P, Garland C W, Nibler J W. Experiments in Physical Chemistry. 5th ed. New York: McGraw-Hill Book company, 1989.

2. 北京大学化学系物理化学教研室. 物理化学实验(第三版). 北京：北京大学出版社，1995：40.

3. Weast R C. CRC Handbook of Chemistry and Physics. Florida: CRC Press, Boca Raton, 1985-1986: 272.

4. 印永嘉. 物理化学简明手册. 北京：高等教育出版社，1988.

5. 朱京，陈卫，金贤德，等. 液体燃烧热和苯共振能的测定. 化学通报，1984,(3)：50.

Experiment 1

Determination of the heat of combustion

Experimental purpose

1. Understand the construction, principles, and usage of the bomb calorimeter, mastering the definition of combustion enthalpy and the relationship between isobaric and isochoric heat effects.

2. Measure the combustion enthalpy of sucrose (or biscuits, cereal bars) using an adiabatic bomb calorimeter.

3. Correct temperature by using Reynolds curve and calculate the temperature changes.

Basical principles

At standard pressure, when a unit molar enthalpy of combustion of combustible substance B completely oxidizes to specified products at temperature T, the standard molar combustion enthalpy change is termed as the standard molar combustion enthalpy (ΔcH). The combustion products are specified as follows: C combusts to $CO_2(g)$, H combusts to $H_2O(l)$, Cl transforms to $HCl(aq)$, S transforms to $SO_2(g)$, N transforms to $N_2(g)$, and metals transform to the free state.

After a chemical change in the system, the heat exchanged by the system, causing the system's temperature to return to the initial state before the reaction, is termed the heat effect of the reaction. If the reaction occurs under constant pressure, the heat is isobaric heat effect (Q_p); if the reaction occurs under constant volume, the heat is isochoric heat effect (Q_V). Generally, the term "heat effect" refers to isobaric heat effect unless specified. Heat effects can be determined as follows: the substance undergoes adiabatic changes in the calorimeter; by measuring the calorimeter's temperature change, the heat released or required to restore to the initial temperature can be calculated. The result is the heat effect during isothermal changes. It is crucial to understand the relationship between isochoric and isobaric heat effects.

According to the first law of thermodynamics

$$\Delta_r H = \Delta_r U + \Delta(pV) \tag{1}$$

when the non-volume work is zero, $Q_V = \Delta_r U$; $Q_p = \Delta_r H$. If the gases involved in the reaction and the gases produced are treated as ideal gases, the relationship between them is given by

$$Q_p = Q_V + \Delta nRT \tag{2}$$

where $\Delta n = \sum n_g(\text{Product}) - \sum n_g(\text{Reactant})$; R is the molar gas constant and its value is 8.314 $J \cdot K^{-1} \cdot mol^{-1}$; T is the thermodynamic temperature during the reaction [$T/K = (273.15 + t)/℃$, t is the environmental temperature, i.e. the temperature of the calorimeter jacket].

There are various types of calorimeter. The adiabatic bomb calorimeter used in this experiment is anenvironmentally isothermal calorimeter. The cross-section of the bomb calorimeter is shown in Fig. 1-1.

The fundamental principle of the bomb calorimeter is the law of conservation of energy. When the sample undergoes complete combustion, the released energy elevates the temperature of the calorimeter and its accompanying accessories, as well as the surrounding medium. Through the measurement of the temperature change before and after the combustion, one can calculate the sample's constant-volume combustion heat. The relationship is expressed as follows

$$-\frac{m(\text{sample})}{M} \cdot Q_V - l \cdot Q_l = [m(\text{water}) \cdot C_{\text{water}} + C_{\text{calorimeter}}] \cdot \Delta T = b \cdot \Delta T \tag{3}$$

where, $m(\text{sample})$ and M are the mass and molar mass of the sample respectively; Q_V is the isochoric heat effect; l is the length of the wire; Q_l is the heat of combustion per unit length of

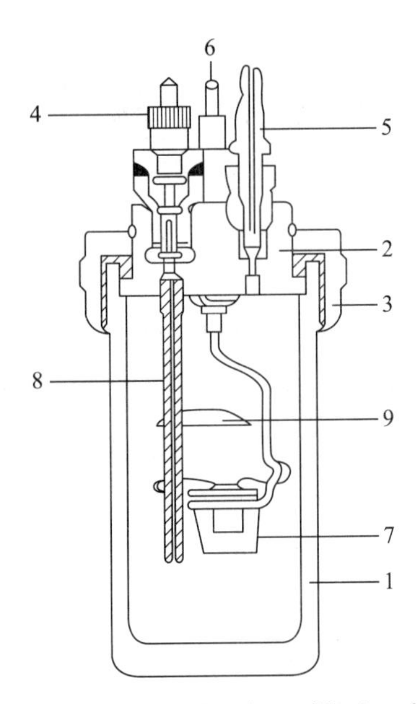

Fig. 1-1　Oxygen-bomb profile drawing

1-Thick-walled cylinder; 2-Spring cap; 3-Nut; 4-Air inlet; 5-Exhaust port; 6-Electrode; 7-Burning cup; 8-Ignition electrode (also an intake pipe); 9-Combustion damper

wire, whose value is $-2.9\ \mathrm{J \cdot cm^{-1}}$; m(water) and C_{water} are the mass and specific heat capacity of water when the measurement medium is water; $C_{\text{calorimeter}}$ is the water equivalent of the calorimeter (i.e. the heat required for the system to increase 1 ℃ except water); ΔT is the change value of water temperature before and after combustion of the sample; b is the heat meter constant [$b=(m_{\text{water}} \cdot C_{\text{water}}+C_{\text{calorimeter}})$, $\mathrm{J \cdot ℃^{-1}}$]. It should be noted that Equation (3) can be used to calculate the isochoric heat effect Q_V value of the sample only if the mass of water used to determine the sample is the same as that used to determine the b value by benzoic acid.

Due to the inevitable heat exchange between the calorimeter and the surrounding environment, it is necessary to use the Reynolds number correction to correct the temperature measurement values. The specific method is as follows: take an appropriate amount of the substance to be tested and estimate that its combustion will raise the water temperature by 1.5 to 2.0 ℃. Pre-adjust the temperature of the water in the calorimeter to be about 1.0 ℃ below the ambient temperature. After completing the experimental operation, plot a series of water temperature *vs.* time, resulting in the curve shown in Fig. 1-2. In this figure, point H indicates the start of combustion; point D is the observed highest temperature. When the ambient temperature corresponds to point J, draw a horizontal line from J intersecting the curve at point I. Draw a line ab perpendicular to the horizontal axis from point I, then extend the FH line and GD line to intersect the line ab at points A' and C', respectively. The temperature difference between A' and C' ($\Delta T = T_{C'} - T_{A'}$) represents the objective value of the temperature increase caused by the sample combustion.

In the case of good adiabatic performance of the calorimeter, a Reynolds number correction chart, as shown in Fig. 1−3, can be obtained. Due to the small heat loss and the significant power of the stirrer, continuous introduction of energy prevents the curve from reaching an extremely high temperature. The temperature increase caused by stirring could be deducted, i. e., the temperature difference between AC' is the corrected ΔT.

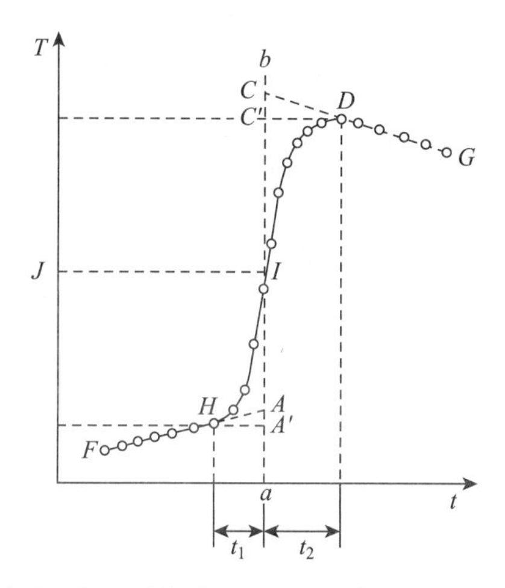

Fig. 1−2 Reynolds Temperature Correction Chart with Slightly Poor Adiabatic Performance

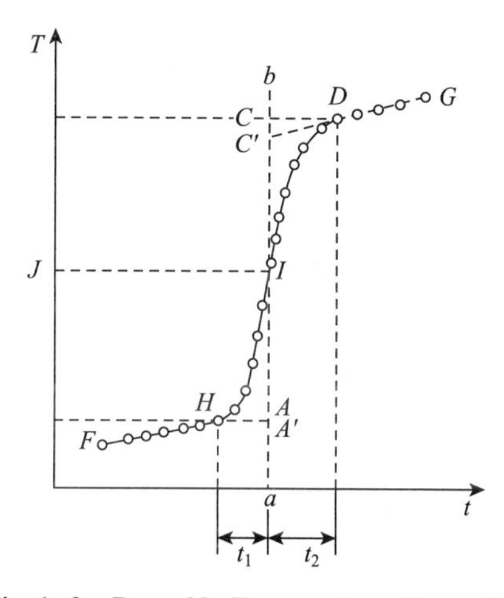

Fig. 1−3 Reynolds Temperature Correction Chart with Good Adiabatic Performance

The XRY−1A bomb calorimeter is used to measure temperature, as shown in Fig. 1−4. The controller panel includes electronic switches for power, stirring, data, end, ignition, and reset, as well as a seven-segment display for complete process operation and temperature display. The left two digits represent the number of temperature measurements, while the right five represents the actual temperature measured. The temperature range of this instrument is 10～35 ℃.

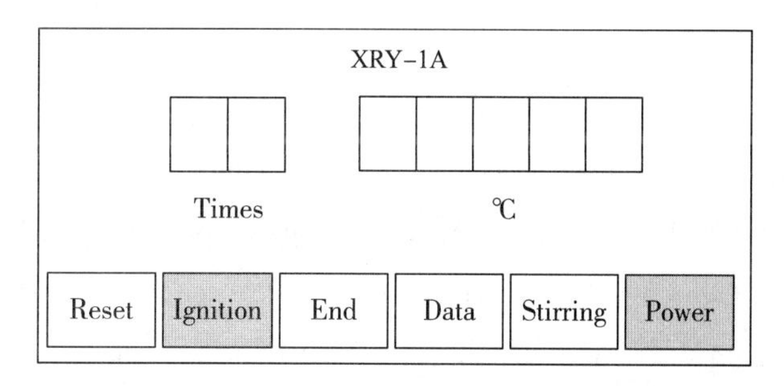

Fig. 1−4 XRY−1A bomb calorimeter controller panel

Upon startup, as long as the "ignition" button is not pressed, the instrument automatically displays temperature data for 100 times, increasing from 00 to 99 every 30 seconds, accompanied by a buzzer. Pressing the "end" button or "reset" button stops the display of temperature measurement times. Pressing the "ignition" button ignites the filament in the bomb calorimeter with an AC voltage of about 24 V, burning the filament, igniting the sample in the crucible, and resetting the number of measurements to zero. Afterwards, the temperature data is displayed one by one on the five-digit display every half minute. When the number of measurements reaches 31, it automatically resets to zero (during the experiment, after ignition, at most 31 numbers can be read, and after reading the 31st number, press the "end" button). Pressing the "end" button resets the instrument to zero, then press the "data" button, and the temperature data after ignition is displayed one by one on the instrument. The experimenter can check the temperature data recorded during the experiment against the displayed data. Each time the experimenter presses the "data" button, the stored temperature data and measurement times are automatically displayed one by one, facilitating the experimenter in checking the temperature records.

Apparatus and software

XRY-1A bomb calorimeter
Oxygen cylinder
Oxygen pressure reducing valve
Pellet press
Plastic beaker
Straightedge
Scissors
Thermometer(0～50 ℃)
Volumetric bottle(1 000 mL)
Electronic analytical balance
Electronic balance
Special wire for ignition
Benzoic acid (AR)
Rock sugar (or biscuits, milk slices)

Experimental procedures

1. Measure the calorimeter constant b

(1) Sample lamination: Use the pellet press to compress about 0.9 g of benzoic acid into a pellet. Use steel ruler to accurately measure 22 cm of cotton thread, 10 cm (timely record the length of the measurement) ignition special wire. Use precision electronic balance to accurately weigh the mass of iron wire and cotton thread, then use the cotton thread that has been weighed to wrap the benzoic acid sheet firmly (cross shape), use the wire that has been weighed to pass between the cotton thread and benzoic acid sheet, use precision electronic balance to accurately weigh its total weight, and get the weight of benzoic acid.

(2) Oxygenation: Tighten the bomb calorimeter, and connect its filling valve to the outlet of the cylinder. Loosen the main valve of the cylinder (rotating counterclockwise) so that the pointer of the high pressure gauge points to about 10 MPa. Slowly tighten the pressure reducer (rotate clockwise), so that oxygen is filled into the oxygen bomb, and the low-pressure gauge reads about 2 MPa. After 3 min, check if the low-pressure gauge pointer has dropped to determine if the bomb calorimeter is leaking. If the pointer does not drop, it indicates that the calorimeter does not leak, and the regulator can be closed, and the copper tube connected to the bomb filling valve can be removed. After all students have filled the oxygen, the main valve of the cylinder must be closed. Because there is residual gas between the main valve and the regulator, the regulator should be tightened again to release the residual gas, and then the regulator should be tightened again to restore the oxygen cylinder gauge to its original state. During the oxygen filling process, if any anomalies are found, the cause should be identified and eliminated.

(3) Measure the temperature during combustion: ① Measure the temperature of the calorimeter jacket using a temperature sensor. Then adjust the water temperature to be about 1.0 ℃ below the jacket temperature (ambient temperature) using a temperature sensor. ② Add precisely measured 2 000 mL of water at a temperature about 1 ℃ below ambient temperature to the water bucket, place the bomb calorimeter in the bucket, and place the bucket in the calorimeter. ③Connect the ignition wire, connect all the circuit wires on the control box, and add accurately measured 1 000 mL of water at about 1 ℃ below room temperature to the water bucket. Cover it with a wooden lid and insert the temperature sensor into the inner bucket. Turn on the power and stirring switch, and the instrument begins to display the inner bucket's water temperature, with a buzzer sounding every 30 seconds. After the inner bucket's water temperature uniformly rises, record the displayed temperature every 30 seconds. When recording the 10th temperature, press the "ignition" button simultaneously (the number of measurements automatically resets to zero), record the temperature data every half minute, and measure about 31 data (tracking records, after the experiment stops, press the "end" button, and the instrument displays "zero," then press the "data" button, and the data after ignition can be displayed on the instrument one by one. The experimenter can check the temperature data recorded during the experiment against the displayed data). ④After checking that the data is correct, stop stirring, remove the temperature sensor, and open the bucket lid (note: take out the sensor first, then open the bucket lid). Remove the bomb calorimeter, loosen the bomb discharge valve to release the gas in the bomb, open the bomb, check if the combustion is

complete (if the combustion is incomplete, there will be a large amount of black ash in the bomb, and it needs to be redone), measure the remaining length of the iron wire and record it. Pour the water in the water bucket into the specified bucket.

2. Determine the Q_V of rock sugar (or biscuits, milk slices)

Weigh 1.4～1.6 g of rock sugar (or biscuits, milk slices) using an electronic balance and measure its combustion heat, following the steps in 1.

Data recording and processing

1. Record the raw data of the measured benzoic acid and rock sugar (or biscuits, milk slices) in the pre-designed table.

2. Use the Reynolds graph to determine ΔT of benzoic acid and substitute it into Equation (3) to obtain the b value.

3. Calculate the Q_V of the rock candy (or biscuits, or milk tablets)

Use the Reynolds graph to determine ΔT of rock sugar (or biscuits, milk slices) and substitute it into the following equation to obtain Q_V.

$$-\frac{m(\text{sample})}{M} \cdot Q_V - l \cdot Q_l = b \cdot \Delta T$$

4. Calculate the Q_p of rock sugar (or biscuits, milk slices)

Calculate Q_p of rock sugar (or biscuits, milk slices) using the Q_V value.

Thinking and discussing

1. Why should the solid samples be pressed into a pellet?
2. Why is the Reynolds number correction used to correct temperature measurements?
3. Discuss the sources of errors in this experiment and how to minimize them.
4. What aspects of the experimental design philosophy are worth emulating?

Literature value

Table 1-1 Literature values of Enthalpy of Combustion for Several Substances

Constant pressure enthalpy of combustion	$kcal \cdot mol^{-1}$	$kJ \cdot mol^{-1}$	$J \cdot g^{-1}$	Determination condition
Benzoic acid	−771.24	−3 226.9	−26 460	$p^{\ominus}$, 20 ℃
Sucrose	−1 348.7	−5 643.0	−16 486	$p^{\ominus}$, 25 ℃
Naphthalene	−1 231.8	−5 153.8	−40 205	$p^{\ominus}$, 25 ℃

Reference materials

1. Shoemaker D P, Garland C W, Nibler J W. Experiments in Physical Chemistry. 5th ed. New York: McGraw-Hill Book company, 1989.

2. Physical Chemistry Teaching and Research Office, Department of Chemistry, Peking University. Physical Chemistry Experiments. (3rd ed). Beijing: Peking University Press, 1995: 40.

3. Weast R C. CRC Handbook of Chemistry and Physics. Florida: CRC Press, Boca Raton, 1985-1986: 272.

4. Yin Y. A Short Course in Physical Chemistry. Beijing: Higher Education Press, 1988.

5. Zhu J, Chen W, Jin X, et al. Measurement of the combustion heat of liquids and the resonance energy of benzene. Chemistry, 1984, (3): 50.

实验2 氨基甲酸铵分解反应平衡常数的测定

实验目的

1. 测定氨基甲酸铵的分解压力,求分解反应的平衡常数和有关热力学函数的变化值。
2. 了解温度对反应平衡常数的影响。
3. 掌握用等压计测定静态平衡压力的方法。

基本原理

氨基甲酸铵(白色不稳定固体)是合成尿素的中间产物,很不稳定,受热易分解,其分解反应为

$$NH_2COONH_4(s) \rightleftharpoons 2NH_3(g) + CO_2(g)$$

该反应是可逆的多相反应,若不将分解产物从系统中移走,则很容易达到平衡。在压力不太大时,气体的逸度近似为1,纯固态物质的活度为1,气体可看成理想气体,则分解反应的标准平衡常数 $K_p^\ominus$ 为

$$K_p^\ominus = \left(\frac{p_{NH_3}}{p^\ominus}\right)^2 \cdot \left(\frac{p_{CO_2}}{p^\ominus}\right) \qquad (1)$$

式中 p_{NH_3}、p_{CO_2} 分别为 NH_3、CO_2 在实验温度下的平衡分压，$p^\ominus = 100$ kPa。设分解反应系统的总压力为 $p_{总}$，因固体氨基甲酸铵的蒸气压力可忽略，故系统的总压为 $p_{总} = p_{NH_3} + p_{CO_2}$。

从氨基甲酸铵分解反应式可知：$p_{NH_3} = \frac{2}{3}p_{总}$；$p_{CO_2} = \frac{1}{3}p_{总}$。

代入式(1)得

$$K_p^\ominus = \left(\frac{2}{3} \times \frac{p_{总}}{p^\ominus}\right)^2 \times \left(\frac{1}{3} \times \frac{p_{总}}{p^\ominus}\right) = \frac{4}{27} \times \left(\frac{p_{总}}{p^\ominus}\right)^3 \tag{2}$$

可见，当系统达到平衡后，只要测量其总压，便可求得实验温度下的标准平衡常数 $K_p^\ominus$。

由 van't Hoff 公式的等压微分式可知温度与平衡常数的关系为

$$\frac{\mathrm{d}\ln K_p^\ominus}{\mathrm{d}T} = \frac{\Delta_r H_m^\ominus}{RT^2} \tag{3}$$

式中 $\Delta_r H_m^\ominus$ 为氨基甲酸铵分解反应的标准摩尔焓变；T 为热力学温度；R 为摩尔气体常数，其值为 8.314 $J \cdot K^{-1} \cdot mol^{-1}$。若温度变化范围不大，$\Delta_r H_m^\ominus$ 可视为常数。对式(3)做不定积分，得

$$\ln K_p^\ominus = -\frac{\Delta_r H_m^\ominus}{RT} + c \tag{4}$$

以 $\ln K_p^\ominus$ 对 $\frac{1}{T}$ 作图得到一直线，其斜率为 $-\frac{\Delta_r H_m^\ominus}{R}$，由此可求得 $\Delta_r H_m^\ominus$。

由某温度下的标准平衡常数 $K_p^\ominus$，可以求算该温度下的标准摩尔反应 Gibbs 函数的变化值 $\Delta_r G_m^\ominus$

$$\Delta_r G_m^\ominus = -RT\ln K_p^\ominus \tag{5}$$

利用实验温度范围内分解反应的平均等压热效应 $\Delta_r H_m^\ominus$ 和某温度下的标准摩尔 Gibbs 自由能变化 $\Delta_r G_m^\ominus$，可近似算出该温度下的标准摩尔熵变 $\Delta_r S_m^\ominus$

$$\Delta_r S_m^\ominus = \frac{\Delta_r H_m^\ominus - \Delta_r G_m^\ominus}{T} \tag{6}$$

仪器和试剂

真空装置 1 套

等压计

储气罐

恒温槽

样品管

三通真空活塞

数字式真空压力计

氨基甲酸铵（自制）

硅油

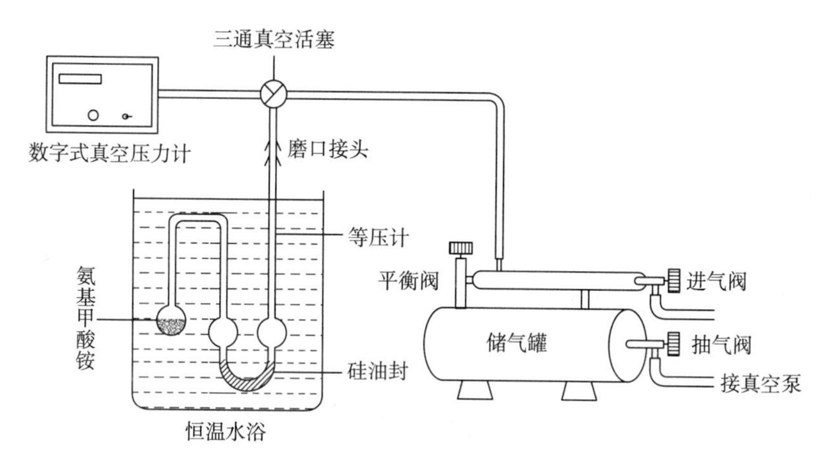

图 2-1 实验装置示意图

实验步骤

1. 调节恒温水浴温度为 35.00 ℃,开动搅拌,打开数字式压力计,记录大气压和室温(实验前后都要记录,数据处理时取平均值)。

2. 抽真空

关闭进气阀,打开抽气阀和平衡阀,开启真空泵。抽气至精密数字压力计读数约为 90 kPa 左右,关闭抽气阀和真空泵。继续抽气 10 min 后,关闭平衡阀。然后慢慢打开进气阀,至油封液面齐平后,关闭进气阀。由于反应并未达到平衡,所以油封液面再次出现落差,故需要反复调节进气阀,直到油封液面齐平并保持 10 min 不变时,可确认反应已达到平衡,记录分解压力和反应温度。

3. 升温至 35.50 ℃,再调节封闭液齐平,5 min 不变时可记录分解压力;间隔 1 min 后,再读 1 次分解压力。

4. 按步骤 3 的操作,依次测出 36.00、36.50 和 37.00 ℃时的分解压力。

5. 测量完毕,打开平衡阀和进气阀,使系统与大气相通。

数据记录与处理

1. 设计合适的表格将实验数据记录其中。

2. 以 $\ln K_p^\ominus$ 对 $\dfrac{1}{T}$ 作图,计算氨基甲酸铵分解反应的等压反应热效应 $\Delta_r H_m^\ominus$。

3. 计算 35.00 ℃时氨基甲酸铵分解反应的标准摩尔 Gibbs 自由能变化 $\Delta_r G_m^\ominus$ 和标准摩尔熵变 $\Delta_r S_m^\ominus$。

思考与讨论

1. 如何检测本实验测量装置是否漏气?

2. 当空气缓慢进入系统时,如放入的空气过多,将有什么现象出现,怎样克服?

3. 实验前为什么要抽净系统中的空气？若空气没有抽净，对测量结果（压力、平衡常数）有何影响？

4. 实验中对等压计中所用的油封液体有何要求？

参考资料

1. Joncich M J, Solka B H, Bower E. J Chem Edu, 1967(44): 598.

2. 东北师范大学，等. 物理化学实验（第二版）. 北京：高等教育出版社，1989：126.

Experiment 2

Determination of equilibrium constants for the decomposition reaction of ammonium carbamate

Experimental purpose

1. Determine the decomposition pressure of ammonium carbamate, calculate the equilibrium constant of the decomposition reaction, and assess the thermodynamic function changes involved.

2. Understand the impact of temperature on the equilibrium constant of the reaction.

3. Master the method of using a manometer to measure static equilibrium pressure.

Basical principles

Ammonium carbamate (white, unstable solid) is an intermediate product in the synthesis of urea, highly unstable and prone to decomposition upon heating. Its decomposition reaction is:

$$NH_2COONH_4(s) \rightleftharpoons 2NH_3(g) + CO_2(g)$$

The decomposition reaction is reversible and multi-phase. If the decomposition product is not removed from the system, equilibrium is easily reached. At relatively low pressure, the fugacity of gases is approximated to 1, the activity of pure solid substances is also 1, and suppose the gases can be treated as ideal gases. The standard equilibrium constant $K_p^\ominus$ for the decomposition reaction is given by

$$K_p^\ominus = \left(\frac{p_{NH_3}}{p^\ominus}\right)^2 \cdot \left(\frac{p_{CO_2}}{p^\ominus}\right) \tag{1}$$

where p_{NH_3} and p_{CO_2} are the equilibrium partial pressures of NH_3 and CO_2 respectively at the experimental temperature, $p^\ominus = 100$ kPa. Let the total pressure of the decomposition reaction system be p_{total}. Since the vapor pressure of solid ammonium carbamate can be neglected, the total pressure of the system is

$$p_{\text{total}} = p_{NH_3} + p_{CO_2}$$

From the decomposition reaction equation of ammonium carbamate, it is known that

$$p_{NH_3} = p_{\text{total}} \text{ ; } p_{NH_3} = \frac{2}{3} p_{\text{total}}$$

Substituting into equation (1) yields:

$$K_p^{\ominus} = \left(\frac{2}{3} \times \frac{p_{\text{total}}}{p^{\ominus}}\right)^2 \times \left(\frac{1}{3} \times \frac{p_{\text{total}}}{p^{\ominus}}\right) = \frac{4}{27} \times \left(\frac{p_{\text{total}}}{p^{\ominus}}\right)^3 \tag{2}$$

It can be seen that when the system reaches equilibrium, the standard equilibrium constant $K_p^{\ominus}$ at the experimental temperature can be found by measuring its total equilibrium pressure.

The relationship between temperature and the equilibrium constant is described by the van't Hoff equation:

$$\frac{\mathrm{d}\ln K_p^{\ominus}}{\mathrm{d}T} = \frac{\Delta_r H_m^{\ominus}}{RT^2} \tag{3}$$

where $\Delta_r H_m^{\ominus}$ is the standard molar enthalpy change for the decomposition of ammonium carbamate, T is the thermodynamic temperature, and R is the molar gas constant with a value of $8.314 \text{ J} \cdot \text{K}^{-1} \cdot \text{mol}^{-1}$. $\Delta_r H_m^{\ominus}$ can be considered as a constant if the temperature does not vary over a wide range. Making the indefinite integral of equation (3), we get

$$\ln K_p^{\ominus} = -\frac{\Delta_r H_m^{\ominus}}{RT} + c \tag{4}$$

Plotting $\ln K_p^{\ominus}$ against $\frac{1}{T}$ yields a line with slope $-\frac{\Delta_r H_m^{\ominus}}{R}$, from which $\Delta_r H_m^{\ominus}$ can be determined.

From the standard equilibrium constant $K_p^{\ominus}$ at a certain temperature, the change in the standard-state free energy of reaction ($\Delta_r G_m^{\ominus}$) at that temperature can be calculated

$$\Delta_r G_m^{\ominus} = -RT\ln K_p^{\ominus} \tag{5}$$

Utilizing the average isobaric heat effect $\Delta_r H_m^{\ominus}$ of the decomposition reaction within the experimental temperature range and the change in standard molar Gibbs free energy $\Delta_r G_m^{\ominus}$ at a specific temperature, the standard molar entropy change $\Delta_r S_m^{\ominus}$ at that temperature can be approximately calculated

$$\Delta_r S_m^{\ominus} = \frac{\Delta_r H_m^{\ominus} - \Delta_r G_m^{\ominus}}{T} \tag{6}$$

Apparatus and reagents

Vacuum apparatus set
Gas tank
Sample tube
Digital vacuum manometer
Silicone oil
Manometer
Thermostatic bath
Three-way vacuum stopcock
Ammonium carbamate

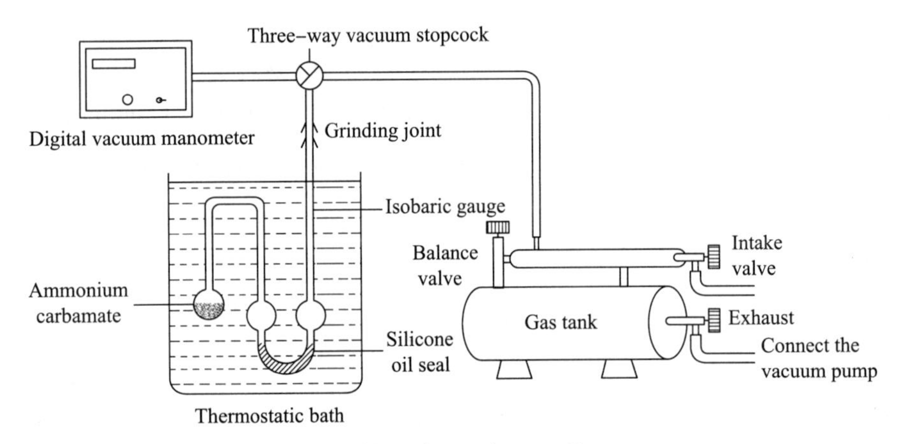

Fig. 2–1 Experimental setup diagram

Experimental procedures

1. Set the thermostat water bath temperature to 35.00 ℃, start stirring, open the digital manometer, and record atmospheric pressure and room temperature (record before and after the experiment, taking the average during data processing).

2. Evacuate

Close the intake valve, open the evacuation valve and balance valve, and start the vacuum pump. Evacuate until the precise digital pressure gauge reads approximately 90 kPa. Close the evacuation valve and the vacuum pump. Continue to evacuate for 10 minutes, then close the balance valve. Slowly open the intake valve until the oil seal liquid level is level, then close the intake valve. Adjust the intake valve repeatedly until the oil seal liquid level is level and remains unchanged for 10 minutes, confirming that the reaction has reached equilibrium. Record the decomposition pressure and reaction temperature.

3. Raise the temperature to 35.50 ℃, read the decomposition pressure after adjusting the sealing liquid level, and record it after 5 minutes. Read the decomposition pressure again 1 minute later.

4. Repeat step 3 for temperatures of 36.00, 36.50, and 37.00 ℃.

5. After measurements are completed, open the balance valve and intake valve to vent the system to the atmosphere.

Data recording and processing

1. Design an appropriate table to record the experimental data.

2. Plot $\ln K_p^\ominus$ against $1/T$ and calculate the isobaric reaction enthalpy change $\Delta_r H_m^\ominus$ for the decomposition of ammonium carbamate.

3. Calculate the standard molar Gibbs free energy change $\Delta_r G_m^\ominus$ and standard molar entropy change $\Delta_r S_m^\ominus$ for the decomposition of ammonium carbamate at 35.00 ℃.

Thinking and discussing

1. How to detect whether there is a gas leak in the measurement apparatus?

2. If too much air is introduced when air slowly enters the system, what phenomenon occurs, and how to overcome it?

3. Why is it necessary to evacuate the air from the system before the experiment? What is the impact if the air is not evacuated on the measurement results (pressure, equilibrium constant)?

4. What are the requirements for the sealing liquid used in the manometer in the experiment?

Reference materials

1. Joncich M J, Solka B H, Bower E. J Chem Edu, 1967(44): 598.

2. Northeast Normal University, et al. Physical Chemistry Experiments (2nd ed). Beijing: Higher Education Press, 1989: 126.

实验 3 凝固点降低法测定葡萄糖的摩尔质量

实验目的

1. 学会使用凝固点降低法测定葡萄糖的摩尔质量。

2. 通过实验加深理解稀溶液的依数性。

基本原理

在一定压力下，固液两相达成平衡时的温度称为液态纯物质的凝固点，固态纯溶剂与溶液呈平衡状态时的温度称为溶液的凝固点。根据稀溶液的依数性，指定溶剂的种类和数量后，向溶剂中加入非挥发性溶质，稀溶液的凝固点低于纯溶剂的凝固点，稀溶液的凝固点的降低值只取决于所含溶质分子的数目。对于稀溶液，凝固点降低值与溶液成分的关系可由式(1)计算

$$\Delta T_f = \frac{R\,(T_f^*)^2}{\Delta_{fus}H_m(A)} \times \frac{n_B}{n_B + n_A} \tag{1}$$

式中 ΔT_f 为凝固点降低值；T_f^* 为纯溶剂的凝固点；$\Delta_{fus}H_m(A)$ 为纯 A 的摩尔凝固焓；n_A 和 n_B 分别为溶剂和溶质的物质的量。当溶液浓度很稀时，$n_B \leqslant n_A$，则

$$\Delta T_f \approx \frac{R\,(T_f^*)^2}{\Delta_{fus}H_m(A)} \times \frac{n_B}{n_A} = \frac{R\,(T_f^*)^2}{\Delta_{fus}H_m(A)} \times \frac{M_A n_B}{m(A)} = \frac{R\,(T_f^*)^2 M_A}{\Delta_{fus}H_m(A)} m_B = k_f m_B \tag{2}$$

式中 M_A 为溶剂的摩尔质量；m_B 为溶质的质量摩尔浓度；k_f 称为质量摩尔凝固点降低常数。

如果已知溶剂的凝固点降低常数 k_f，并测得此溶液的凝固点降低值 ΔT_f，以及溶剂和溶质的质量 $m(A)$、$m(B)$，则溶质的摩尔质量由下式求得

$$M_B = k_f \frac{m(B)}{\Delta T_f m(A)} \tag{3}$$

应该注意，如果溶质在溶液中有解离、缔合、溶剂化和配合物形成等情况时，不能简单地运用公式(3)计算溶质的摩尔质量。显然，溶液凝固点降低法可用于溶液热力学性质的研究，例如电解质的电离度、溶质的缔合度、溶剂的渗透系数和活度系数以及无机化合物的结晶水数等。

图 3-1 是水的步冷曲线和葡萄糖溶液的步冷曲线。通过水的步冷曲线可以看出：将凝固点管插入-3 ℃的寒剂中以后，系统温度逐渐降低(A-B)；当温度降低至高于溶剂粗测凝固点 0.5 ℃（B 点处）时，将凝固点管从寒剂中取出放入空气套管中，温度缓慢下降(B-C)；在温度低于溶剂粗测凝固点 0.2 ℃左右时，加入一颗小冰粒作为晶种，促使溶剂结晶，由于结晶放出凝固热，使系统温度回升(C-D)；当固液两相平衡共存时，温度不再改变，此时的平衡温度即为溶剂的凝固点(D-E)。因此，在测量纯水的凝固点时，当贝克曼温度计的温度不变时，记录该温度，此温度即为纯水的凝固点。

葡萄糖溶液的步冷曲线可以看到类似的现象。将凝固点管插入-3 ℃的寒剂中以后，系统温度逐渐降低(a-b)；当温度降低至高于溶液的粗测凝固点 0.5 ℃(b 点处)时，将凝固点管放入空气套管中，温度缓慢下降(b-c)至过冷；在温度低于溶液粗测凝固点 0.2 ℃左右时，加入晶种，促使溶液析出溶剂晶体，系统温度回升(c-d)。温度达到最高(d 点)后，系统温度又开始下降(d-e)。其原因可根据稀溶液的依数性解释：随着溶剂的析出，溶液的浓度逐渐增大，凝固点逐渐降低。因此，可将温度回升后到达的最高温度作为溶液的凝固点。

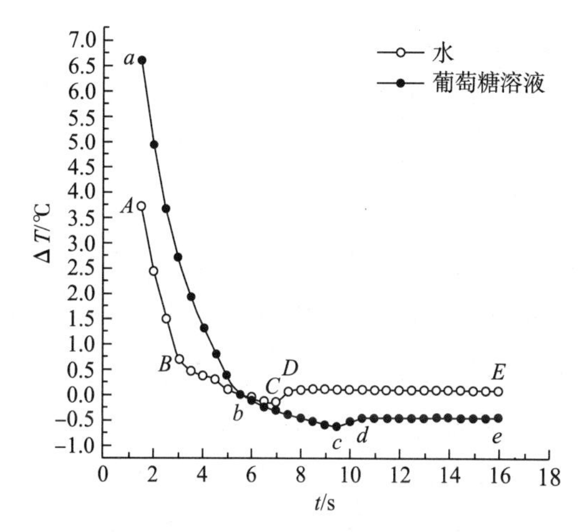

图 3-1 水和葡萄糖溶液的步冷曲线

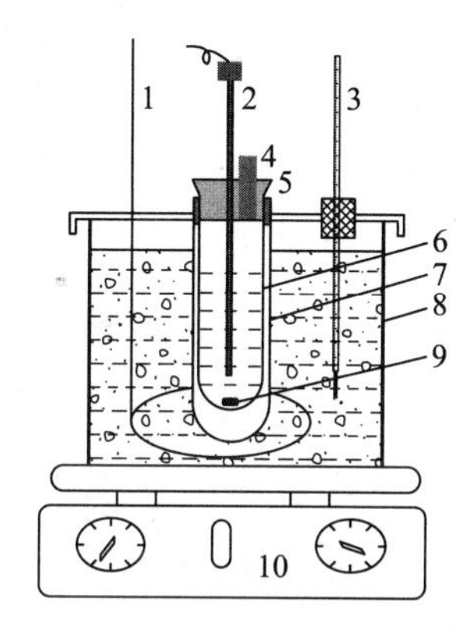

图 3-2 凝固点降低法测摩尔质量改进装置图

（1.寒剂搅拌器；2.贝克曼温度计；3.温度计；4.小胶塞；5.双口塞；6.凝固点管；7.空气套管；8.浴槽；9.磁子；10. 磁力搅拌器）

实验过程中应注意准确观察读取最高温度（*d* 点），因为稀溶液经过冷温度回升到最高温度后，系统温度下降很缓慢。

仪器和试剂

SWC-LG 凝固点实验装置	SWC-ⅡC 数字式贝克曼温度计
电子分析天平	温度计
移液管（25 mL）	葡萄糖（分析纯）
蒸馏水	NaCl

实验步骤

1. 葡萄糖的脱水处理

将一定质量葡萄糖放入 105 ℃ 烘箱中，烘干 8 h，以去掉结晶水。将烘干后的葡萄糖放入磨口瓶中密封保存，待用。用分析天平准确称量 1.2 g 左右葡萄糖，用于摩尔质量的测定。

2. 溶剂凝固点的测定

打开凝固点实验装置和数字式贝克曼温度计电源，进行预热，以保证实验数据的稳定。用食盐、冰、水调节寒剂温度为-3 ℃。将洁净干燥的空气套管放入寒剂中，用胶塞塞在管口，使管内温度降低。用温度计测量蒸馏水的温度并记录，之后用移液管向清洁、干燥的凝固点管内加入 25.00 mL 蒸馏水，并将清洗干净的磁子放入凝固点管内，将凝固点管也放入寒剂中，调节磁子的搅拌速度，使磁子在凝固点管底部快速转动，充分搅拌溶液。

将贝克曼温度计探头用酒精擦洗干净后，插入凝固点管并塞紧胶塞，使贝克曼温度计探头固定在凝固点管的中央并靠近凝固点管底部位置。观察凝固点管中贝克曼温度计的温度变化，当温度达到最低点后，又开始回升，回升到最高点后又开始下降，然后趋于平衡。记录平衡时的温度，即为蒸馏水的粗测凝固点。

取出凝固点管，使管中固体完全融化，再将凝固点管直接插入寒剂中，使溶剂较快冷却，当溶剂温度降至高于粗测凝固点 0.5 ℃时，取出凝固点管，迅速擦干后放入空气套管中，使水温均匀而缓慢降低。当温度降到低于近似凝固点 0.2 ℃时，通过胶塞上的大孔（不用时用小胶塞塞紧）向溶剂中加入一粒小绿豆粒大小的冰粒作为晶核，促使固体析出。仔细观察温度回升后贝克曼温度计的变化，直至稳定，此即为水的凝固点，记录该温度。重复上述操作 3 次，记录每次所测的纯水的凝固点，且保证水的凝固点之间相差在±0.003 ℃以内。

3. 溶液凝固点的测定

取出凝固点管，使管中的冰完全融化，加入已知质量的葡萄糖样品，按溶剂凝固点的测定方法进行测定，不同的是，溶液凝固点是取过冷后温度回升所达到的最高温度。重复测定 3 次，取其平均值。保证葡萄糖的凝固点之间相差在±0.003 ℃以内。

数据记录与处理

1. 将所得数据列表记录。

2. 根据书后附录水的密度公式计算 25 mL 水的质量。

3. 计算葡萄糖的摩尔质量。

4. 计算葡萄糖摩尔质量测定结果的相对误差。

思考与讨论

1. 利用凝固点降低这一稀溶液的依数性可以解决哪些实际问题？

2. 如何用凝固点降低这一稀溶液的依数性解释寒剂的控温作用？

3. 控制溶液的过冷深度都有哪些方法？

4. 溶液过冷太甚对实验结果有何影响？

参考资料

1. 傅献彩，侯文华. 物理化学（第六版）上册. 北京：高等教育出版社，2022.

2. 孙越，刘懿. 冯春梁. 介绍一个绿色物理化学实验——凝固点降低法测定葡萄糖的摩尔质量. 大学化学，2007，22(04)：44-46.

Experiment 3

Determination of molar mass of glucose by freezing point depression

Experimental purpose

1. Learn the use of the freezing point depression method to determine the molar mass of glucose.

2. Deepen understanding of the colligative properties of dilute solutions through experimentation.

Basical principles

The freezing point is the temperature of a liquid at which it changes its state from liquid to solid at a certain pressure. The temperature at which a solid pure solvent and a solution are in equilibrium is known as the freezing point of the solution. According to the colligative properties of dilute solutions, when a non-volatile solute is added to a solvent, the freezing point of the dilute solution is lower than that of the pure solvent, and the depression of the freezing point depends only on the number of solute molecules present. For dilute solutions, the relationship between the depression of the freezing point and the composition of the solution can be calculated using Equation (1).

$$\Delta T_{\mathrm{f}} = \frac{R\left(T_{\mathrm{f}}^{*}\right)^{2}}{\Delta_{\mathrm{fus}}H_{\mathrm{m}}(\mathrm{A})} \times \frac{n_{\mathrm{B}}}{n_{\mathrm{B}} + n_{\mathrm{A}}} \tag{1}$$

where, ΔT_{f} is the freezing point depression; T_{f}^{*} is the freezing point of pure solvent; $\Delta_{\mathrm{fus}}H_{\mathrm{m}}(\mathrm{A})$ represents the molar enthalpy of fusion of the solvent A; n_{A} and n_{B} are the amounts of substances in the solvent and solute respectively. When the solution concentration is very dilute, $n_{\mathrm{A}} \leqslant n_{\mathrm{B}}$, the equation simplifies to

$$\Delta T_{\mathrm{f}} \approx \frac{R\left(T_{\mathrm{f}}^{*}\right)^{2}}{\Delta_{\mathrm{fus}}H_{\mathrm{m}}(\mathrm{A})} \times \frac{n_{\mathrm{B}}}{n_{\mathrm{A}}} = \frac{R\left(T_{\mathrm{f}}^{*}\right)^{2}}{\Delta_{\mathrm{fus}}H_{\mathrm{m}}(\mathrm{A})} \times \frac{M_{\mathrm{A}}n_{\mathrm{B}}}{m(\mathrm{A})} = \frac{R\left(T_{\mathrm{f}}^{*}\right)^{2}M_{\mathrm{A}}}{\Delta_{\mathrm{fus}}H_{\mathrm{m}}(\mathrm{A})}m_{\mathrm{B}} = k_{\mathrm{f}}m_{\mathrm{B}} \tag{2}$$

where, M_{A} is the molar mass of solvent; m_{B} is the molality of solute; K_{f} is the cryoscopic constant, relates molality to freezing point depression (which is a colligative property).

If the cryoscopic constant k_{f} of the solvent is known, and the measured freezing point depression value ΔT_{f} of the solution, along with the masses m (A) and m (B) of the solvent and solute are available, the molar mass of the solute can be calculated using the following formula

$$M_B = k_f \frac{m(B)}{\Delta T_f m(A)} \tag{3}$$

It is important to note that when the solute undergoes dissociation, association, or forms complexes in the solution, Equation (3) cannot be applied directly.

The freezing point depression method can be used to study thermodynamic properties of solutions, such as the ionization degree of electrolytes, association degree of solutes, permeability coefficients, activity coefficients, and the number of crystalline water molecules in inorganic compounds.

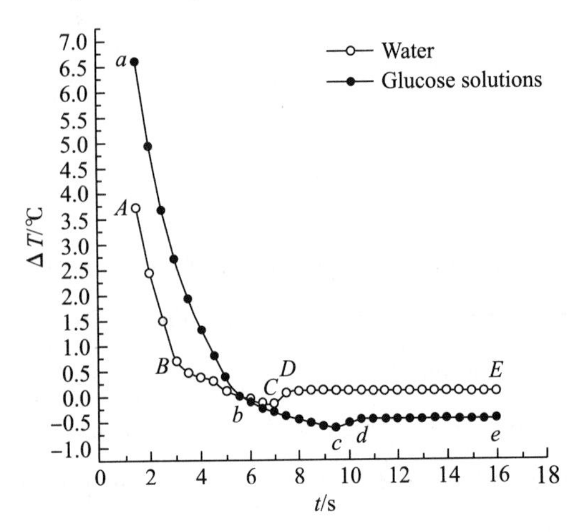

Fig. 3-1 Step-cooling curves of water and glucose solutions

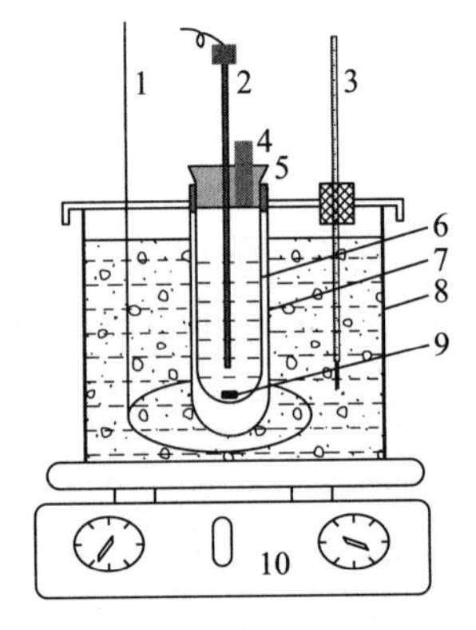

Fig. 3-2 Molar quality improvement device of freezing point depression method

(1. Cold mixer; 2. Beckmann thermometer; 3. Thermometer; 4. Small rubber plug; 5. Double mouth plug; 6. Freezing point tube; 7. Air sleeve; 8. Bath; 9. Magnetons; 10. Magnetic stirrer)

Fig. 3-1 illustrates the step-cooling curves for water and a glucose solution. In the cooling curve for water, upon inserting the freezing point tube into the cold agent at −3 ℃, the system temperature gradually decreases ($A-B$). When the temperature drops to 0.5 ℃ above the approximate freezing point of the solvent (point B), the freezing point tube is removed from the cold agent and placed into the air sleeve, resulting in a slow temperature decrease ($B-C$). When the temperature falls below the freezing point of the solvent, approximately 0.2 ℃, a small ice grain is introduced as a crystal seed to facilitate solvent crystallization. As a result of the crystallization releasing latent heat, the system temperature rises ($C-D$). At the point of coexistence between the solid and liquid phases, the temperature remains constant, representing the equilibrium temperature and the freezing point of the solvent ($D-E$). Consequently, when determining the freezing point of water, the temperature is recorded when the Beckmann thermometer remains steady.

A similar behavior is observed in the step-cooling curve of the glucose solution. After placing the freezing point tube into the cold agent at −3 ℃, the system temperature gradually decreases (*a*−*b*). Upon reaching a temperature 0.5 ℃ above the initially measured freezing point of the solution (point *b*), the freezing point tube is transferred to the air sleeve, leading to a slow temperature decrease (*b*−*c*) and supercooling. When the temperature drops below the freezing point of the solution, approximately 0.2 ℃, crystal seeds are added to induce the precipitation of solvent crystals, resulting in a system temperature rise (*c*−*d*). After reaching the highest temperature (point *d*), the system temperature begins to decline again (*d*−*e*). This phenomenon can be explained by the colligative property of dilute solutions: with the precipitation of the solvent, the solution concentration gradually increases, leading to a gradual decrease in the freezing point. Therefore, the highest temperature attained after the temperature rise can be considered the freezing point of the solution.

Throughout the experiment, meticulous attention should be paid to accurately observe and record the highest temperature (point *d*). This is crucial as following the rise in temperature of the dilute solution, the system temperature decreases very slowly.

Apparatus and reagents

SWC-LG freezing pointexperimental apparatus

SWC-ⅡC digital Beckmann thermometer

Electronic analytical balance

Thermometer

Pipette (25mL)

Glucose (AR)

Distilled water

NaCl

Experimental Procedures

1. Dehydration of glucose

Place a certain mass of glucose in a 105 ℃ oven for 8 hours to remove crystalline water. Seal the dried glucose in a ground-mouth bottle for later use. Weigh approximately 1.2 g of dried glucose with an analytical balance for molar mass determination.

2. Determination of solvent freezing point

Turn on the freezing point experimental apparatus and the digital Beckmann thermometer for preheating to ensure the stability of the test data. Adjust the temperature of the cold agent to −3 ℃ using salt, ice and water. Insert a clean and dry air sleeve into the cold agent, and seal it with a stopper to lower the temperature inside the tube.

Next, measure the temperature of distilled water using the thermometer. Subsequently, use a pipette to add 25.00 mL of distilled water into the freezing point tube. Simultaneously, place the cleaned magneton into the freezing point tube and insert the tube into the cold agent. Adjust the stirring speed of the magneton to facilitate rapid rotation at the tube's bottom, ensuring through stirring of the solution.

After cleaning the Beckman thermometer probe with alcohol, insert the freezing point tube and seal it with glue to fix the Beckman thermometer probe in the center and near the tube's bottom. Observe the temperature changes of the Beckmann thermometer in the freezing point tube. Record the temperature at the lowest point, marking the subsequent rise, fall, and equilibrium. This recorded temperature represents the approximate freezing point of distilled water.

Remove the freezing point tube, cover it with your hand to induce complete melting of the solid contents, and then promptly reinsert it into the cold agent for rapid solvent cooling. Retrieve the tube when the solvent temperature is 0.5 ℃ higher than the freezing point on the coarse side. Quickly dry the tube and place it into the air sleeve to achieve uniform but gradual cooling of the water temperature.

When the temperature drops to approximately 0.2 ℃ below the estimated freezing point, introduce a small ice grain (similar in size to a mung bean) into the solvent through the large hole on the rubber plug. This acts as a crystal nucleus, promoting solid precipitation. Thoroughly observe the Beckmann thermometer's changes after the temperature rise until stabilization, which indicates the freezing point of water, and record this temperature.

Repeat the above procedure three times, recording the freezing point of pure water each time, and ensure that the differences between the freezing points of water are within ±0.003 ℃.

3. Determination of the solution freezing point

Retrieve the freezing point tube, ensuring the ice is completely melted. Add the glucose sample with a known mass to the tube and proceed with the measurement using the solvent freezing point method. However, note that the freezing point of the solution is identified as the highest temperature reached after taking into account supercooling. Repeat this measurement three times and calculate the average value. Ensure that the discrepancies in the glucose freezing point do not exceed ±0.003 ℃.

Data recording and processing

1. Record the obtained data.

2. Calculate the mass of 25 mL of water using the density formula given in the Appendix.

3. Calculate the molar mass of the glucose.

4. Calculate the relative error of the molar mass determination.

Thinking and discussing

1. What practical problems can be solved by using the colligative property of freezing point depression?

2. How can the colligative property of freezing point depression explain the temperature control effect of the freezing agent?

3. What methods can be used to control the degree of supercooling of a solution?

4. How does excessive supercooling affect experimental results?

Reference materials

1. Fu X, Hou W. Physical Chemistry (6th ed, Volume I). Beijing: Higher Education Press, 2022.

2. Sun Y, Liu Y. A green physical chemistry experiment: Determination of the molar mass of glucose using the freezing point depression method. University Chemistry, 2007, (04): 44-46.

实验 4　双液系的气-液平衡相图

实验目的

1. 了解阿贝折光仪的原理及使用方法,学会由折光率(折射率)确定二元液体的组成。

2. 测定环己烷-无水乙醇二组分系统气-液平衡的相关数据,并绘出其沸点-组成相图。

3. 由相图确定环己烷-无水乙醇二组分系统的恒沸点及恒沸混合物的组成。

基本原理

研究多相系统的状态如何随温度、压力和浓度等变量的改变而发生变化,并用图形来表示系统状态的变化,这种图就叫相图。两种纯液体组分混合构成的二组分系统称为双液系。若两个纯液体组分能够按任意比例互相混溶,则称其为完全互溶双液系。若两个纯液体只能在一定比例范围互相混溶为一相,其他比例范围内为两相,则称其为部分互溶双液系。环己烷-无水乙醇二组分系统是完全互溶双液系。

当混合物的蒸气压等于外压时,混合物开始沸腾。此时的温度即为该混合物的沸点。在一定的外压下,纯液体的沸点有其确定值。但双液系的沸点不仅与外压有关,还与两种液体的相对含量有关。根据相律

$$f = C - \Phi + 2$$

二组分系统($C=2$)的自由度最多为3,即系统的状态可以由三个独立变量来决定,这三个独立变量通常为温度、压力和组成。这样,绘制二组分系统的状态图需要用具有三个坐标的立体图来表示。若温度、压力、组成中的任意一个变量为常量,那么二组分系统的状态图就可以用二维图形来描述。例如,在一定温度下,可以绘出系统压力 p 和组分 x 的关系图($p-x$ 图);如系统的压力确定,则可绘出温度 T 和组成 x 的关系图($T-x$ 图);若系统的组成确定,则可以绘制出系统温度 T 和系统压力 p 的关系图($T-p$ 图)。常用的是 $p-x$ 和 $T-x$ 图。二组分完全互溶双液系的 $T-x$ 图的示意图如图 4-1 至图 4-3。

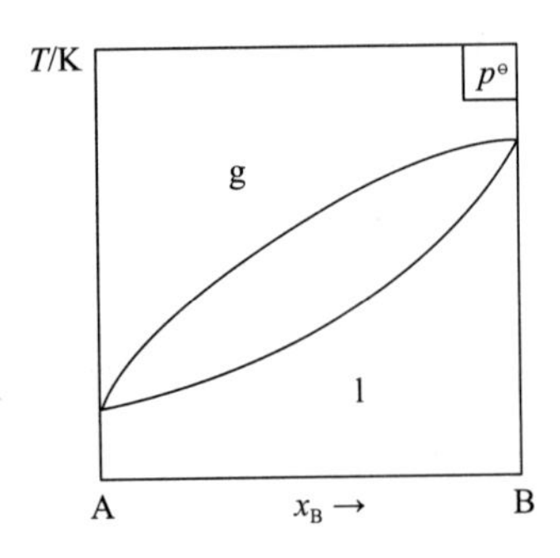

图 4-1　理想完全互溶(或偏差不大)相图

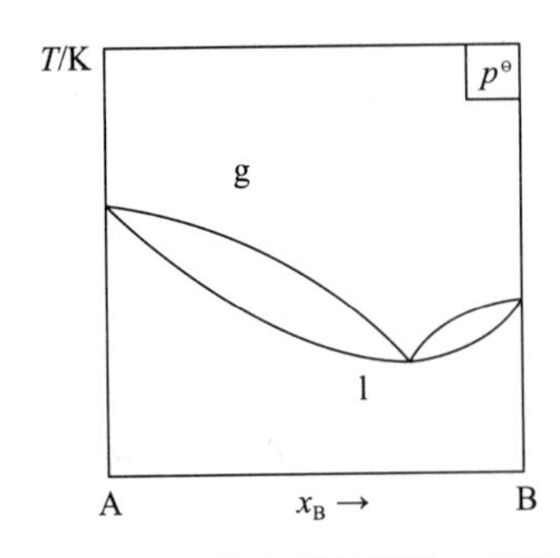

图 4-2　具有最低恒沸点相图(正偏差很大)

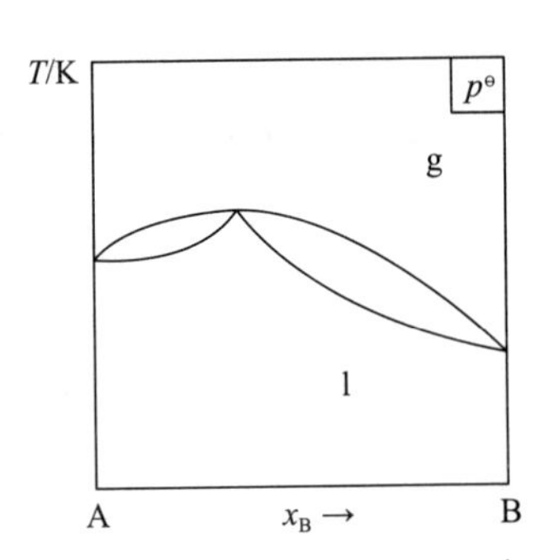

图 4-3　具有最高恒沸点相图(负偏差很大)

本实验所测的环己烷-无水乙醇二组分系统的 $T-x$ 图是一个典型的具有最低恒沸点的相图(属于正偏差较大的非理想完全互溶双液系)。

实验装置如图 4-4 所示。这是一只带回流冷凝管的长颈圆底烧瓶。冷凝管底部有一半球形小槽,用以收集冷凝下来的气相样品。

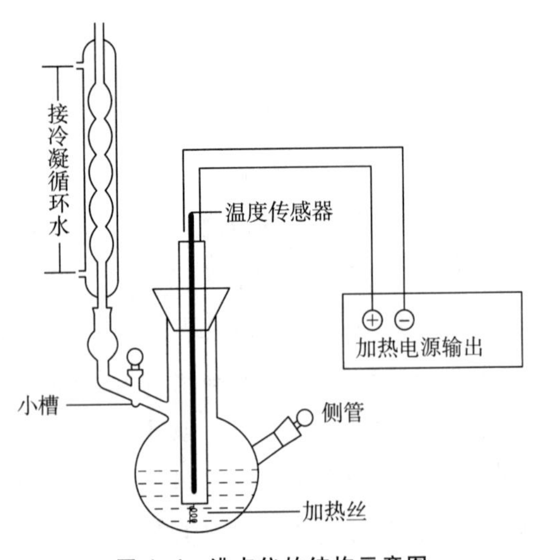

图 4-4　沸点仪的结构示意图

折光率是物质的一个重要物理常数,从折光率可以定量地分析溶液的成分,检验物质的纯度；另外还可以求算物质摩尔折射度、分子的偶极矩及测定分子结构等。由于本实验选用的环己烷和无水乙醇的折光率相差颇大，而测定折光率又只需要少量样品，因此可以测定一系列不同配比的环己烷-无水乙醇溶液的沸点和折光率，在组成-折光率工作曲线上查出所测折光率对应的组成，就可绘制环己烷-无水乙醇二组分系统的 $T-x$ 相图。压力不同时，其 $T-x$ 相图将略有差异。表 4-1 为 $p^{\ominus}$下乙醇-环己烷溶液的恒沸点、组成的数据。[引自：①Advances in Chemistry. Series 116 (Compiled by Horsley L H). Azeotropic Data-Ⅲ. Washington D C: American Chemical Society, 1973: 136. ②Advances in Chemistry. Series 6 (Compiled by Horsley L H). Azeotropic Data. Washington D C: American Chemical Society: 62.]

表 4-1 在 $p^{\ominus}$下乙醇-环己烷溶液的恒沸点、组成数据

沸点/℃	乙醇质量分数/%	$x_{环己烷}$
64.9	40	/
64.8	29.2	0.570
64.8	31.4	0.545
64.9	30.5	0.555

在 $p^{\ominus}$、25 ℃时，环己烷-无水乙醇系统的折光率-组成关系如表 4-2 所示。[引自：Timmermans J (Ed). The Physico-Chemical Constants of Binary Systems in Concentrated Solutions. London: Interscience Publishers, 1959(2): 36.]在实际实验时，若使用的试剂与表 4-2 中试剂的批次不同，需要测试实际环己烷折光率，从而利用实测值对表中数据进行校正。

表 4-2 25 ℃时环己烷-无水乙醇系统的折光率-组成关系

$x_{无水乙醇}$	$x_{环己烷}$	n_D^{25}
1.0000	0.0000	1.35935
0.8992	0.1008	1.36867
0.7948	0.2052	1.37766
0.7089	0.2911	1.38412
0.5941	0.4059	1.39216
0.4983	0.5017	1.39836
0.4016	0.5984	1.40342
0.2987	0.7013	1.40890
0.2050	0.7950	1.41356
0.1030	0.8970	1.41855
0.0000	1.0000	1.42338

折光率可以采用阿贝折光仪进行测量。在使用阿贝折光仪进行测量时,需要用的样品量少,数滴液体即可进行测量;测量方法简便,读数准确,重复性好,无需特殊光源设备,普通日光或其他光源即可;棱镜的夹层可通恒温水,保持所需的恒定温度。

阿贝折光仪是根据光的全反射原理设计的,即利用全反射临界的测定方法测定未知物质的折光率,其外形如图 4-5 所示(详细的阿贝折光仪光学原理,请参看仪器说明书或有关资料)。

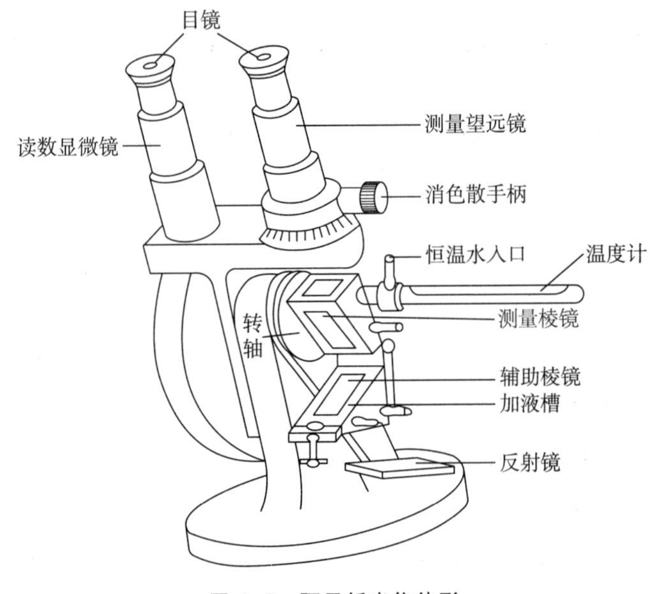

图 4-5　阿贝折光仪外形

使用阿贝折光仪时,先用橡皮管将仪器上测量棱镜和辅助棱镜上保温夹套的进水口与超级恒温槽串接起来(确保连接可靠),恒温温度以折光仪上的温度计读数为准,一般选用(20.0 ± 0.1) ℃或(25.0 ± 0.1) ℃。然后,打开折射棱镜部件,移去擦镜纸。检查上、下棱镜表面,用滴管滴加少量丙酮(或无水乙醇)清洗镜面,用洗耳球将镜面吹干或用擦镜纸轻轻吸干镜面。(注意:用滴管时勿使管尖碰触镜面;测完样品后必须仔细清洁两个镜面,但切勿用滤纸擦拭。)接下来,滴加 1～2 滴试样于棱镜的毛玻璃面上,锁紧棱镜。调节反射镜使入射光线达最强,通过目镜观察视场,同时旋转调节手轮,使明暗分界线落在交叉线视场中,如从目镜中看到视场是暗的,可将调节手轮逆时针旋转;如是明亮的,则顺时针旋转。明亮区域在视场的顶部。在明亮视场下旋转目镜,使视场中的交叉线最清晰。因光源为白光,故在交叉线处有时呈现彩色,旋转消色散手柄使彩色消失,使视场中明暗两部分具有良好的反差,明暗分界线具有最小的色散,明暗清晰,再转动棱镜使明暗界线正好与目镜中的十字线交点重合(图 4-6),这时从读数显微镜即可读出被测物的折光率 D(每次测定时,两个棱镜都要啮紧,防止两棱镜所夹的液层成劈状,影响数据重复性)。为了数据的准确,必须按上述步骤测定 3 次样品,取其平均值。在测量结束后,必须用少量丙酮(或无水乙醇)和擦镜纸清洗镜面。合上折射镜部件前须在两个棱镜之间放一张擦镜纸。在测量过程中需要注意的是,折光仪要放置在不被日光直接照射或靠近热的光源(如电灯泡)的位置,以免影响测定温度。

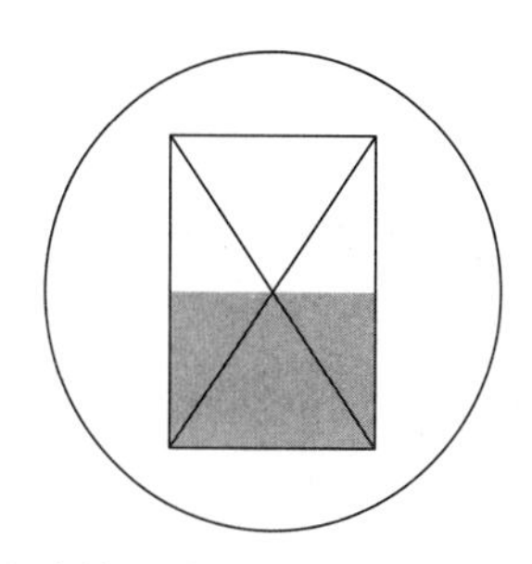

图 4-6　准确的明暗分界线与交叉线位置示意图

仪器和试剂

沸点测定仪　　数字阿贝(Abbe)折光仪(棱镜等温)

超级恒温槽　　调压变压器

容量瓶(100 mL)　　玻璃漏斗(直径 5 cm)

滴管　　丙酮(分析纯)

环己烷(分析纯)　　无水乙醇(分析纯)

实验步骤

1. 配制溶液(用 100 mL 容量瓶)

按表 4-3 所列数据配制 8 瓶环己烷-无水乙醇溶液各 100 mL。

表 4-3　配制 100 mL 环己烷-无水乙醇溶液所需环己烷的体积

瓶号	1	2	3	4	5	6	7	8
$V_{环}$/mL	10	23	31	45	70	93	96	98

2. 沸点的测定

将样品加到沸点仪中,并使传感器和加热丝浸入溶液内。打开冷凝水,打开电源开关调节“加热电源调节”旋钮,逐渐加大电压使溶液慢慢开始沸腾,蒸气在冷凝管中回流的高度保持 1.5 cm 左右。当温度恒定后(系统达到平衡),记下温度读数(沸点),停止加热。用湿毛巾将气相冷凝液的凹槽进行冷却,用装有冰水的 250 mL 烧杯套在烧瓶下方,使溶液快速冷却,以防溶液挥发,给组成测定带来误差。

需要注意的是,为加速达到气液平衡,可倾斜蒸馏瓶,将凹槽中的气相冷凝液倾回蒸馏瓶中,重复 3 次 (加热时间不宜太长,以免物质挥发),每次倾入之前看一下温度读数

(沸点),3 次读数基本不变,可认为系统达到平衡。

3. 折光率的测定

将控制折光仪温度的恒温槽温度设为(25.0 ± 0.2)℃。用滴管取冷却的气相样品,测定其折光率。每个样品(加一次样)读 3 次数,取平均值(相邻读数不能相差 0.0002)。用另一支滴管取液相溶液,测其折光率,每个样品(加一次样)读 3 次数取平均值(相邻读数不能相差 0.0002)。

取样和测量动作要迅速。每次取样前可用洗耳球吹干滴管,将其干燥。每测完一个样品,阿贝折光仪毛玻璃面也要用洗耳球吹干(注意:一定要先测气相,后测液相)。

气相、液相折光率都测完后,将沸点仪中的溶液倒入原来装样品的容量瓶中(尽量倒干净,但不必吹干),以便循环使用。换下一个样品,重复 2、3 步操作。8 个样品全部测完,经指导老师查合格后,方可结束实验。

数据记录与处理

1. 工作曲线的绘制

实验前,要根据文献中报道的 25 ℃时环己烷-无水乙醇系统的折光率、组成数据(列于表 4-2 中)绘制“折光率-组成”工作曲线。

2. 实验数据的记录和处理

结合实验的基本原理、步骤,合理地设计数据表,将实验数据及处理结果列入表中。

3. T-x 相图的绘制

在“折光率-组成”的工作曲线上,根据折光率的数值找到对应的组成,并绘出环己烷-无水乙醇的 T-x 相图,由相图读出最低恒沸点和恒沸混合物的组成。

思考与讨论

1. 在测定恒沸点时,若溶液过热或出现分馏现象,那么绘制出的相图的形状会发生什么变化?

2. 为什么工业上常用 95% 的乙醇? 只用精馏含水酒精的方法是否可能获得无水乙醇? 如何获得无水乙醇? (可参阅:Buckingham J. Dictionary of Organic Compounds. 5th ed. Chapman and Hall, 1982: 2486)。

3. 试设计其他方法测定气-液两相组成,并与本实验描述的方法进行对比,讨论两者的优缺点。

4. 实验所用样品循环使用对实验结果是否有影响?

5. 每测完一个样品要将溶液倒回原瓶中,倾倒后蒸馏瓶底部的少量残留液对下一样品的测定是否有影响?

6. 讨论本实验误差的主要来源。

参考资料

1. 傅献彩，侯文华. 物理化学(第六版)上册. 北京：高等教育出版社，2022.

2. Daniel F, Alberty R A, Williams J W, et al. Experimental Physical Chemistry. 7th ed. New York: McGraw-Hill, Inc, 1970: 61.

3. 杨晓晔，杨志明. 对双液系气液平衡相图实验的一点意见——该不该对沸点进行压力校正. 贵州师范大学学报：自然科学版，1995(2)：34.

Experiment 4

Binary gas-liquid phase diagram

Experimental purpose

1. Understand the principles and usage of the Abbe refractometer, and learn how to determine the composition of binary liquids by refractive index.

2. Determine the gas-liquid equilibrium data of the cyclohexane-anhydrous ethanol binary system and plot its boiling point-composition phase diagram.

3. Determine the constant boiling point and composition of the azeotropic mixture in the cyclohexane-anhydrous ethanol binary system using the phase diagram.

Basical principles

The study of how the state of a multiphase system changes with variables such as temperature, pressure, and concentration is represented graphically using phase diagrams. A binary system composed of two pure liquid components is referred to as a binary system. If two pure liquid components can mix in any proportion, it is called a completely miscible binary system. The cyclohexane-anhydrous ethanol binary system is a completely miscible binary system.

The boiling point of a mixture is reached when its vapor pressure equals the external pressure. The boiling point of a pure liquid has a constant value under a specific external pressure. However, in a binary system, the boiling point depends not only on external pressure but also on the relative content of the two liquids. According to phase law

$$f = C - \Phi + 2$$

For a binary system ($C=2$), the maximum degrees of freedom (f) are 3, meaning the system's state can be determined by three independent variables—usually temperature, pressure, and composition. Therefore, a three-coordinate system is needed to represent the state of a binary system. Commonly used representations are p-x (pressure-composition) and T-x (temperature-composition) diagrams. The T-x phase diagram of the two-component fully miscible two-liquid system is as follows

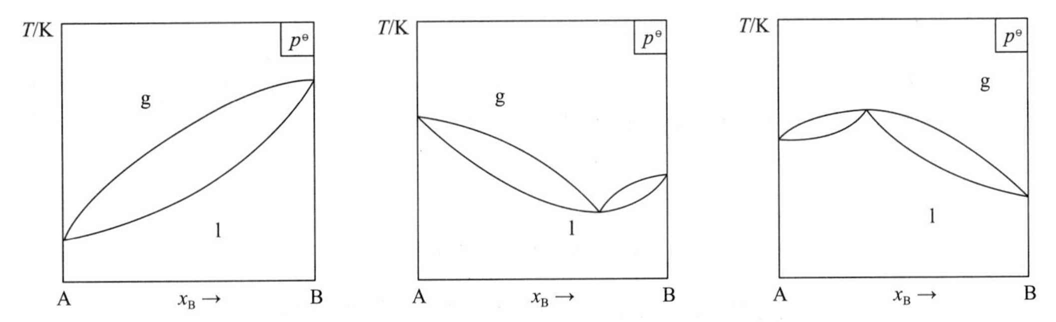

Fig. 4-1 Ideal perfectly miscible (or slightly deviated) phase diagram

Fig. 4-2 Phase diagram with lowest constant boiling point (positive deviation is large)

Fig. 4-3 Phase diagram with the highest constant boiling point (large negative deviation)

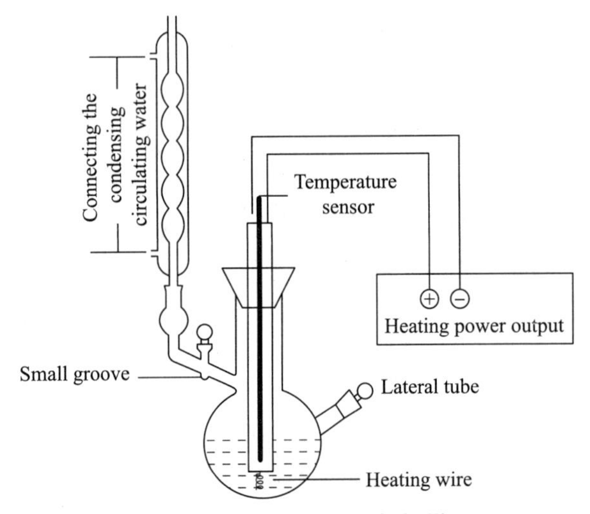

Fig. 4-4 Structure scheme of ebulliometer

The temperature-composition phase diagram of the cyclohexane-anhydrous ethanol two-component system, as measured in this experiment, represents a typical phase diagram exhibiting the lowest constant boiling point. This system is characterized as a non-ideal completely miscible two-liquid system with a significant positive deviation.

Fig. 4-4 illustrates the experimental setup, featuring a long-necked round-bottomed flask equipped with a reflux condensing tube. A half-spherical chamber at the base of the condensing tube serves to collect the condensed gas-phase samples.

In this experiment, cyclohexane and anhydrous ethanol were chosen, and despite their notable difference in refractive indices, only a limited number of samples were required to determine the boiling point and refractive index of various cyclohexane-anhydrous ethanol solutions. The corresponding compositions are then identified on the composition-refraction working curve, enabling the construction of the temperature-composition phase diagram for the gas-liquid system. It's worth noting that variations in pressure will introduce slight differences in the phase diagram of the two-liquid system. Table 4-1 presents the azeotropic data for ethanol-cyclohexane solutions at constant boiling points under atmospheric pressure. [referenced from: ①Advances in Chemistry. Series 116 (Compiled by Horsley L H). Azeotropic Data - Ⅲ. Washington D C: American Chemical Society, 1973: 136. ②Advances in Chemistry. Series 6 (Compiled by Horsley L H). Azeotropic Data. Washington D C: American Chemical Society. 62.]

Table 4-1 Azeotropic data for ethanol-cyclohexane solutions at constant boiling points at 101.325 kPa

Boiling point/℃	Ethanol mass fraction /%	$x_{\text{cyclohexane}}$
64.9	40.0	/
64.8	29.2	0.570
64.8	31.4	0.545
64.9	30.5	0.555

Table 4-2 The relationship between the refractive index and composition of the cyclohexane-anhydrous ethanol system at 25 ℃

$x_{\text{ethyl alcohol}}$	$x_{\text{cyclohexane}}$	n_D^{25}
1.0000	0.0000	1.35935
0.8992	0.1008	1.36867
0.7948	0.2052	1.37766
0.7089	0.2911	1.38412
0.5941	0.4059	1.39216
0.4983	0.5017	1.39836
0.4016	0.5984	1.40342
0.2987	0.7013	1.40890
0.2050	0.7950	1.41356
0.1030	0.8970	1.41855
0.0000	1.0000	1.42338

Cite: Timmermans J (Ed). The Physico-Chemical Constants of Binary Systems in Concentrated Solutions. London: Interscience Publishers, 1959(2): 36.

The refractive index stands as a crucial physical constant for substances. This parameter allows for a quantitative analysis of solution composition, providing a means to assess substance purity. Furthermore, it facilitates the calculation of molar refraction, dipole moment, and molecular structure. The Abbe refractometer is a versatile optical instrument commonly employed in physical and chemical experiments due to its efficiency and simplicity.

(Ⅰ) Structure of the Abbe refractometer

The Abbe refractometer is designed based on the principle of total light reflection, wherein the refractive index of unknown substances is measured by determining the critical angle for total reflection. The instrument's structure is depicted in Fig. 4-5. For a detailed understanding of the optical principles behind the Abbe refractometer, please consult the instrument manual or relevant materials.

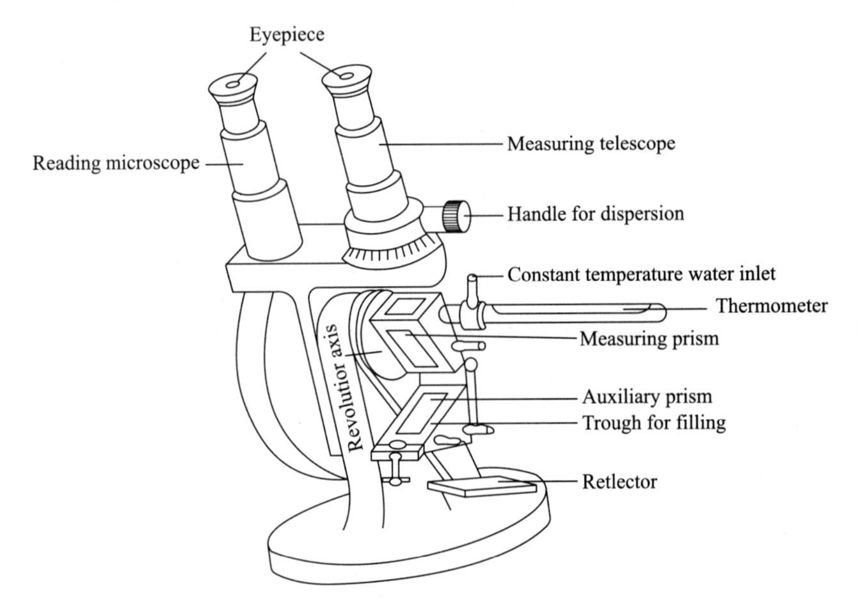

Fig. 4-5 Outline of Abbe refractometer

(Ⅱ) Usage methods and precautions

1. Connect the water inlet of the measuring prism on the instrument and the insulation jacket on the auxiliary prism in series with the super constant temperature groove using a rubber tube (ensure a secure connection). Maintain a constant temperature as indicated by the thermometer reading on the refractometer, typically set at (20.0±0.1)℃ or (25.0±0.1)℃.

2. Open the refracting prism component and remove the protective paper. Inspect the surfaces of the upper and lower prisms, clean them with a small amount of acetone (or anhydrous alcohol) using a dropper, and dry the surfaces with an ear ball or gently blot them with mirror wiping paper. (Note: Avoid allowing the dropper tip to touch the prism surface;

After testing the sample, ensure both prisms are meticulously cleaned, refraining from using filter paper for wiping.)

3. Apply 1 ~ 2 drops of the sample onto the ground glass surface of the prism to secure it.

4. Adjust the reflector to maximize incident light, observe the field of view through the eyepiece, and rotate the adjustment handwheel to align the terminator with the cross-line field of view. If the field of view appears dark, rotate the adjustment handwheel counter clockwise; if it is bright, rotate it clockwise. Bright areas should be at the top of the field of view. Rotate the eyepiece under the bright field of view to ensure the clearest crossing lines. As the light source is white, occasional color may appear at the crossing lines. Rotate the achromatic handle to eliminate the color, ensuring good contrast between light and dark areas with minimal dispersion for a clear chiaroscuro. Then, rotate the prism to align the chiaroscuro with the cross line in the eyepiece. At this point, the refractive index D of the measured object can be read from the microscope on the scale (ensure both prisms are tightened each time during measurement to prevent the liquid layer between them from splitting, affecting data repeatability). For accurate results, measure the sample three times and calculate the average.

5. After measurement, use a small amount of acetone (or anhydrous alcohol) and wiping paper to clean the prism. Place a sheet of mirror wiping paper between the two prisms before closing the refractor component.

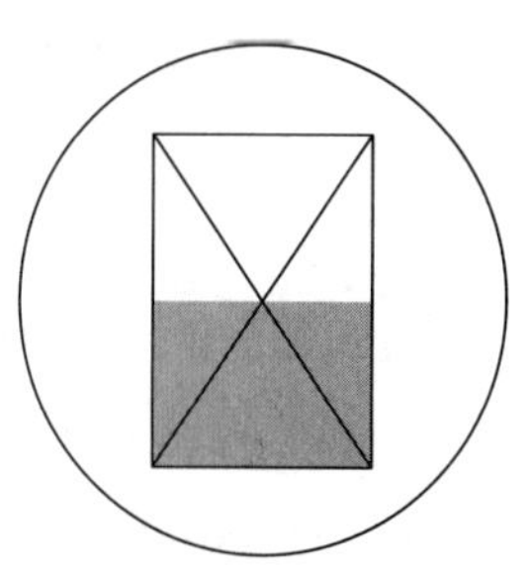

Fig. 4-6 Exact location of the terminator and crossover lines

6. The refractometer should not be exposed directly to sunlight or placed near heat sources such as light bulbs to avoid interference with temperature measurements.

Apparatus and reagents

Ebulliometer	Abbe refractometer (Prism constant temperature)
Super thermostatic bath	Voltage regulating transformer
Volumetric bottle (100 mL)	Glass funnel (diameter 5 cm)
Dropper	Cyclohexane (AR)
Anhydrous ethanol (AR)	Acetone (AR)

Experimental Procedures

1. Prepare solution (100 mL volumetric bottle)

Prepare eight bottles of cyclohexane-anhydrous ethanol solutions, each containing 100 mL, according to the quantities listed in Table 4-3.

Table 4-3 Volume of cyclohexane required to prepare 100 mL cyclohexan-anhydrous ethanol solution

No.	1	2	3	4	5	6	7	8
$V_{cyclohexane}$/mL	10	23	31	45	70	93	96	98

2. Measure the boiling point

Add the sample to the boilometer, ensuring the sensor and heating wire are immersed. Start the condensate water, turn on the power switch, and adjust the "heating power adjustment" knob, gradually increasing the voltage (not exceeding 12 V).

Allow the solution to slowly boil until a vapor height of approximately 1.5 cm in the condensate pipe is reached. Record the constant temperature (boiling point) when the system is in equilibrium.

Stop heating, cool the condensate groove with a wet towel, and quickly cool the solution with a 250 mL beaker of ice water under the flask to prevent errors in composition determination due tovolatility.

Note: Tilt the distillation bottle to aid gas-liquid balance, pouring condensate in the gas phase back into the distillation bottle. Repeat three times, ensuring stable temperature readings.

3. Measure refractive index

Set the constant temperature tank controlling the refractometer at (25.0±0.2)℃.

Measure the refractive index of the gas condensate and liquid solution using separate droppers. Take three readings for each sample, averaging them (adjacent readings should not differ by 0.0002). The refraction of liquid solution is measured with another dropper. Each sample (plus one sample) is read 3 times to take the average value (adjacent readings cannot differ by 0.0002).

Dry the dropper before each sample, and after testing, dry the frosted glass surface of the Abbe refractometer.

Pour the solution from the boilometer back into the original volume bottle for recycling.

Data recording and processing

1. Draw a work curve

Draw a work curve for "refraction ratio-composition" before the experiment using literature values listed in Table 4-2.

2. Experimental data recording and processing

Reasonable design of data recording table, experimental data and processing results included in the table.

3. Draw the $T-x$ phase diagram

Draw the $T-x$ phase diagram of cyclohexane-anhydrous ethanol, locating the lowest constant boiling point and the composition of the azeotropic mixture.

Thinking and discussing

1. How would overheating or fractional distillation affect the shape of the phase diagram?

2. Why is 95% ethanol commonly used in industries? Is it possible to obtain anhydrous ethanol by distilling aqueous ethanol? How can anhydrous ethanol be obtained? (See: Buckingham J. Dictionary of Organic Compounds. 5th ed. Chapman and Hall, 1982: 2486).

3. Propose alternative methods to determine gas-liquid phase compositions and compare their advantages and disadvantages with the described method.

4. Discuss the potential impact of recycling the same samples on the experimental results.

5. Could the residual liquid left at the bottom of the distillation flask after pouring the solution back into the bottle affect the measurement of the next sample?

6. Identify and discuss the main sources of error in this experiment.

Reference materials

1. Fu X, Hou W. Physical Chemistry (6th ed, Volume I). Beijing: Higher Education Press, 2022.

2. Daniel F, Alberty R A, Williams J W, et al. Experimental Physical Chemistry. 7th ed. New York: McGraw-Hill, Inc, 1970: 61.

3. Yang X, Yang Z. Some opinions on the gas-liquid equilibrium phase diagram experiment for binary systems-should boiling points be pressure-corrected? Journal of Guizhou Normal University(Natural Science), 1995(2): 34.

实验 5 $KCl-HCl-H_2O$ 三组分系统相图的绘制

实验目的

1. 掌握相律和用等边三角坐标表示三组相图的方法。

2. 了解湿固相法的原理,学会确定溶液中纯固相组成点的方法。

3. 绘制 $KCl-HCl-H_2O$ 三组分系统相图。

基本原理

对于三组分系统($C=3$),根据相律$f=C-\Phi+2$,系统最多可能有四个自由度(温度、压力、两个浓度项),用三维空间的立体模型已不足以表示这种相图。保持温度(或压力、或一个浓度项)不变,则其条件自由度$f^{*}=4-\Phi$,可用立体模型表示其相图。当处于等温等压条件时,其条件自由度$f^{**}=3-\Phi$。系统最大条件自由度$f_{max}^{**}=3-1=2$,因此,最多有两个浓度变量,可用平面图表示系统状态和组成之间的关系。通常在平面图上用等边三角形来表示各组分的浓度。如图 5-1 所示,等边三角形的三个顶点分别表示纯组分 A、B 和 C,三条边 AB、BC、CA 分别表示 A 和 B、B 和 C、C 和 A 所组成的二组分系统,三角形内任何一点都表示三组分系统。三角形内任一点 O,引平行于各边的平行线 a、b 和 c,根据几何学的知识可知,a、b 和 c 的长度之和应等于三角形一边之长,即 $a+b+c=AB=BC=CA=1$。因此,O 点的组成可由这些平行线在各边上的截距 a'、b'和 c' 来表示。通常是沿着逆时针的方向在三角形的三边上标出 A、B 和 C 三个组分的质量分数。即从 O 点作与 BC 的平行线,在 AC 线上的长度 a',即为 A 的质量分数 w_A;从 O 点作与 AC 的平行线,在 AB 线上的长度 b',即为 B 的质量分数 w_B;从 O 点作与 AB 的平行线,在 BC 线上的长度 c',即为 C 的质量分数 w_C。

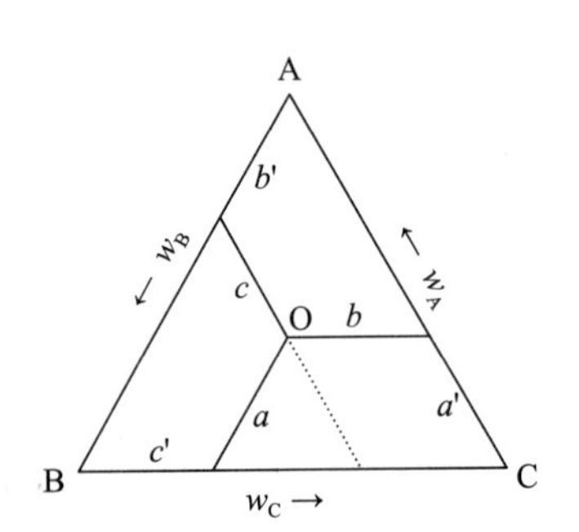

图 5-1 等边三角形法表示三元相图

由 KCl、HCl、H_2O 形成的三组分系统,当 HCl 的含量不太高时,HCl 完全溶于水形成盐酸溶液。该溶液与 KCl 有共同的负离子 Cl^-,所以当饱和的 KCl 水溶液中加入盐酸时,由于同离子效应,使 KCl 的溶解度降低。本实验即研究在不同浓度的盐酸溶液中 KCl 的溶解度,并通过此实验了解水盐系统的相图绘制方法和一般性质。

为了分析平衡系统各相的成分,可以采取分离各相方法,例如对于液体可以用分液漏斗来分离。由于固体上总会带有一些母液很难分离干净,而且有些固相极易风化潮解,不能离开母液而稳定存在,因此固相的分离比较困难。这时,常常采用不用分离母液而确定固相组成的湿固相法。这一方法是根据带有饱和溶液的固相的组成点，必定处于饱和溶液的组成点和纯固相的组成点的连接线上，因此同时分析几对饱和溶液和湿固相的成分，将它们连成直线，这些直线的交点即为纯固相成分。本实验就是采用这种方法求取固相组成的。

仪器和试剂

100 mL 磨口锥形瓶	50 mL 磨口锥形瓶
2 mL 移液管	恒温槽
KCl(分析纯)	$AgNO_3$(分析纯)
NaOH(分析纯)	盐酸

实验步骤

1. 配制 0.1 mol · L^{-1} $AgNO_3$、0.1 mol · L^{-1} NaOH 以及浓度分别为 1、2、4、6、8 和 12 mol · L^{-1} 的盐酸溶液。

2. 在 6 个洗净的 100 mL 磨口锥形瓶中，分别注入 25 mL 浓度为 1、2、4、6、8 mol · L^{-1} 的盐酸溶液，剩下一个加 25 mL 煮沸后放冷的蒸馏水。

3. 在每个锥形瓶中加入约 10 g 的 KCl 固体，然后将每个锥形瓶置于约 30 ℃的水浴中，不断摇荡 5 min 后，取出置于 25 ℃的恒温水浴中，继续摇荡 3 min，然后在恒温水浴中继续恒温，静置 5 min，待溶液澄清后，用滴管在每个锥形瓶中取饱和溶液约 0.5 g，放入已知质量的 50 mL 磨口锥形瓶中(或用称量瓶也可)，于分析天平上称量，记录每个样品的质量。

4. 取饱和溶液样品的同时，用玻璃勺取湿固相约 0.2～0.3 g 样品于另一已知质量的称量瓶中，用分析天平称其质量。在取样时应注意下述问题：

(1) 系统的温度不能改变，因此不要将锥形瓶离开恒温水槽。

(2) 取样时固相可以带有母液，但饱和溶液不能带有固相。因此，取样时要特别小心谨慎，等固相完全下沉以后再进行取样。

(3) 取样的滴管的温度应比系统的温度高些，以免饱和溶液在移液管中析出结晶，引起误差。为此，取样滴管最好先预热一下。但滴管温度也不能太高，一方面避免改变系统的温度，另一方面防止水分蒸发改变浓度。

5. 将已称过质量的湿固相样品用约 50 mL 的蒸馏水洗到 250 mL 锥形瓶中，进行滴定分析。先以酚酞作指示剂，用 0.1 mol · L^{-1} NaOH 溶液滴定样品中的酸量，至终点后，记下滴定用去的 NaOH 溶液的体积。然后再向锥形瓶中滴入 1～2 滴稀 HNO_3 溶液，使系统带微酸性。最后以 K_2CrO_4 作为指示剂，用 $AgNO_3$ 溶液滴定样品中 Cl^-的浓度，记下所用

$AgNO_3$ 的浓度及消耗的体积。

数据记录与处理

1. 根据本实验的基本原理和实验步骤，设计数据记录表并填入数据。

2. 用下列公式计算每个饱和溶液样品及湿固相样品中 HCl、KCl 和 H_2O 的质量分数，并用表列出。

$$w(\mathrm{HCl})/(\%) = (c_1V_1 \times 36.5)/(m \times 1\,000)$$

$$w(\mathrm{KCl})/(\%) = (c_2V_2 - c_1V_1) \times 74.56/(m \times 1\,000)$$

$$w(\mathrm{H_2O})/(\%) = 100 - w(\mathrm{HCl}) - w(\mathrm{KCl})$$

式中：c_1 为滴定时所用 NaOH 的浓度（$mol \cdot L^{-1}$），V_1 为滴定时所消耗 NaOH 的体积（mL）；c_2 为 $AgNO_3$ 的浓度（$mol \cdot L^{-1}$），V_2 为 $AgNO_3$ 的体积（mL）；m 为样品质量（g）；74.56 及 36.5 分别为 KCl 和 HCl 的摩尔质量。

3. 查阅 25 ℃时 KCl 在水中的溶解度，并将其换算成质量分数。

4. 将第 2～3 步所得的结果标记在三角相图上，并将各个饱和溶液的组成点连成一饱和溶解度曲线，同时将饱和溶液的组成点与其成平衡的湿固相的组成点作连接线，将各连接线延长于一点，交点即为固相成分。

5. 标明相图中各相区的成分和各组相区的意义。

思考与讨论

1. 为什么根据系统由清变浑的现象即可测定相界？

2. 本实验中根据什么原理求出 KCl-HCl-H_2O 系统的连接线？

参考资料

1. Daniels F, et al. Experimental Physical Chemistry. New York: McGraw-Hill Book Company, 1975: 128.

2. Seidell A. Solubilities of Inorganic and Metal Organic Compounds, Vol. I. William F. Link, D. Vam Nostrand Company, 1958.

3. B. Б. Коган. СпрабоуНик no Pactbopимостй. T. I. Ⅱ, M. JI., A H CCCP, 1961.

4. 傅献彩，侯文华. 物理化学（第六版）上册. 北京：高等教育出版社，2022.

5. 顾菡珍，叶于浦. 相平衡和相图基础. 北京大学出版社，1991.

6. 北京大学化学学院物理化学实验教学组. 物理化学实验（第 4 版）. 北京大学出版社，2019.

Experiment 5

Construction of the phase diagram for the KCl–HCl–H_2O ternary system

Experimental purpose

1. Master the phase laws and the method of representing ternary phase diagrams using equilateral triangular coordinates.

2. Understand the principles of the wet solid-phase method and learn to determine the composition points of pure solid phases in a solution.

3. Construct the phase diagram for the KCl–HCl–H_2O ternary system.

Basical principles

For a ternary system ($C=3$), according to the phase rule $f=C-\Phi+2$, the system can have a maximum of four degrees of freedom (temperature, pressure, two concentration terms). Using a three-dimensional model is not sufficient to represent such a phase diagram. When keeping temperature (or pressure, or one concentration term) constant, the degrees of freedom $f^{*}=4-\Phi$ can be represented using a three-dimensional model. Under isothermal and isobaric conditions, the degrees of freedom $f^{**}=3-\Phi$. The maximum degrees of freedom $f^{**}_{max}=3-1=2$, hence, there can be at most two concentration variables, and a plane graph can be used to represent the relationship between system states and compositions. Equilateral triangles are commonly used on the plane graph to represent concentrations of each component.

As shown in Fig. 5–1, the three vertices of the equilateral triangle represent pure components A, B and C, and the three edges AB, BC, CA represent binary systems of A and B, B and C, C and A. Any point inside the triangle represents the ternary system. For any point O within the triangle, parallel lines a, b and c are drawn perpendicular to the edges, and the lengths $a+b+c$ should equal the length of one side of the triangle ($a+b+c=AB=BC=CA=1$). Therefore, the composition of point O can be represented by the intercepts a', b' and c' of these parallel lines on the edges. Usually, the mass fractions of components A, B and C are marked counterclockwise along the edges of the triangle. That is, starting from point O, draw parallel lines to the edges BC, AC, and AB. The length a' on line AC represents the mass fraction w_A of component A. Similarly, the length b' on line AB represents the mass fraction w_B of component B, and the length c' on line BC represents the mass fraction w_C of component C.

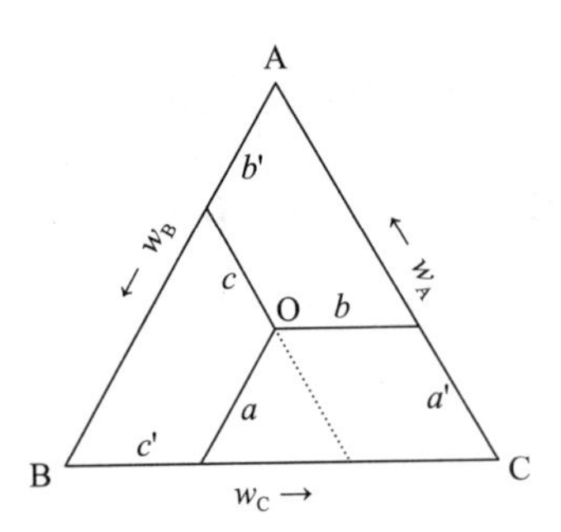

Fig. 5-1 Equilateral triangle for representing ternary phase diagrams

For the ternary system formed by KCl, HCl and H_2O, when the HCl content is not too high, HCl completely dissolves in water to form hydrochloric acid solution. This solution shares the common negative ion Cl^- with KCl, so when hydrochloric acid is added to a saturated KCl water solution, the solubility of KCl decreases due to the common ion effect. This experiment studies the solubility of KCl in hydrochloric acid solutions of different concentrations, aiming to understand the phase diagram construction method and general properties of aqueous salt systems.

To analyze the composition of equilibrium system phases, methods such as separation funnels can be used for liquid-liquid separations. However, separating solid phases can be challenging due to difficulty in completely removing the mother liquor and some solid phases being prone to weathering. In such cases, the wet solid-phase method is employed. This method determines the composition of the solid phase based on the connecting line between the composition points of the saturated solution and the pure solid phase. The points of intersection are the composition of the pure solid phase. This experiment employs this approach to ascertain the composition of the solid phase.

Apparatus and reagents

100 mL ground-mouth conical flasks

50 mL ground-mouth conical flasks

2 mL pipette

Thermostatic bath

KCl (AR)

$AgNO_3$ (AR)

NaOH (AR)

Hydrochloric acid

Experimental procedures

1. Prepare 0.1 mol · L^{-1} $AgNO_3$, 0.1 mol · L^{-1} NaOH, and hydrochloric acid solutions with concentrations of 1, 2, 4, 6, 8 and 12 mol · L^{-1}.

2. Pour 25 mL of hydrochloric acid solution with concentrations of 1, 2, 4, 6, 8 mol · L^{-1}

into 6 clean 100 mL conical flasks. Pour 25 mL of distilled water into one flask.

3. Add approximately 10 g of KCl solid to each flask, then place each flask in a water bath at approximately 30 ℃. Shake continuously for 5 minutes, then transfer to a 25 ℃ constant temperature water bath. Continue shaking for 3 minutes, then let the solutions stand in the constant temperature water bath for 5 minutes until the solutions clarify. Use a dropper to transfer approximately 0.5 g of saturated solution from each flask into a known mass 50 mL ground-mouth conical flask (or weighing bottle) and record the mass of each sample.

4. While taking the saturated solution samples, use a glass spoon to transfer approximately 0.2～0.3 g of wet solid phase to another known mass weighing bottle. Weigh the sample using an analytical balance. Pay attention to the following:

(1) Do not remove the conical flask from the constant temperature water bath to keep the system temperature constant.

(2) Solid phase may contain mother liquor, but saturated solution should not contain solid phase. Be cautious while sampling; wait until the solid phase completely settles before sampling.

(3) The temperature of the sampling dropper should be slightly higher than the system temperature to avoid crystallization of the saturated solution in the pipette, causing errors. It is advisable to preheat the sampling pipette. However, the pipette temperature should not be too high to avoid changing the system temperature and prevent evaporation affecting the concentration.

5. Wash the wet solid phase sample, weighed previously, into a 250 mL conical flask with approximately 50 mL of distilled water and perform titration analysis. First, titrate the acidity of the sample with 0.1 mol · L^{-1} NaOH solution using phenolphthalein as an indicator. Record the volume of NaOH solution consumed at the endpoint. Then add 1～2 drops of dilute HNO_3 solution to the flask to make the system slightly acidic. Finally, titrate the chloride ion concentration in the sample with $AgNO_3$ solution using K_2CrO_4 as an indicator. Record the concentration of $AgNO_3$ used and the volume consumed.

Data recording and processing

1. Design a data recording table based on the basic principles and experimental procedures and fill in the data.

2. Use the following formulas to calculate the mass fractions of HCl, KCl and H_2O in each saturated solution sample and wet solid phase sample. List the results in a table.

$$w(\mathrm{HCl})/(\%) = (c_1 V_1 \times 36.5)/(m \times 1\ 000)$$

$$w(KCl)/(\%) = (c_2V_2 - c_1V_1) \times 74.56/(m \times 1\ 000)$$

$$w(H_2O)/(\%) = 100 - w(HCl) - w(KCl)$$

where c_1 is the concentration of NaOH used during titration ($mol \cdot L^{-1}$), V_1 is the volume of NaOH consumed during titration (mL); c_2 is the concentration of $AgNO_3$($mol \cdot L^{-1}$), V_2 is the volume of $AgNO_3$(mL); m is the sample mass (g); 74.56 and 36.5 are the molar masses of KCl and HCl, respectively.

3. Check the solubility of KCl in water at 25 ℃ and convert it into mass fractions.

4. Mark the results obtained in steps 2 and 3 on the ternary phase diagram. Connect the composition points of each saturated solution to form a solubility curve, and connect the composition points of the saturated solution to the composition points of the coexisting wet solid phase. Extend all connection lines to a point, and the intersection is the composition of the solid phase.

5. Indicate the compositions of different phases in the phase diagram and explain the significance of each phase region.

Thinking and discussing

1. Why can the phase boundary be determined based on the phenomenon of the system changing from clear to turbid?

2. According to what principle are the tie lines determined for the KCl-HCl-H_2O system in this experiment?

Reference materials

1. Daniels F, et al. Experimental Physical Chemistry. New York: McGraw-Hill Book Company, 1975: 128.

2. Seidell A. Solubilities of Inorganic and Metal Organic Compounds, Vol. I, William F. Link, D. Vam Nostrand Company, 1958.

3. В.Б. Коган. СправоуНик по Растворимостй. Т. I. Ⅱ, М. JI., А Н СССР, 1961.

4. Fu X, Hou W. Physical Chemistry (6th ed, Volume I). Beijing: Higher Education Press, 2022.

5. Gu H, Ye Y. Fundamentals of Phase Equilibria and Phase Diagrams. Peking University Press, 1991.

6. Physical Chemistry Experiment Teaching Group, College of Chemistry, Peking University. Physical Chemistry Experiments (4th ed). Peking University Press, 2019.

实验6　蔗糖水解反应速率常数的测定

实验目的

1. 用旋光法测定蔗糖水解反应的速率常数和半衰期。

2. 了解旋光仪的测量原理和使用方法。

3. 明确一级反应的特点。

基本原理

蔗糖水解转化为葡萄糖和果糖的反应为

$$C_{12}H_{22}O_{11}(\text{蔗糖}) + H_2O \xrightarrow{H^+} C_6H_{12}O_6(\text{葡萄糖}) + C_6H_{12}O_6(\text{果糖})$$

该反应的速率方程是由 Wilhelmy 在 1850 年建立的，其速率方程如式(1)所示，其中 c 为 t 时刻反应物(蔗糖)的浓度，k 为反应速率常数。速率方程中没有水的浓度项，是由于在反应系统中水是大量存在的，水分子的消耗相对于水的浓度来说是微不足道的，可以近似认为水的浓度在反应前后不发生变化。这类反应称为准一级反应。

$$r = -\frac{\mathrm{d}c}{\mathrm{d}t} = kc \tag{1}$$

由于蔗糖水解反应在纯水中进行时反应速率极慢，而在 H^+离子催化作用下可以明显加快反应速率，因此研究了酸催化下的蔗糖水解反应，H^+离子可看作催化剂，其浓度在反应前后不变。同样，反应中 H_2O 和 H^+的浓度基本上不变，该反应的反应速率只与蔗糖浓度成正比，此反应可看作一级反应，其反应速率方程仍可写成式(1)的形式，由于 H_2O 和 H^+的浓度可看作是不变的，已将它们归入速率常数中。在酸性溶液中蔗糖的水解反应仍然是准一级反应。

将式(1)积分得

$$\ln\left(\frac{c_0}{c}\right) = kt$$

$$\ln c = -kt + \ln c_0 \tag{2}$$

式中 c_0 是反应物(蔗糖)的初始浓度。若以 $\ln(c/\mathrm{mol\cdot dm^{-3}})$ 对 t 作图，可得一直线，由直线斜率即可求得反应速率常数 k。

在本反应中，无论是反应物还是产物均具有旋光性，且旋光能力不同，故可用系统反应过程中旋光度的变化来跟踪反应进程。测量旋光度所用的仪器称为旋光仪。所测旋光度的大小与系统中所含旋光物质的旋光能力、溶剂性质、溶液的浓度、样品管长度、光源波长以及温度等均有关系。在其他条件均固定时，旋光度 α 与物质浓度呈直线关系，即 $\alpha =$

βc。比例常数 β 与物质的旋光度、溶剂性质、溶液浓度、样品管长度、温度等均有关。

物质的旋光能力可用比旋光度 $[\alpha]_D^{20}$ 来度量。反应物蔗糖是右旋性物质，比旋光度 $[\alpha]_D^{20}=66.6°$；生成物中葡萄糖也是右旋性物质，比旋光度 $[\alpha]_D^{20}=52.5°$；果糖是左旋性物质，比旋光度 $[\alpha]_D^{20}=-91.9°$。由于生成物中果糖的左旋性比葡萄糖的右旋性大，因此当水解作用进行时，右旋角不断减小，到反应终了时，系统将变成左旋。

设最初的旋光度为 α_0，最后的旋光度为 α_∞，则

$$\alpha_0=\beta_{反}c_0 \quad (蔗糖尚未转化,t=0) \tag{3}$$

$$\alpha_\infty=\beta_{生}c_0 \quad (蔗糖全部转化,t=\infty) \tag{4}$$

式中，$\beta_{反}$、$\beta_{生}$ 分别为反应物与生成物的比例常数；c_0 为反应物的初始浓度，亦即生成物最后的浓度。当时间为 t 时，蔗糖浓度为 c，旋光度为 α_t，则

$$\alpha_t=\beta_{反}c+\beta_{生}(c_0-c) \tag{5}$$

由式(3)和(4)得

$$c_0=\frac{\alpha_0-\alpha_\infty}{\beta_{反}-\beta_{生}}=K(\alpha_0-\alpha_\infty) \tag{6}$$

由式(4)和(5)得

$$c=\frac{\alpha_t-\alpha_\infty}{\beta_{反}-\beta_{生}}=K(\alpha_t-\alpha_\infty) \tag{7}$$

将式(6)和式(7)代入式(2)得

$$\ln\left(\frac{\alpha_0-\alpha_\infty}{\alpha_t-\alpha_\infty}\right)=kt \tag{8}$$

或

$$\ln(\alpha_t-\alpha_\infty)=-kt+\ln(\alpha_0-\alpha_\infty) \tag{9}$$

以 $\ln(\alpha_t-\alpha_\infty)$ 对 t 作图可得一直线，由直线斜率即可求得反应速率常数 k。反应物浓度消耗一半所需要的时间称为半衰期，用 $t_{1/2}$ 表示。将 $c=\frac{1}{2}c_0$ 代入式(2)，得

$$t_{1/2}=\frac{1}{k}\ln\frac{c_0}{\frac{1}{2}c_0}=\frac{\ln 2}{k} \tag{10}$$

上式说明一级反应的半衰期只与反应速率常数 k 有关，而与反应物的初始浓度无关。这是一级反应的一个特点。

仪器和试剂

旋光仪　　磨口锥形瓶(100 mL)

秒表　　量筒(100 mL)

移液管(25 mL)　　细口瓶(500 mL，公用)

水浴装置 盐酸(分析纯)

蔗糖(分析纯)

实验步骤

1. 用蒸馏水练习样品管的装样技术,熟悉旋光仪的使用及读数(仪器使用前需要预热10 min)。

2. 配制 0.8 和 0.4 $mol \cdot L^{-1}$蔗糖溶液各 500 mL,配制 3.0 和 1.5 $mol \cdot L^{-1}$ HCl 溶液各500 mL。

3. 用移液管吸取 25 mL 0.8 $mol \cdot L^{-1}$的蔗糖溶液至 100 mL 碘量瓶中,再用另一支移液管吸取 25 mL 3.0 $mol \cdot L^{-1}$的 HCl 溶液至装有蔗糖溶液的碘量瓶中,HCl 流出一半时开始计时,摇动碘量瓶使溶液混合均匀。迅速用少量反应液润洗样品管 2 次,然后将反应液装满样品管,盖好玻璃片,旋紧套盖(此时样品管内不能有气泡存在)。将样品管外面的液体擦净,放入旋光仪内,测旋光度。第 1 个数据最好在 2～4 min 时读出,前 10 min 内每 2 min读一次数,11～30 mim 每隔 3～5 min 读一次数。

4. 将装有剩余反应液的碘量瓶置于 50～60 ℃水浴中加热 30 min(可与前述 α_t 的测试同时),然后取出碘量瓶使溶液冷却至实验温度,再测此溶液的旋光度,即为 α_∞ 值。注意水浴温度不可过高,否则将产生副反应,溶液颜色变黄。加热过程亦应避免溶液蒸发影响 α_∞ 的测定。

5. 将蔗糖溶液换成 0.4 $mol \cdot L^{-1}$,使其与 3 $mol \cdot L^{-1}$的 HCl 溶液反应,按前述步骤 3 及4 的操作测定 α_t 和 α_∞ 的值。

6. 利用 0.8 $mol \cdot L^{-1}$蔗糖溶液与 1.5 $mol \cdot L^{-1}$ HCl 溶液反应,仍按前述步骤 3 及 4 的操作测定 α_t 和 α_∞ 的值。

需要注意的是,由于反应液酸性较强,对旋光仪有很强腐蚀性,因此实验结束后,必须立即将所用仪器擦拭干净。

数据记录与处理

1. 将实验记录的数据及处理结果列表表示。

2. 以 $\ln(\alpha_t-\alpha_\infty)$ 对 t 作图求速率常数 k 和半衰期 $t_{1/2}$。

思考与讨论

1. 通过实验结果分析蔗糖水解是否是一级反应。

2. 不同浓度盐酸对反应速率及速率常数是否有影响,有何影响,为什么?

3. 不同浓度蔗糖对反应速率及速率常数是否有影响,有何影响,为什么?

4. 如何判断某一旋光物质是左旋还是右旋?

参考资料

1. 伏洛勃约夫，等. 物理化学实验. 北京：高等教育出版社，1954：129.

2. Daniels F, Alberty R A, Williams J W, et al. Experimental Physical Chemistry. 7th ed. New York: McGraw-Hill, Inc, 1975: 149.

3. 印永嘉，李大珍. 物理化学简明教程. 北京：高等教育出版社，1980：445.

Experiment 6

Determination of the rate constant for sucrose hydrolysis reaction

Experimental purpose

1. Determine the rate constant and half-life of the sucrose hydrolysis reaction using the polarimetry method.

2. Understand the measurement principles and usage of a polarimeter.

3. Clarify the characteristics of a first-order reaction.

Basical principles

The reaction of sucrose hydrolysis converting into glucose and fructose, is represented as

$$C_{12}H_{22}O_{11}(\text{sucrose}) + H_2O \xrightarrow{H^+} C_6H_{12}O_6(\text{glucose}) + C_6H_{12}O_6(\text{fructose})$$

The rate equation for this reaction was established by Wilhelmy in 1850. The rate equation is given by Equation (1), where c is the concentration of the reactant (sucrose) at time t, and k is the reaction rate constant. The rate equation doesn't include the concentration term for water since it's assumed to be constant due to its abundance in the reaction system. This type of reaction is referred to as a pseudo-first-order reaction.

The hydrolysis of sucrose in pure water is extremely slow, but in the presence of H^+ ions, it significantly accelerates. Therefore, people studies the acid-catalyzed sucrose hydrolysis reaction, where H^+ ions act as catalysts. The reaction is still considered pseudo-first-order, as the concentrations of H_2O and H^+ are assumed to be constant.

The reaction rate is very slow when the reaction is carried out in pure water, but the reaction rate can be obviously accelerated under the catalysis of H^+ ions. Because there is a large amount of water in the reaction, it can be approximately considered that the concentration of water does not change before and after the reaction, and H^+ ion is the catalyst, then its concentration does not change before and after the reaction. Therefore, the reaction rate is only proportional to the sucrose concentration, and this reaction can be regarded as a first-order reaction, and its reaction rate equation is

$$r = -\frac{dc}{dt} = kc \tag{1}$$

Integration of the above equation gives

$$\ln\left(\frac{c_0}{c}\right) = kt$$

$$\ln c = -kt + \ln c_0 \tag{2}$$

where c_0 is the initial concentration of the reactant (sucrose). Plotting $\ln(c/\text{mol} \cdot \text{dm}^{-3})$ against t yields a straight line, and the slope of this line provides the reaction rate constant k.

In this reaction, both reactants and products exhibit optical rotation, with different specific rotations. The change in optical rotation during the reaction can be used to monitor the progress. The instrument used to measure optical rotation is called a polarimeter. The optical rotation α is related to the concentration c by the equation $\alpha = \beta c$, where β is a proportionality constant dependent on the specific rotation, solvent properties, solution concentration, sample tube length, wavelength of light source and temperature.

The optical rotatory power of substances is measured by specific rotation $[\alpha]_D^{20}$. The reactant sucrose is a right-handed substance with specific rotation $[\alpha]_D^{20} = 66.6°$, the product glucose is also a right-handed substance with specific rotation $[\alpha]_D^{20} = 52.5°$, and the fructose is a left-handed substance with specific rotation $[\alpha]_D^{20} = -91.9°$. As the reaction progresses, the system changes from right-handed to left-handed rotation due to the greater left-handed rotation of fructose compared to the right-handed rotation of glucose.

The initial and final optical rotations (α_0 and α_∞) are related to the initial concentration of the reactant (c_0) by Equations (3) and (4).

$$\alpha_0 = \beta_r c_0 \quad (\text{Sucrose not yet converted, } t=0) \tag{3}$$

$$\alpha_\infty = \beta_p c_0 \quad (\text{Total conversion of sucrose, } t=\infty) \tag{4}$$

where β_r and β_p are the proportional constants of reactants and products respectively; c_0 is the initial concentration of reactants, that is, the final concentration of products. When the time is t, the sucrose concentration is c, the optical rotation is α_t, then

$$\alpha_t = \beta_r c + \beta_p(c_0 - c) \tag{5}$$

From Equations (3) and (4)

$$c_0 = \frac{\alpha_0 - \alpha_\infty}{\beta_r - \beta_p} = K(\alpha_0 - \alpha_\infty) \tag{6}$$

From Equations (4) and (5)

$$c = \frac{\alpha_t - \alpha_\infty}{\beta_r - \beta_p} = K(\alpha_t - \alpha_\infty) \tag{7}$$

Substituting Equations (6) and (7) into Equation (2), we get

$$\ln\left(\frac{\alpha_0 - \alpha_\infty}{\alpha_t - \alpha_\infty}\right) = kt \tag{8}$$

or

$$\ln(\alpha_t - \alpha_\infty) = -kt + \ln(\alpha_0 - \alpha_\infty) \tag{9}$$

Plotting $\ln(\alpha_t - \alpha_\infty)$ against t gives a straight line, and the slope of this line provides the reaction rate constant k. The half-life ($t_{1/2}$) of a first-order reaction is only dependent on the reaction rate constant k and is independent of the initial concentration of the reactant, as shown in Equation (10).

$$t_{1/2} = \frac{1}{k}\ln\frac{c_0}{\frac{1}{2}c_0} = \frac{\ln 2}{k} \tag{10}$$

Equation (10) shows that the half-life of the first-order reaction is only related to the reaction rate constant k, but has nothing to do with the initial concentration of reactants. This is a characteristic of the first-order reaction.

Apparatus and reagents

Polarimeter
Stopwatch
Pipette (25 mL)
Water bath
Sucrose (AR)
Grinding conical flask (100 mL)
Graduated cylinder (100 mL)
Narrow-mouth bottle (500 mL, shared)
Hydrochloric acid (AR)

Experimental procedures

1. Practice sample tube loading technique with distilled water, familiarize with the usage and reading of the polarimeter (preheat the instrument for 10 minutes before use).

2. Prepare 500 mL 0.8 mol · L^{-1} and 500 mL 0.4 mol · L^{-1} sucrose solutions. Prepare 500 mL 3.0 mol · L^{-1} and 500 mL 1.5 mol · L^{-1} HCl solutions.

3. Pipette 25 mL of 0.8 mol · L^{-1} sucrose solution into a 100 mL volumetric flask. Then, using another pipette, add 25 mL of 3.0 mol · L^{-1} HCl solution to the flask containing the sucrose solution. Start timing when half of the HCl has been added, shake the flask to mix the

solution uniformly. Quickly rinse the sample tube with a small amount of the reaction solution twice, then fill the sample tube with the reaction solution, cover it with a glass plate, tighten the cover (there should be no bubbles inside the sample tube at this point). Wipe off any liquid on the outside of the sample tube, place it in the polarimeter, and measure the optical rotation. The first data point is best read at 2～4 minutes, then read every 2 minutes within the first 10 minutes, and every 3～5 minutes from 11～30 minutes.

4. Place the remaining reaction solution in the volumetric flask in a 50～60 ℃ water bath and heat for 30 minutes (this can be done simultaneously with the testing of α_t described above). Then, take out the volumetric flask, cool the solution to the experimental temperature, and measure the optical rotation, which is the α_∞ value. Note that the water bath temperature should not be too high, or else side reactions may occur, leading to yellowing of the solution. Avoid evaporation during the heating process, as it may affect the determination of α_∞.

5. Replace the sucrose solution with a 0.4 mol · L^{-1} solution and react it with 3.0 mol · L^{-1} HCl solution, measuring α_t and α_∞ values as described as steps 3 and 4.

6. Repeat the procedure using a 0.8 mol · L^{-1} sucrose solution and 1.5 mol · L^{-1} HCl solution, measuring α_t and α_∞ values as described as steps 3 and 4.

It is important to note that the reaction solution is highly acidic and corrosive to the polarimeter. Therefore, immediately wipe the used equipment clean after completing the experiment.

Data recording and processing

1. List the recorded experimental data and results.

2. Plot $\ln(\alpha_t-\alpha_\infty)$ against t to determine the rate constant k and half-life $t_{1/2}$.

Thinking and discussing

1. Analyze the experimental results to determine if sucrose hydrolysis is a first-order reaction.

2. Investigate whether different concentrations of hydrochloric acid affect the reaction rate and rate constant, and discuss the reasons for any observed effects.

3. Explore the impact of different sucrose concentrations on the reaction rate and rate constant, providing explanations.

4. Discuss how to determine if a specific optically active substance is levorotatory or dextrorotatory.

Reference materials

1. Volobyov, et al. Physical Chemistry Experiments. Beijing: Higher Education Press, 1954: 129.

2. Daniels F, Alberty R A, Williams J W, et al. Experimental Physical Chemistry. 7th ed. New York: McGraw-Hill, Inc, 1975: 149.

3. Yin Y, Li D. Concise Tutorial on Physical Chemistry. Beijing: Higher Education Press, 1980: 445.

实验7 乙酸乙酯皂化反应速率常数的测定

实验目的

1. 深入理解电导法测定乙酸乙酯皂化反应的速率常数的基本设计思想。
2. 用电导法测定乙酸乙酯皂化反应的速率常数并计算反应的活化能。
3. 了解二级反应的特点。
4. 掌握电导率仪的使用方法。

基本原理

乙酸乙酯皂化反应是二级反应,两种反应物的起始浓度可以相等,也可以不相等。若以 a、b 代表乙酸乙酯和碱(NaOH)的起始浓度,x 为在 t 时刻的生成物浓度,当两种反应物的起始浓度不相等时,反应物浓度和生成物浓度随时间的变化可表示为

$$CH_3COOC_2H_5 + OH^- = CH_3COO^- + C_2H_5OH$$

	$CH_3COOC_2H_5$	OH^-	CH_3COO^-	C_2H_5OH
$t=0$	a	b	0	0
$t=t$	$a-x$	$b-x$	x	x
$t=\infty$	$a-b$	0	b	b

该反应的速率方程式为

$$\frac{dx}{dt} = k(a-x)(b-x) \tag{1}$$

式中 k 为反应速率常数

$$\frac{dx}{(a-x)(b-x)} = k dt \tag{2}$$

积分式(2),得

$$\frac{1}{a-b}\ln\frac{b(a-x)}{a(b-x)} = kt \quad (设\ a>b) \tag{3}$$

由实验测得不同时间 t 时的 x 值,则可依式(3)计算出不同 t 的 k 值。如果 k 值为常数,就可证明反应是二级反应。通常由 $\ln\frac{(a-x)}{(b-x)}$ 对 t 图,若所得的是一条直线,就证明反应是二级反应,其直线的斜率为 k 值。

不同时刻下生成物的浓度可用化学分析法测定(例如分析反应液中的 OH^- 浓度),也可以用物理化学分析法测定(如测量电导率)。本实验用电导率仪测定反应系统的电导率(κ)值,跟踪反应进程,进而测定其反应速率常数。

本实验的基本设计思想为:

(1) 反应系统是稀水溶液,可以假定 CH_3COONa 全部电离。溶液中参与导电的离子有 Na^+、OH^- 和 CH_3COO^- 等,而 Na^+ 在反应前后浓度不变,溶液中 OH^- 离子的导电能力远远大于 CH_3COO^- 离子(即反应物与生成物的电导率值相差很大)。因此,随着反应的进行,OH^- 离子的浓度不断降低,溶液的电导率值也就随之下降。

(2) 在稀溶液中,可以认为每种强电解质的电导率与其浓度成正比,而且溶液的总电导率等于组成溶液的电解质的电导率之和。

依据上述两点,对乙酸乙酯皂化反应来说,反应物与生成物只有 $NaOH$ 和 CH_3COONa 是强电解质。在一定浓度范围内,可以认为系统电导率值的减少量与 CH_3COONa 的浓度 x 成正比,即

$$t=0:\quad x=\beta(\kappa_0-\kappa_t) \tag{4}$$

$$t=\infty:\quad b=\beta(\kappa_0-\kappa_\infty) \tag{5}$$

式中 κ_0 和 κ_t 分别为溶液起始时和时间为 t 时的电导率值,κ_∞ 为反应终了时的电导率值,β 为比例常数。由式(4)、(5)可求出 x

$$x=\frac{b(\kappa_0-\kappa_t)}{\kappa_0-\kappa_\infty} \tag{6}$$

将式(6)代入式(3)得

$$\ln\left(\frac{a}{b}\cdot\frac{\kappa_0-\kappa_\infty}{\kappa_t-\kappa_\infty}-\frac{\kappa_0-\kappa_t}{\kappa_t-\kappa_\infty}\right)=(a-b)kt+\ln\frac{a}{b} \tag{7}$$

以 $\ln\left(\frac{a}{b}\cdot\frac{\kappa_0-\kappa_\infty}{\kappa_t-\kappa_\infty}-\frac{\kappa_0-\kappa_t}{\kappa_t-\kappa_\infty}\right)$ 对 t 作图,可得到一条直线,由直线斜率可以求出速率常数 k。

反应速率常数 k 与温度 T/K 的关系一般符合阿仑尼乌斯公式,即

$$\frac{\mathrm{d}\ln k}{\mathrm{d}T}=\frac{E_a}{RT^2} \tag{8}$$

当表观活化能 E_a 为常数的时候,对式(8)做不定积分和定积分,分别得

$$\ln k = -\frac{E_a}{RT} + C \tag{9}$$

$$\ln \frac{k_2}{k_1} = \frac{E_a}{R}\left(\frac{1}{T_1} - \frac{1}{T_2}\right) \tag{10}$$

式中 C 为积分常数。显然，在不同的温度下测定速率常数 k，由 $\ln k$ 对 $1/T$ 作图，应得一直线，由直线的斜率就可算出 E_a 的值。或者测两个温度下的速率常数代入式(10)，可以计算出活化能。

仪器和试剂

数字式电导率仪
恒温槽
铂黑电极
水浴加热箱
短颈容量瓶(200 mL)
秒表
粗试管 1 只
细试管 2 只
移液管(25 mL) 3 支
三角瓶(250 mL) 3 只
NaOH(分析纯)
乙酸乙酯(分析纯)
酚酞试剂
草酸(分析纯)
蒸馏水

实验步骤

1. 配制溶液

(1) 乙酸乙酯溶液的配制(浓度约为 0.02 mol · L^{-1})

将 200 mL 的短颈容量瓶洗净，加少量二次蒸馏水(少于 1/3)，放在电子天平上去皮，滴加乙酸乙酯 0.35 g 左右，准确称其质量并计算浓度。

注意：在称量过程中容量瓶的盖子要盖好。

(2) NaOH 溶液的配制(浓度约为 0.02 mol · L^{-1})

用台秤称 1 g NaOH，用二次蒸馏水快速洗涤两次后，再加二次蒸馏水将 NaOH 溶解并把溶液转移至 500 mL 塑料瓶中，用草酸标定其浓度。(用电子天平准确称取 3 份草酸分别放入 3 个三角瓶中，每份约 0.025 g。)

2. 电导率仪的调节

以 SLDS-Ⅰ数显电导率仪为例，来说明电导率仪的使用方法。

(1) 将电极插头插入电极插座（插头、插座上的定位销对准后，按下插头顶部即可），接通仪器电源，仪器处于校准状态，校准指示灯亮。让仪器预热 15 min。

(2) 将“温度补偿”旋钮的标志线置于 25 ℃位置(无补偿作用)。

(3) 调节常数旋钮，使仪器所显示值为所用电极的常数标称值。若电极常数为 0.92，则调“常数”旋钮使显示 9 200；若常数为 1.10，则调“常数”旋钮使显示 11 000；以此类推。

（4）按“测量/转换”键，使仪器处于测量状态（测量指示灯亮），待显示值稳定后，该显示数值即为被测液体在该温度下的电导率值。

（5）测量中，若显示屏显示为“OUL”，表示被测值超出量程范围，应置于高一档量程来测量，若读数很小，就置于低一档量程，以提高精度。

（6）测量高电导的溶液，若被测溶液的电导率高于 20 mS · cm^{-1}时，应选用 DJS-10 电极，此时量程范围可扩大到 200 mS · cm^{-1}（20 mS · cm^{-1}档可测至 200 mS · cm^{-1}，2 mS · cm^{-1}档可测至 20 mS · cm^{-1}，但显示数须乘 10）。测量纯水或高纯水的电导率，宜选 0.01 常数的电极，被测值 = 显示数×0.01；也可用 DJS-0.1 电极，被测值 = 显示数×0.1；被测液的电导低于 30 μS · cm^{-1}，宜选用 DJS-I 光亮电极；电导率高于 30 μS · cm^{-1}，应选用 DJS-1 铂黑电极。

表 7-1　电导率范围及对应电极常数推荐表

电导率范围/（μS · cm^{-1}）	电阻率范围/（Ω · cm）	推荐使用电极常数/cm^{-1}
0.05～2	20 M～500 k	0.01，0.1
2～200	500 k～5 k	0.1，1.0
200～2 000	5 k～500	1.0
2 000～20 000	500～50	1.0，10
20 000～2×10^5	50～5	10

（7）仪器可长时间连续使用，可用输出讯号（0～10 mV）外接记录仪进行连续监测，也可选配 RS232C 串口，由电脑显示监测。

3. κ_0 的测量

用移液管准确移取 25.0 mL NaOH 溶液放到干燥的细试管中，再用另一只移液管准确移取 25.0 mL 二次蒸馏水至该试管中，将溶液混合均匀。用滴管吸少量混合液淋洗电极 2 次，并把电极放入该试管中。将试管放到恒温槽中（控制实验所用 2 个恒温槽的温度分别为 25 ℃和 30 ℃）恒温 10 min，然后测定 κ_0，读 2 次数（间隔 1 min）取平均值。

4. κ_t 的测量

用移液管移取 25.0 mL NaOH 溶液放到干燥的粗试管中，用另一支移液管移取 25.0 mL 乙酸乙酯溶液放到干燥的细试管中。将装有溶液的两支试管盖上胶塞并放到恒温槽中恒温 10 min。取出两试管，用毛巾迅速擦干试管外壁，并迅速将乙酸乙酯倒入 NaOH 中（溶液倒入一半时开启秒表计时），将两个试管中的溶液迅速来回倒几次，使反应液混合均匀，将混合液全部倒入粗试管中，再将粗试管放入恒温槽中。用滴管吸少量反应液淋洗电极，将电极插入粗试管中（注意电极事先要恒温），塞紧胶塞，最好在第 3 min 内读取第一个数据。实验中两名同学交替读数，记录自己的读数，共测 40 min。

第 1 名同学读数时间为:3,5,7,9,12,15,18,21,26,30,34,38 min。

第 2 名同学读数时间为:4,6,8,10,13,16,19,24,28,32,36,40 min。

5. κ_∞ 的测量

测定 κ_t 后,将粗试管置于 50～60 ℃的水浴箱中,加热 30 min。取出后用自来水冷却并放入恒温槽中恒温 10 min,测定 κ_∞,读取 3 次数据(间隔 1 min)取其平均值。

测完 κ_∞,停止实验。清洗所用的玻璃仪器,将铂黑电极浸入蒸馏水中。

数据记录与处理

1. 设计表格将数据列入表中

(1) 将相关常量列表表示。

(2) 将实验数据及处理结果填入表格中,并用电脑进行数据处理。

2. 以 $\ln\left(\frac{a}{b}\cdot\frac{\kappa_0-\kappa_\infty}{\kappa_t-\kappa_\infty}-\frac{\kappa_0-\kappa_t}{\kappa_t-\kappa_\infty}\right)$ 对 t 作图,求速率常数 k。

3. 计算实验活化能。

思考与讨论

1. 配制乙酸乙酯溶液时,为什么在容量瓶中要事先加入适量的蒸馏水?

2. 测 κ_0 时,25 mL NaOH 溶液为何要加 25 mL 的二次蒸馏水?

3. 为什么乙酸乙酯与 NaOH 溶液的浓度不能太大?

4. 若乙酸乙酯与 NaOH 溶液的起始浓度相等,应如何计算 k 值?

参考资料

1. Daniels F, Alberty R A, Williams J W, et al. Experimental Physical Chemistry. 7th ed. New York: McGraw-Hill, Inc, 1975: 144.

2. 傅献彩, 侯文华. 物理化学(第六版)下册. 北京: 高等教育出版社, 2022.

Experiment 7

Determination of the rate constant for the saponification reaction of ethyl acetate

Experimental purpose

1. Gain a deep understanding of the basic design principles of the electrical conductivity method for determining the rate of ethyl acetate saponification.

2. Use the electrical conductivity method to determine the rate constant of the ethyl acetate saponification reaction and calculate the activation energy of the reaction.

3. Understand characteristics of a second-order reaction.

4. Master the use of the conductivity meter.

Basical principles

The saponification reaction of ethyl acetate is a second-order reaction. The initial concentrations of the two reactants, ethyl acetate and sodium hydroxide (NaOH), can be equal or unequal. If the initial concentrations of ethyl acetate and NaOH are denoted by a and b, and x as the concentration of the product at time t, when the initial concentrations are unequal, the concentration of reactants and the concentration of the product over time can be expressed as

$$CH_3COOC_2H_5 + OH^- = CH_3COO^- + C_2H_5OH$$

$t = 0$	a	b	0	0
$t = t$	$a - x$	$b - x$	x	x
$t = \infty$	$a - b$	0	b	b

The rate equation of the reaction is

$$\frac{dx}{dt} = k(a - x)(b - x) \tag{1}$$

where k is the reaction rate constant

$$\frac{dx}{(a - x)(b - x)} = k dt \tag{2}$$

Integrating Equation (2), we get

$$\frac{1}{a - b} \ln \frac{b(a - x)}{a(b - x)} = kt \quad (\text{if } a > b) \tag{3}$$

By experimentally measuring the values of x at different times t, the value of k at different times t can be calculated according to Equation (3). If k is constant, it proves that the reaction is a second-order reaction. When plotting $\ln \frac{(a - x)}{(b - x)}$ against t, a straight line indicates a second-order reaction, and the slope of the line is the value of k.

The concentration of the product at different times can be determined by chemical analysis (e.g., analyzing the concentration of OH^- in the reaction solution) or by physical-chemical analysis (e.g., measuring conductivity). This experiment uses a conductivity meter to measure the conductivity (κ) of the reaction system, track the progress of the reaction, and determine the rate constant.

The basic design idea of this experiment is:

(1) The reaction system is a dilute aqueous solution, and it can be assumed that CH_3COONa is completely ionized. The ions participating in the conductivity are Na^+, OH^- and CH_3COO^-. Since Na^+ concentration remains constant before and after the reaction, the conductivity of OH^- ions is much greater than that of CH_3COO^- ions (i.e., the conductivity values of reactants and products differ significantly). Therefore, as the reaction proceeds, the concentration of OH^- ions continuously decreases, and the conductivity of the solution also decreases.

(2) In dilute solutions, it can be assumed that the conductivity of each strong electrolyte is proportional to its concentration, and the total conductivity of the solution is equal to the sum of the conductivity of the electrolytes in the solution.

Based on these two points, for the saponification reaction of ethyl acetate, only NaOH and CH_3COONa are strong electrolytes. Within a certain concentration range, the decrease in the conductivity value of the system is proportional to the concentration x of CH_3COONa. Therefore

$$t = 0: \quad x = \beta(\kappa_0 - \kappa_t) \tag{4}$$

$$t = \infty: \quad b = \beta(\kappa_0 - \kappa_\infty) \tag{5}$$

where κ_0 and κ_t are the initial and time t conductivity values of the solution, κ_∞ is the final conductivity value of the reaction, and β is a proportionality constant. By Equations (4) and (5), x can be determined

$$x = \frac{b(\kappa_0 - \kappa_t)}{\kappa_0 - \kappa_\infty} \tag{6}$$

Substituting Equation (6) into Equation (3), we get

$$\ln\left(\frac{a}{b}\cdot\frac{\kappa_0 - \kappa_\infty}{\kappa_t - \kappa_\infty} - \frac{\kappa_0 - \kappa_t}{\kappa_t - \kappa_\infty}\right) = (a - b)kt + \ln\frac{a}{b} \tag{7}$$

Plotting $\ln\left(\frac{a}{b}\cdot\frac{\kappa_0 - \kappa_\infty}{\kappa_t - \kappa_\infty} - \frac{\kappa_0 - \kappa_t}{\kappa_t - \kappa_\infty}\right)$ against t should result in a straight line, and the slope of the line can be used to calculate the rate constant k.

The relationship between the rate constant k and the temperature T/K in general follows the Arrhenius equation

$$\frac{\mathrm{d}\ln k}{\mathrm{d}T} = \frac{E_a}{RT^2} \tag{8}$$

When the apparent activation energy E_a is constant, integrating Equation (8) with indefinite and definite integrals yields

$$\ln k = -\frac{E_a}{RT} + C \tag{9}$$

$$\ln \frac{k_2}{k_1} = \frac{E_a}{R}\left(\frac{1}{T_1} - \frac{1}{T_2}\right) \tag{10}$$

where C is the integration constant. Obviously, measuring the rate constant k at different temperatures, plotting $\ln k$ against $1/T$ should yield a straight line, and the slope of the line can be used to calculate the value of E_a. Alternatively, measuring the rate constants at two temperatures and substituting them into Equation (10) can calculate the activation energy.

Apparatus and reagents

Digital conductivity meter
Thermostat
Platinum black electrode
Water bath heating box
Short neck volumetric flask (200 mL)
Stopwatch
Coarse test tube (1)
Fine test tube (2)
Pipettes (25 mL) (3)
Triangular flasks (250 mL) (3)
NaOH (AR)
Ethyl acetate (AR)
Phenolphthalein reagent
Oxalic acid (AR)
Distilled water

Experimental procedures

1. Preparation of solutions

(1) Prepare of ethyl acetate solution (concentration approximately 0.02 $mol \cdot L^{-1}$):

Clean a 200 mL short-necked volumetric flask, add a small amount of distilled water (less than 1/3), place it on an electronic balance, peel off, add approximately 0.35 g of ethyl acetate, accurately weigh its mass, and calculate the concentration.

Note: The flask cap should be closed during the weighing process.

(2) Prepare NaOH solution (concentration approximately 0.02 $mol \cdot L^{-1}$):

Weigh 1 g of NaOH using a balance, wash it twice with distilled water, then dissolve NaOH in distilled water and transfer the solution to a 500 mL plastic bottle, calibrate its concentration using oxalic acid. (Use an electronic balance to accurately weigh three portions of oxalic acid and place them in three triangular flasks, each weighing about 0.025 g.)

2. Adjustment of the conductivity meter:

Using the SLDS-I digital conductivity meter as an example, explain the use of the conductivity meter.

(1) Insert the electrode plug into the electrode socket (after aligning the plug and socket, press the top of the plug). Turn on the instrument power, and the instrument is in the calibration

state with the calibration indicator light on. Let the instrument preheat for 15 minutes.

(2) Set the "temperature compensation" knob indicator line to the 25 ℃ position (no compensation effect).

(3) Adjust the constant knob to make the displayed value of the instrument equal to the nominal constant value of the electrode in use. If the electrode constant is 0.92, adjust the "constant" knob to make the display show 9 200. If the constant is 1.10, adjust the "constant" knob to make the display show 11 000, and so on.

(4) Press the "measurement/conversion" key to put the instrument in measurement state (measurement indicator light on). After the displayed value stabilizes, the displayed value is the conductivity value of the measured liquid at that temperature.

(5) During measurement, if the display shows "OUL", it means that the measured value exceeds the range and should be measured in a higher range. If the reading is very small, switch to a lower range to improve accuracy.

(6) For measuring the conductivity of high-conductivity solutions, if the conductivity of the measured solution is higher than 20 $mS \cdot cm^{-1}$, use the DJS-10 electrode. At this time, the range can be expanded to 200 $mS \cdot cm^{-1}$ (the 20 $mS \cdot cm^{-1}$ range can be measured up to 200 $mS \cdot cm^{-1}$, the 2 $mS \cdot cm^{-1}$ range can be measured up to 20 $mS \cdot cm^{-1}$, but the displayed value must be multiplied by 10). For measuring the conductivity of pure water or high-purity water, it is advisable to use an electrode with a constant of 0.01. The measured value is equal to displayed value×0.01; alternatively, use the DJS-0.1 electrode, the measured value is equal to displayed value×0.1; for solutions with a conductivity lower than 30 $\mu S \cdot cm^{-1}$, use the DJS-I bright electrode; for a conductivity higher than 30 $\mu S \cdot cm^{-1}$, use the DJS-1 platinum black electrode.

Table 7-1 Recommended electrode constants for different conductivity ranges

Conductivity range/($\mu S \cdot cm^{-1}$)	Resistivity range/($\Omega \cdot cm$)	Electrode constants recommended/cm^{-1}
0.05～2	20 M～500 k	0.01, 0.1
2～200	500 k～5 k	0.1, 1.0
200～2 000	5 k～500	1.0
2 000～20 000	500～50	1.0, 10
20 000～2×10^5	50～5	10

(7) The instrument can be used continuously for an extended period. It allows external connection to a data logger with an output signal range of 0～10 mV for continuous monitoring. Additionally, it is equipped with an RS232C serial port, enabling connection to a computer for real-time display and monitoring.

3. Measurement of κ_0

Use a pipette to accurately transfer 25.0 mL of NaOH solution to a dry fine test tube, and use another pipette to accurately transfer 25.0 mL of distilled water into a same test tube. Mix the solution thoroughly. Use a dropper to rinse the electrode with a small amount of mixed solution twice, and place the electrode in the test tube. Place the test tube in a constant temperature bath (control the temperatures of the two constant temperature baths used in the experiment to 25 ℃ and 30 ℃, respectively) for 10 minutes, then measure κ_0, and read the values twice (with a 1-minute interval) and take the average.

4. Measurement of κ_t

Use a pipette to transfer 25.0 mL of NaOH solution to a dry coarse test tube and use another pipette to transfer 25.0 mL of ethyl acetate solution to a dry fine test tube. Cover both tubes with rubber stoppers and place them in a constant temperature bath for 10 minutes. Remove both tubes, quickly dry the outer walls of the tubes with a towel, quickly pour ethyl acetate into NaOH (start timing when half of the solution is poured), quickly mix the solutions by pouring them back and forth several times, pour all of the mixed solution into the coarse test tube, and then place the coarse test tube in a constant temperature bath. Use a dropper to take a small amount of reaction solution to rinse the electrode, insert the electrode into the coarse test tube (note that the electrode should be preheated), seal the rubber stopper tightly, and preferably read the first set of data within 3 minutes. In the experiment, two students alternately read the data and recorded their readings for a total of 40 minutes.

Student 1 read times: 3, 5, 7, 9, 12, 15, 18, 21, 26, 30, 34 and 38 min.

Student 2 read times: 4, 6, 8, 10, 13, 16, 19, 24, 28, 32, 36 and 40 min.

5. Measurement of κ_∞

After measuring κ_t, place the coarse test tube in a water bath at 50～60 ℃ and heat for 30 minutes. After taking it out, cool it with tap water and place it in a constant temperature bath for 10 minutes. Measure κ_∞ and read the data three times (with a 1-minute interval) and take the average.

After measuring κ_∞, stop the experiment. Clean the glassware used and immerse the platinum black electrode in distilled water.

Data recording and processing

1. Design a table to record the data.

(1) List relevant constants.

(2) Fill in the experimental data and processing results in the table and use a computer for data processing.

2. Plot $\ln\left(\frac{a}{b}\cdot\frac{\kappa_0-\kappa_\infty}{\kappa_t-\kappa_\infty}-\frac{\kappa_0-\kappa_t}{\kappa_t-\kappa_\infty}\right)$ against t to obtain the rate constant k.

3. Calculate the experimental activation energy.

Thinking and discussing

1. Why is it necessary to add a suitable amount of distilled water to the volumetric flask when preparing the ethyl acetate solution?

2. Why is it necessary to add 25 mL of distilled water to 25 mL of NaOH solution when measuring κ_0?

3. Why can't the concentrations of ethyl acetate and NaOH solution be too large?

4. If the initial concentrations of ethyl acetate and NaOH solutions are equal, how should k value be calculated?

References materials

1. Daniels F, Alberty R A, Williams J W, et al. Experimental Physical Chemistry. 7th ed. New York: McGraw-Hill, Inc, 1975: 144.

2. Fu X, Hou W. Physical Chemistry (6th ed, Volume Ⅱ). Beijing: Higher Education Press, 2022.

实验 8 原电池电动势的测定和相关热力学函数的计算

实验目的

1. 掌握对消法测定电池电动势的原理及电位差计的使用方法。

2. 利用电位差计测定一些电池的电动势。

3. 掌握测定电极电势的方法。

基本原理

在电化学中对可逆电池的研究,为揭示化学能转化为电能的限度、改善电池性能提供了理论依据,也为采用可逆电池的原理研究热力学问题提供了强有力的工具。电池电动势的准确测量,能够为许多热力学数据(如平衡电势、活度系数、解离常数、溶解度、配合常数等)的获得提供很大帮助。因此,精确地测量某一电池的电动势,在物理化学研究中具有重要意义。

1. 电位差计(对消法或补偿法)测电池电动势的基本原理

对于在平衡状态或无限接近于平衡状态的情况下工作的可逆电池,在等温等压条件下,当系统发生变化时,系统 Gibbs 自由能的减少等于对外所做的最大非膨胀功,即 $\Delta_r G = W_{f,max}$。若非膨胀功只有电功且反应进度 $\xi = 1$ mol 时,$\Delta_r G_m = -zEF$。式中 z 是按所写的电极反应,在反应进度为 1 mol 时,反应式中电子的计量系数,量纲为 1;E 是可逆电池的电动势,单位为伏特;F 为法拉第常数。

在电池中,化学能转变为电能的过程是以热力学可逆方式进行的,则该电池称为可逆电池。可逆电池必须满足以下三个条件,缺一不可。首先,电池反应是可逆的;其次,电池必须在接近平衡状态的情况下工作,即无论是放电还是充电过程,通过电池的电流十分微小;最后,除此之外,电池不存在任何不可逆的液体接界。

为了使电池反应在接近热力学可逆条件下进行,必须没有电流通过电池。为了达到这个目的,一般采用电位差计测量电池的电动势。电位差计在物理化学实验中的应用非常广泛,主要用以测定电动势、校正各种电表;其次作为输出可变的精密稳压电源,可应用在极谱分析、电流滴定等实验中;再次,有些电位差计(如学生型)中的滑线电阻可单独用作电桥桥臂,供精密测量电阻时应用。

电位差计是利用对消法进行电势测量的仪器,其原理是用一个大小相等、方向相反的外加电势对抗待测电池,这样待测电池中没有电流通过,外加电势的大小即等于待测电池的电动势。对消法简单原理如图 8-1 所示。它由工作回路、标准回路和测量回路组成。

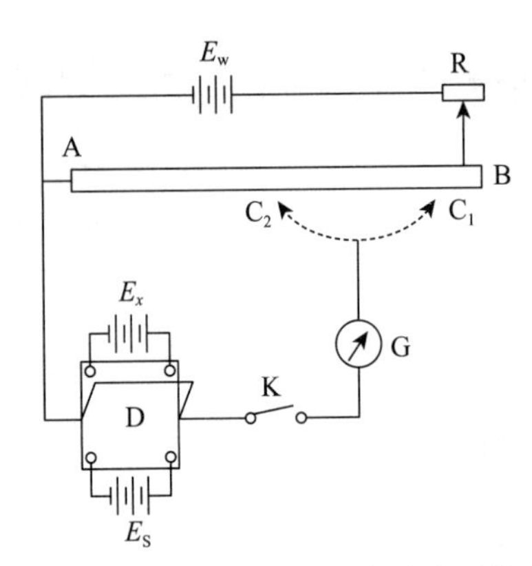

图 8-1　电位差计（对消法）测电池电动势示意图

（1）工作回路

AB 为均匀滑线电阻，与可变电阻 R 及工作电源 E_w 构成回路，调节可变电阻 R，可使流过回路的电流成为某一定值 I_0，这样 AB 上有一定的电位降产生。工作电源 E_w 可用电池串联或稳压电源，其输出电压必须大于待测电池的电动势。

（2）标准回路

E_s 为标准电池，C 是可在 AB 上移动的触点，D 是双向电钥，K 是电键，KC 间有一灵敏度很高的检流计 G。当 D 扳向 E_s 一方时，标准回路接通，若标准电池 E_s 的电动势为 1.01865 V，则先将 C 点移动到 AB 上标记 1.01865 V 的 C_1 处，闭合 K，迅速调节 R 直至 G 中无电流通过。这时 E_s 的电动势与 AC_1 之间的电位降大小相等、方向相反而对消。

（3）测量回路

当 D 换向 E_x 的一方时，在保持工作电流不变的情况下，闭合 K，将 C 在 AB 上迅速移动到 C_2 点，使 G 中无电流通过，这时 E_x 的电动势与 AC_2 间的电位降大小相等，方向相反而对消，于是 C_2 点所标记的电位降为 E_x 的电动势。由于使用过程中工作电池的电压会有所变化，要求每次测量前均须重新校准工作回路的电流。

实际应用的电位差计，滑线电阻由一系列标准电阻串联而成，工作电流总是标定为一固定数值 I_0，使电位差计总是在统一的 I_0 下达到平衡，从而可以将待测电动势的数值直接标度在各段电阻上（即标在仪器面板上），这样就可直接读取电动势的值。

2. 电极电势的计算

原电池是由两个相对独立的电极所组成的，每一个电极相当于一个半电池。若知道了两个电极的电极电势，那么就可以求出由它们所组成的电池的电动势，$E=\varphi_+-\varphi_-$。

以 $Zn(s) \mid ZnSO_4(a_1) \parallel CuSO_4(a_2) \mid Cu(s)$ 电池为例，其电极反应和电池反应分别为

负极反应：$Zn(s)-2e^- \longrightarrow Zn^{2+}(a_1)$

正极反应：$Cu^{2+}(a_2)+2e^- \longrightarrow Cu(s)$

电池反应：$Zn(s)+Cu^{2+}(a_2) \longrightarrow Cu(s)+Zn^{2+}(a_1)$

电极反应和电池反应的 Nernst 方程为

$$\varphi_{Zn^{2+}|Zn} = \varphi^{\ominus}_{Zn^{2+}|Zn} - \frac{RT}{2F}\ln\frac{a_{Zn}}{a_{Zn^{2+}}} \tag{1}$$

$$\varphi_{Cu^{2+}|Cu} = \varphi^{\ominus}_{Cu^{2+}|Cu} - \frac{RT}{2F}\ln\frac{a_{Cu}}{a_{Cu^{2+}}} \tag{2}$$

$$E = \varphi_{Cu^{2+}|Cu} - \varphi_{Zn^{2+}|Zn} = \varphi^{\ominus}_{Cu^{2+}|Cu} - \varphi^{\ominus}_{Zn^{2+}|Zn} - \frac{RT}{2F}\ln\frac{a_{Cu}a_{Zn^{2+}}}{a_{Cu^{2+}}a_{Zn}} = E^{\ominus} - \frac{RT}{2F}\ln\frac{a_{Cu}a_{Zn^{2+}}}{a_{Cu^{2+}}a_{Zn}}$$

$$= E^{\ominus} - \frac{RT}{2F}\ln\frac{a_{Zn^{2+}}}{a_{Cu^{2+}}} \tag{3}$$

到目前为止,无论是从实验上还是从理论上都没有办法计算出某个电极的电极电势,而只能测出由两个电极组成的电池的总电动势。根据 IUPAC(国际纯粹化学与应用化学联合会)的规定,采用标准氢电极作为标准电极,并将其电极电势规定为零,其他电极的电极电势值是与标准氢电极比较而得到的相对值。由于使用标准氢电极条件要求苛刻,因此在实际中常用电势稳定、容易制备、使用方便的二级标准电极(参比电极)来代替标准氢电极。甘汞电极、银-氯化银电极就是常用的二级标准电极,它们的电极电势可以与标准氢电极相比而精确测出,在物理化学手册中可以查到。

本实验用锌电极、铜电极分别与饱和甘汞电极或银-氯化银电极构成原电池,测量电池的电动势,进而获得铜、锌的电极电势。

必须指出,电极电势大小,不仅与电极种类、溶液浓度有关,而且与温度有关。本实验是在实验温度下测得的电极电势 φ_T,由式(1)或式(2)可计算 $\varphi^{\ominus}_T$,为了方便起见,可采用下式求出 298 K 时的标准电极电势 $\varphi^{\ominus}_{298}$

$$\varphi^{\ominus}_T = \varphi^{\ominus}_{298} + \alpha(T - 298) + \frac{1}{2}\beta(T - 298)^2$$

式中 α、β 为电池电极的温度系数。对铜-锌电池来说:

铜电极 (Cu^{2+} | Cu),$\alpha = -1.6\times10^{-5}$ V · K^{-1},$\beta = 0$

锌电极 [Zn^{2+} | Zn(Hg)],$\alpha = 1.0\times10^{-4}$ V · K^{-1},$\beta = 6.2\times10^{-7}$ V · K^{-2}

仪器和试剂

UJ-25 型电位差计　　检流计
标准电池　　电极管
工作电池(3V)　　电线若干
铜、锌电极　　饱和甘汞电极
银-氯化银电极　　硫酸铜(分析纯)
氯化钾(分析纯)　　硫酸锌(分析纯)

实验步骤

1. 铜、锌电极的制备

将铜、锌电极分别用抛光粉抛光后，用蒸馏水冲洗干净，再用少量待测液淋洗。

将洗好后的电极插入装好电解质溶液的电极管中，注意电极管中不能有气泡，旋紧塞子。

2. 电池电动势的测量

（1）组装待测电池

① Zn | $ZnSO_4$(0.1000 mol · L^{-1}) ‖ KCl(饱和) | AgCl | Ag

② Zn | $ZnSO_4$(0.1000 mol · L^{-1}) ‖ KCl(饱和) | Hg_2Cl_2 | Hg

③ Hg | Hg_2Cl_2 | KCl(饱和) ‖ $CuSO_4$(0.1000 mol · L^{-1}) | Cu

④ Ag | AgCl | KCl(饱和) ‖ $CuSO_4$(0.1000 mol · L^{-1}) | Cu

⑤ Zn | $ZnSO_4$(0.1000 mol · L^{-1}) ‖ $CuSO_4$(0.1000 mol · L^{-1}) | Cu

⑥ Cu | $CuSO_4$(0.0100 mol · L^{-1}) ‖ $CuSO_4$(0.1000 mol · L^{-1}) | Cu

参考图 8-2 和图 8-3 的电池①、电池⑤ 组装上述六个电池。

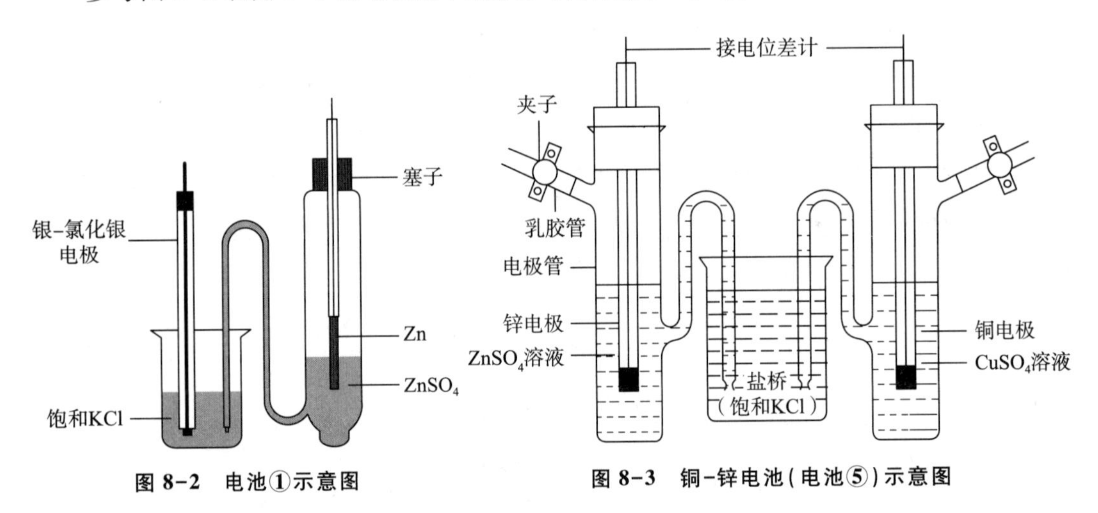

图 8-2　电池①示意图

图 8-3　铜-锌电池（电池⑤）示意图

（2）用电位差计测量电池电动势

① 本实验使用 UJ-25 型电位差计，其面板布局如图 8-4 所示。使用时先将有关的外部线路如工作电池、检流计、标准电池和待测电池等连接好。切不可将标准电池倒置或摇动。

② 接通电源，调节好检流计光点的零位。

③ 将换向开关扳向“N”（校正），调节标准电池温度补偿旋钮，使其读数值与标准电池的电动势的数值一致（注意标准电池电动势的数值受温度影响会发生变动，调节前应先计算实验温度下标准电池电动势的准确数值）。

断续按下粗按钮（当按下粗按钮时，检流计光点在一小格范围内摆动才能按细按钮。

注意按键时间不能超过 1 s)，视检流计光点的偏转情况，调节可变电阻 R（粗、中、细、微）使检流计光点指示零位。

④ 将换向开关扳向“X_1”(若待测电池接于未知 2，则扳向“X_2”)，根据理论计算出待测电池的电动势，将各档测量旋钮(六个大旋钮)预置在合适的位置。轻按粗按钮(当按下粗按钮时，检流计光点在一小格范围内摆动才能按细按钮。注意按键时间不能超过 1 s)，根据检流计光点偏转情况旋转各测量档旋钮，至检流计光点指示零位，此时电位差计各测量档小孔示数的总和，即为被测电池的电动势。

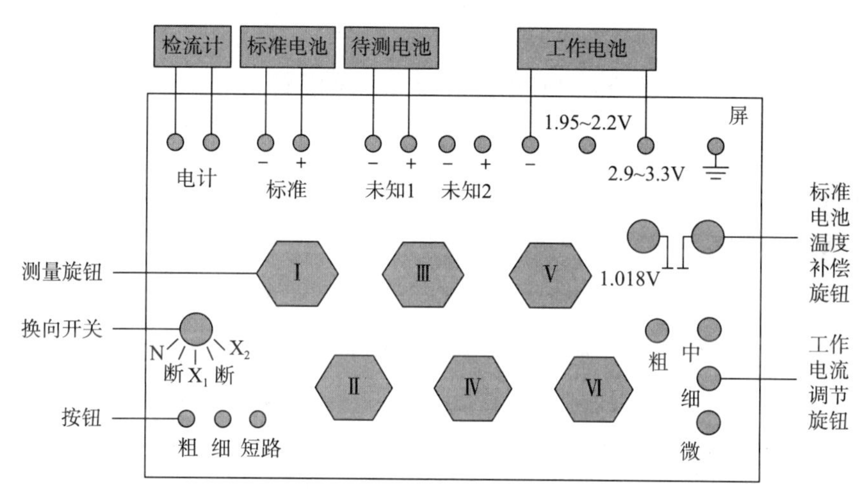

图 8-4　UJ-25 型电位差计面板

注意，每次测量前都要用标准电池对电位差计进行标定，否则，工作电池电压不稳或温度的变化会导致测量结果不准确。组成电池需稳定 15 min 才能读数，读三次数，所读数的偏差小于±0.5 mV，取其平均值。按粗细按钮时，要轻按，按键时间不能超过 1 s。

数据记录与处理

1. 列表表示所测电池的电动势测定值。

2. 根据饱和甘汞电极的电极电势温度校正公式，计算实验温度下的电极电势

$$\Phi_{SCE}/V = 0.2415 - 7.61 \times 10^{-4}(T - 298)$$

3. 根据测定的电池电动势，分别计算铜、锌电极的 φ_T，$\varphi_T^{\ominus}$，$\varphi_{298}^{\ominus}$。

4. 根据有关公式计算铜-锌电池的 $E_{理论}$(活度系数查表)，并与 $E_{实际}$ 进行比较。

实验注意事项

1. 标准电池：(1) 使用温度 4～40 ℃；(2) 正确连接正、负极；(3) 标准电池不能倒置；(4) 不能直接用万用电表测量其电动势；(5) 标准电池不能做电源使用。

2. 测试时必须先按“粗”按钮观察检流计光点是否为零，当检流计光点为零以后再按下“细”按钮观察检流计光点是否为零。无论在校正还是测量过程中，都不能将电键长时

间地按下（以免电极被极化），而应轻轻按下，迅速观察检流计的情况，然后放开按钮，调整相应旋钮后再按按钮检查，反复调整达到目的。在没有调准的情况下，长时间地按下按钮（甚至锁定）会造成标准电池或待测电池长期放电，将损坏标准电池，或导致测量误差很大。

3. 测量过程中若出现检流计受到冲击，应迅速按下“短路”按钮以保护检流计。

4. 实验中的废液、废物不能直接倒入下水道，而应倒入废液桶，以便集中处理。

思考与讨论

1. 为什么不能用伏特计测量电池电动势？

2. 对消法测量电池电动势的主要原理是什么？

3. 应用电位差计测量电动势过程中，若检流计光点总是朝向一个方向偏转，可能是什么原因？

参考资料

1. 武汉大学化学与分子科学学院实验中心. 物理化学实验. 武汉：武汉大学出版社，2012：63–71.

2. 复旦大学，等. 物理化学实验. 北京：高等教育出版社，2004：68–73.

3. 傅献彩，侯文华. 物理化学（第六版）上册. 北京：高等教育出版社，2022.

4. 玉占君，冯春梁，等，物理化学实验. 化学工业出版社，2014.

Experiment 8

Determination of electromotive force of galvanic cell and calculation of thermodynamic functions

Experimental purpose

1. Master the principle of potentiometric determination of the cell electromotive force and the use of a potentiometer.

2. Use the potentiometer to determine the electromotive force of certain cells.

3. Master the methods for determining electrode potentials.

Basical principles

The study of reversible cells in electrochemistry provides a theoretical basis for understanding the limits of chemical energy conversion into electrical energy and improving battery performance. The accurate measurement of cell electromotive force can greatly assist in obtaining thermodynamic data (such as equilibrium potential, activity coefficients, dissociation constants, solubility, complex formation constants, etc.) Therefore, precise measurement of the electromotive force of a cell is of great significance in physical chemistry research.

1. Basic principle of potentiometer (compensation method) for measuring cell electromotive force

For a reversible cell working in a state of equilibrium or approaching equilibrium under isothermal and isobaric conditions, the decrease in the system's Gibbs free energy during a change equals the maximum non-expansion work done on the surroundings, i.e., $\Delta_r G = W_{f,max}$. If the non-expansion work is only electrical work, and the reaction progress $\xi = 1$ mol, then $\Delta_r G_m = -zEF$. Here, z is the stoichiometric coefficient of electrons in the electrode reaction, and F is the Faraday constant.

To achieve near-equilibrium conditions in a cell reaction, it is necessary to prevent current flow through the cell. The potentiometer is commonly used to measure the cell electromotive force. The potentiometer is widely applied in physical chemistry experiments, mainly for determining electromotive force, calibrating various meters, serving as a variable precision stabilized power supply in experiments like polarography and coulometric titration, and providing a precise bridge arm for resistance measurements.

The potentiometer uses the compensation method, where an opposing potential of equal magnitude and opposite direction is applied to the cell under test, ensuring no current flows through the cell. This opposing potential is then equal to the electromotive force of the cell under test. Refer to Fig. 8-1 for a schematic diagram of the potentiometer (compensation method) for measuring cell electromotive force. It consists of three circuits: working circuit, standard circuit and measurement circuit.

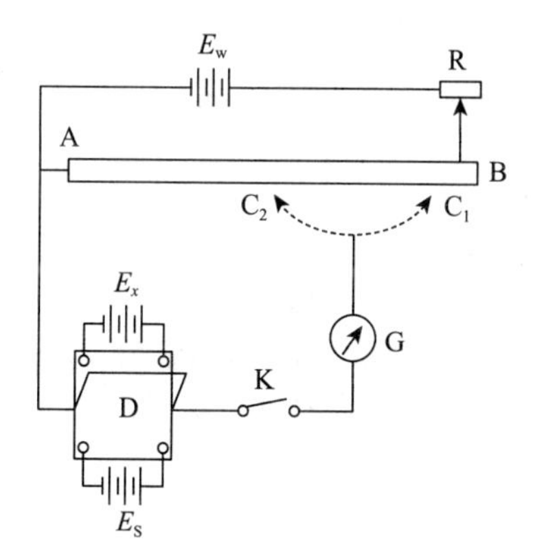

Fig. 8-1 Schematic diagram of potentiometer (compensation method) for measuring cell electromotive force

(1) Working circuit

AB is a uniform sliding wire resistor connected to a variable resistor R and a working power source E_w. Adjusting the variable resistor R sets the current through the circuit to a constant value I_0, creating a potential drop on AB. The working power source E_w can be a battery or a regulated power supply with an output voltage greater than the electromotive force of the cell under test.

(2) Standard circuit

E_n is the Weston standard cell, C is a movable contact on AB, D is a double-throw switch, K is a key, and G is a sensitive galvanometer. When D is switched towards E_s, the standard circuit is connected. If the electromotive force of the standard cell E_s is 1.01865 V, move C to mark point C_1 on AB corresponding to 1.01865 V. Close K, quickly adjust R until G shows no current passing through. At this point, the potential drop across E_s is equal in magnitude but opposite in direction to that across AC_1, canceling each other out.

(3) Measurement Circuit

When D is switched towards E_x, maintain the working current constant, close K, and quickly move C on AB to point C_2. Adjust R until G indicates zero current. At this point, the potential drop across E_x is equal in magnitude but opposite in direction to that across AC_2, canceling each other out. The potential drop at point C_2 represents the electromotive force of the cell under test.

For practical potentiometers, the sliding wire resistor is composed of a series of standard resistors. The working current is always set to a fixed value I_0, ensuring that the potentiometer always reaches equilibrium under the same I_0. The electromotive force of the cell under test can be directly read from the instrument panel by scaling it on various sections of the resistor.

2. Calculation of electrode potentials

A galvanic cell consists of two relatively independent electrodes, each equivalent to a half-cell. Knowing the electrode potentials of the two electrodes allows for the calculation of the cell's electromotive force

$$E=\varphi_+-\varphi_-$$

Taking the $Zn(s)\,|\,ZnSO_4(\alpha_1)\,\|\,CuSO_4(\alpha_2)\,|\,Cu(s)$ cell as an example, the electrode reactions and cell reaction are as follows:

Anode reaction: $Zn(s)-2e^- \longrightarrow Zn^{2+}(a_1)$

Cathode reaction: $Cu^{2+}(a_2)+2e^- \longrightarrow Cu(s)$

Cell reaction: $Zn(s)+Cu^{2+}(a_2) \longrightarrow Cu(s)+Zn^{2+}(a_1)$

The Nernst equation for electrode and cell reactions is given by

$$\varphi_{Zn^{2+}|Zn} = \varphi^{\ominus}_{Zn^{2+}|Zn} - \frac{RT}{2F}\ln\frac{a_{Zn}}{a_{Zn^{2+}}} \tag{1}$$

$$\varphi_{Cu^{2+}|Cu} = \varphi^{\ominus}_{Cu^{2+}|Cu} - \frac{RT}{2F}\ln\frac{a_{Cu}}{a_{Cu^{2+}}} \tag{2}$$

$$E = \varphi_{Cu^{2+}|Cu} - \varphi_{Zn^{2+}|Zn} = \varphi^{\ominus}_{Cu^{2+}|Cu} - \varphi^{\ominus}_{Zn^{2+}|Zn} - \frac{RT}{2F}\ln\frac{a_{Cu}a_{Zn^{2+}}}{a_{Cu^{2+}}a_{Zn}} = E^{\ominus} - \frac{RT}{2F}\ln\frac{a_{Cu}a_{Zn^{2+}}}{a_{Cu^{2+}}a_{Zn}}$$

$$= E^{\ominus} - \frac{RT}{2F}\ln\frac{a_{Zn^{2+}}}{a_{Cu^{2+}}} \tag{3}$$

So far, there has been no experimental or theoretical method to directly calculate the electrode potential of a single electrode. Instead, the total electromotive force (EMF) of a cell comprising two electrodes can be measured. According to IUPAC standards, the standard hydrogen electrode is used as a reference electrode, with its electrode potential defined as zero. The electrode potential values of other electrodes are then determined relative to the standard hydrogen electrode. Due to the demanding conditions of using a standard hydrogen electrode, practical applications often substitute it with secondary standard electrodes, such as the calomel electrode and silver-chloride electrode, which offer stability, ease of preparation, and convenience of use. Detailed information on their electrode potentials can be found in physical chemistry handbooks.

In this experiment, zinc and copper electrodes are combined with either a saturated calomel electrode or a silver-chloride electrode to form the original cell. The cell's EMF is measured to determine the electrode potentials of copper and zinc.

It is essential to note that the electrode potential not only depends on the type of electrode and solution concentration but also varies with temperature. The experiment measures the electrode potential φ_T at the experimental temperature. The standard electrode potential $\varphi^{\ominus}_T$ at 298 K can be calculated using the following formula

$$\varphi^{\ominus}_T = \varphi^{\ominus}_{298} + \alpha(T - 298) + \frac{1}{2}\beta(T - 298)^2$$

where α and β are the temperature coefficient of the electrode. For the copper-zinc cell

The copper electrode($Cu^{2+}|Cu$), $\alpha = -1.6\times10^{-5}\ V\cdot K^{-1}$, $\beta = 0$

The zinc electrode [$Zn^{2+}|Zn(Hg)$], $\alpha = 1.0\times10^{-4}\ V\cdot K^{-1}$, $\beta = 6.2\times10^{-7}\ V\cdot K^{-2}$

Apparatus and reagents

UJ-25 potentiometer
Galvanometer
Standard cell
Electrode tube
Working battery (3V)
wires
Copper and zinc electrodes
Saturated calomel electrode
Silver chloride electrode
$CuSO_4$(AR)
KCl(AR)
$ZnSO_4$(AR)

Experimental procedures

1. Preparation of copper and zinc electrodes

Polish copper and zinc electrodes with polishing powder, rinse with distilled water, and rinse with a small amount of the test solution.

Insert the cleaned electrodes into the electrode tubes containing the electrolyte solution, ensuring no air bubbles, and tighten the stopper.

2. Measurement of the cell electromotive potential

(1) Assemble the cell to be tested

① $Zn(s) | ZnSO_4(0.1000\ mol \cdot L^{-1}) \| KCl(Saturation) | AgCl(s) | Ag(s)$

② $Zn(s) | ZnSO_4(0.1000\ mol \cdot L^{-1}) \| KCl(Saturation) | Hg_2Cl_2(s) | Hg(l)$

③ $Hg(s) | Hg_2Cl_2 | KCl(Saturation) \| CuSO_4(0.1000\ mol \cdot L^{-1}) | Cu(s)$

④ $Ag(s) | AgCl(s) | KCl(Saturation) \| CuSO_4(0.1000\ mol \cdot L^{-1}) | Cu(s)$

⑤ $Zn(s) | ZnSO_4(0.1000\ mol \cdot L^{-1}) \| CuSO_4(0.1000\ mol \cdot L^{-1}) | Cu(s)$

⑥ $Cu(s) | CuSO_4(0.1000\ mol \cdot L^{-1}) \| CuSO_4(0.1000\ mol \cdot L^{-1}) | Cu(s)$

Taking galvanic cell ① as an example: using saturated KCl solution as a salt bridge, and the battery to be tested is assembled according to Fig. 8-2. Fig. 8-3 shows the schematic diagram of the ⑤ Cu-Zn galvanic cell.

(2) Use a potentiometer to measure the electromotive force of the cell

① The UJ-25 type potential difference meter is used in this experiment. Its panel layout is shown in Fig. 8-4. When in use, connect the relevant external circuits such as the working battery, ammeter, standard battery, and the battery under test. Be sure not to invert or shake the standard battery.

② Turn on the power and adjust the zero position of the ammeter light spot.

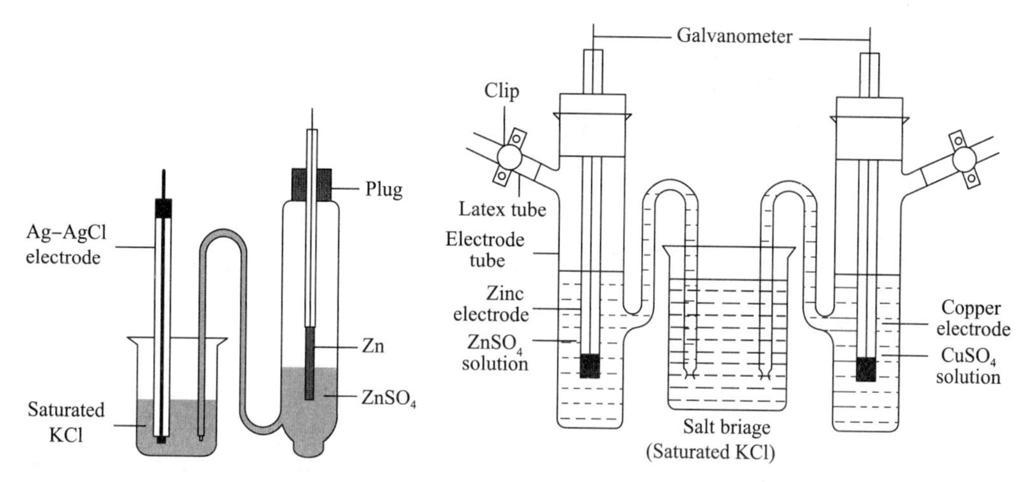

Fig. 8-2 Cell diagram

Fig. 8-3 Cu-Zn galvanic cell

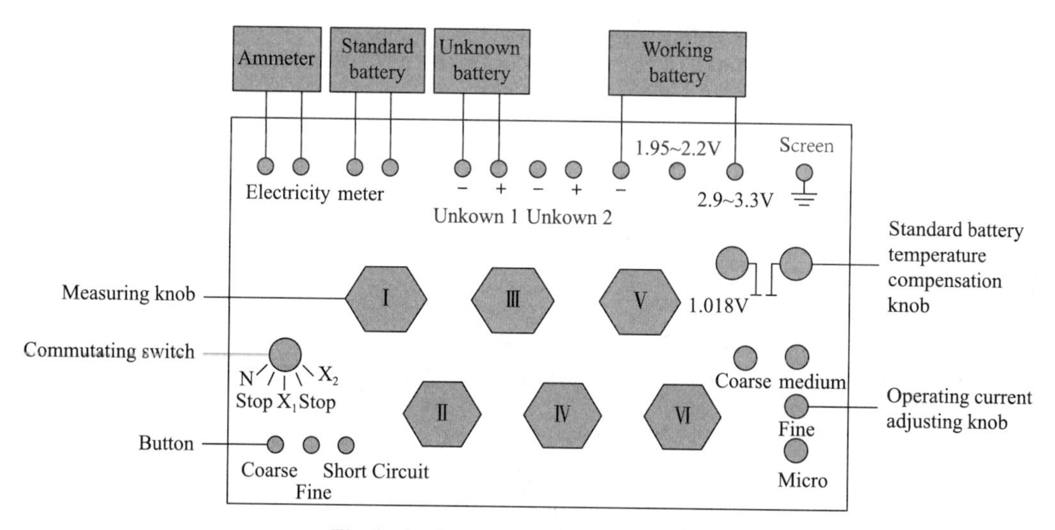

Fig. 8-4 Schematic diagram of UJ-25

③ Turn the switch to "N" (calibration), adjust the temperature compensation knob of the standard battery to make its reading consistent with the electromotive force value of the standard battery (note that the electromotive force value of the standard battery is subject to temperature changes, and the accurate value of the standard battery electromotive force at the experimental temperature should be calculated before adjustment).

Continuously press the coarse button (when the coarse button is pressed, the light spot of the ammeter swings within a small range before pressing the fine button. Note that the button press time should not exceed 1s). Depending on the deflection of the ammeter light spot, adjust the variable resistor R (coarse, medium, fine, micro) to make the ammeter light spot indicate zero.

④ Turn the switch to "X_1" (if the battery under test is connected to the unknown 2, then switch to "X_2"), calculate the theoretical electromotive force of the battery under test, preset the measuring knobs (six large knobs) to appropriate positions. Gently press the coarse button (when the coarse button is pressed, the light spot of the ammeter swings within a small range before pressing the fine button. Note that the button press time should not exceed 1s) and rotate the measuring knobs according to the deflection of the ammeter light spot. When the ammeter light spot indicates zero, the sum of the small hole readings of the potentiometer at each measuring position is the electromotive force of the battery under test.

Note that the potentiometer should be calibrated with a standard battery before each measurement. Otherwise, the measurement results will be inaccurate due to the unstable voltage of the working battery or the change of temperature. The battery needs to be stable for 15 min to read. Read three times, the reading deviation should be less than ± 0.5 mV, take its average. When pressing the thickness button, press it gently for no more than 1 s.

Data recording and processing

1. List represents the EMF determination value of the measured battery.

2. According to the temperature correction formula of SCE electrode potential, the electrode potential at experimental temperature is calculated

$$\Phi_{SCE}/V = 0.2415 - 7.61 \times 10^{-4}(T - 298)$$

3. According to the measured electromotive force of the battery, φ_T, $\varphi_T^\ominus$, $\varphi_{298}^\ominus$ of the copper and zinc electrodes are calculated respectively.

4. Calculate the theoretical electromotive force of the copper-zinc cell, considering activity coefficients and compare it with the experimental electromotive force.

Experimental precautions

1. Standard battery: (1) Use within the temperature range of 4 to 40 ℃. (2) Ensure that connection of positive and negative poles is correct. (3) Do not invert the standard cell. (4) Do not directly measure its electromotive force with a multimeter. (5) The standard cell should not be used as a power source.

2. During the test, press the "coarse" button first to observe whether the ammeter's light spot is zero. Only after the light spot is zero should you press the "fine" button and check if the light spot remains at zero. Whether during calibration or measurement, avoid pressing the button for an extended period (to prevent electrode polarization). Press the button lightly, quickly observe the ammeter's status, release the button, adjust the corresponding knob, and press the button again to check. Repeat adjustments until the desired outcome is achieved. Prolonging

button press without proper calibration can lead to long-term discharge of the standard cell or the test cell, causing damage to the standard cell and significant measurement errors.

3. If the ammeter experiences a shock during the measurement process, quickly press the "short circuit" button to protect the ammeter.

4. Waste liquids and materials from the experiment should not be directly poured into the sewer but should be disposed of in a waste liquid container for centralized processing.

Thinking and discussing

1. Why is a voltmeter not suitable for measuring cell EMF?

2. What is the main principle of the compensation method for measuring cell EMF?

3. If the ammeter consistently deflects in one direction during the potentiometer measurement, what could be the possible reasons?

Reference materials

1. Wuhan University, College of Chemistry and Molecular Sciences, Experimental Center. Physical Chemistry Experiment. Wuhan: Wuhan University Press, 2012: 63-71.

2. Fudan University, et al. Physical Chemistry Experiment. Beijing: Higher Education Press, 2004: 68-73.

3. Fu X, Hou W. Physical Chemistry (6th ed, Volume Ⅰ). Beijing: Higher Education Press, 2022.

4. Yu Z, Feng C, et al. Physical Chemistry Experiments. Beijing: Chemical Industry Press, 2014.

实验 9 电势-pH 曲线的测定与应用

实验目的

1. 测定 Fe^{3+}/Fe^{2+}-EDTA 配位系统在不同 pH 条件下的电极电势，绘制电势-pH 曲线。

2. 据测定的电势-pH 曲线设计较合适的脱硫条件，并进行实验研究。

基本原理

1. Fe^{3+}/Fe^{2+}-EDTA 系统的电势-pH 曲线

电极电势的数值反映了物质氧化还原能力，可以由其判断电化学反应是否能够发生。

对于有 H^+或 OH^-参加的电极反应,它们的电极电势数值与溶液的 pH 有关。对于这样的系统,有必要考察其电极电势与 pH 之间的关系,从而对电极反应得到一个比较完整、清晰的认识。把一些有 H^+或 OH^-参加的电极电势与溶液的 pH 绘制成图,这样就绘制出了电势-pH 曲线,也称电势-pH 图。图 9-1 为 Fe^{3+}/Fe^{2+}-EDTA 和 S/H_2S 系统的电势与 pH 的关系示意图。

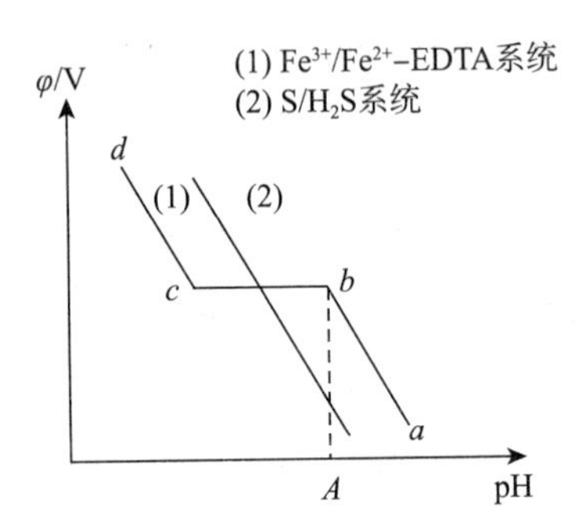

图 9-1　Fe^{3+}/Fe^{2+}-EDTA 和 S/H_2S 系统的电势与 pH 的关系

对于 Fe^{3+}/Fe^{2+}-EDTA 系统,pH 不同,形成的配合物也有所差异。假定 EDTA 的酸根离子为 Y^{4-},下面将 pH 分成 3 个区间来讨论 Fe^{3+}/Fe^{2+}-EDTA 系统电极电势的变化。

(1) 在高 pH (图 9-1 中的 ab 区间) 时,溶液的配合物为 $Fe(OH)Y^{2-}$和 FeY^{2-},其电极反应为

$$Fe(OH)Y^{2-}+e^- \Longrightarrow FeY^{2-}+OH^-$$

根据电极反应的能斯特(Nernst)方程,其电极电势为

$$\varphi = \varphi^{\ominus} - \frac{RT}{F}\ln \frac{a_{FeY^{2-}} \cdot a_{OH^-}}{a_{Fe(OH)Y^{2-}}} \tag{1}$$

式中 $\varphi^{\ominus}$为标准电极电势,a 为活度。

a 与活度因子 γ 和质量摩尔浓度 m 的关系为

$$a = \gamma \cdot \frac{m}{m^{\ominus}} \tag{2}$$

同时考虑到在稀溶液中水的活度积 K_w,按照 pH 的定义,则式(1)可改写为

$$\varphi = \varphi^{\ominus} - \frac{RT}{F}\ln \frac{\gamma_{FeY^{2-}} \cdot K_w}{\gamma_{Fe(OH)Y^{2-}}} - \frac{RT}{F}\ln \frac{m_{FeY^{2-}}}{m_{Fe(OH)Y^{2-}}} - \frac{2.303RT}{F}\text{pH} \tag{3}$$

令 $b_1 = \frac{RT}{F}\ln \frac{\gamma_{FeY^{2-}} \cdot K_w}{\gamma_{Fe(OH)Y^{2-}}}$, 在溶液离子强度和温度一定时,$b_1$ 为常数。则

$$\varphi = (\varphi^{\ominus} - b_1) - \frac{RT}{F}\ln \frac{m_{FeY^{2-}}}{m_{Fe(OH)Y^{2-}}} - \frac{2.303RT}{F}\text{pH} \tag{4}$$

当 EDTA 过量时,系统中的配合物的浓度可近似地看作溶液中铁离子的浓度,即

$m_{FeY^{2-}} \approx m_{Fe^{2+}}$，$m_{Fe(OH)Y^{2-}} \approx m_{Fe^{3+}}$。当 $m_{Fe^{3+}}$ 与 $m_{Fe^{2+}}$ 比例一定时，φ 与 pH 呈线性关系，即图 9-1 中的 ab 段。

（2）在特定的 pH 范围内，Fe^{2+}、Fe^{3+} 与 EDTA 生成稳定的配合物 FeY^{2-} 和 FeY^{-}，其电极反应为

$$FeY^{-} + e^{-} \xlongequal{\quad} FeY^{2-}$$

电极电势表达式为

$$\varphi = \varphi^{\ominus} - \frac{RT}{F}\ln\frac{a_{FeY^{2-}}}{a_{FeY^{-}}} = \varphi^{\ominus} - \frac{RT}{F}\ln\frac{\gamma_{FeY^{2-}}}{\gamma_{FeY^{-}}} - \frac{RT}{F}\ln\frac{m_{FeY^{2-}}}{m_{FeY^{-}}}$$

$$= (\varphi^{\ominus} - b_2) - \frac{RT}{F}\ln\frac{m_{FeY^{2-}}}{m_{FeY^{-}}} \tag{5}$$

令 $b_2 = \frac{RT}{F}\ln\frac{\gamma_{FeY^{2-}}}{\gamma_{FeY^{-}}}$，当温度和离子强度一定时，$b_2$ 为常数。在此 pH 范围内，系统的电极电势只与 $\frac{m_{FeY^{2-}}}{m_{FeY^{-}}}$ 的比值有关，或者说只与配制溶液时 $\frac{m_{Fe^{2+}}}{m_{Fe^{3+}}}$ 的比值有关。曲线中出现平台区（如图 9-1 中的 bc 段）。

（3）在低 pH 时，系统的电极反应为

$$FeY^{-} + H^{+} + e^{-} \xlongequal{\quad} FeHY^{-}$$

同理可求得

$$\varphi = \varphi^{\ominus} - \frac{RT}{F}\ln\frac{a_{FeHY^{-}}}{a_{FeY^{-}}} = \varphi^{\ominus} - \frac{RT}{F}\ln\frac{\gamma_{FeHY^{-}}}{\gamma_{FeY^{-}}} - \frac{RT}{F}\ln\frac{m_{FeHY^{-}}}{m_{FeY^{-}}}$$

$$= (\varphi^{\ominus} - b_3) - \frac{RT}{F}\ln\frac{m_{FeHY^{-}}}{m_{FeY^{-}}} - \frac{2.303RT}{F}\mathrm{pH} \tag{6}$$

令 $b_3 = \frac{RT}{F}\ln\frac{\gamma_{FeHY^{-}}}{\gamma_{FeY^{-}}}$，当温度和离子强度一定时，$b_3$ 为常数，在 $\frac{m_{Fe^{2+}}}{m_{Fe^{3+}}}$ 不变时，φ 与 pH 呈线性关系（即图 9-1 中 cd 段）。

由此可见，用惰性金属（如 Pt）作导体，将 Fe^{3+}/Fe^{2+}-EDTA 系统组成电极，与另一参比电极（如饱和甘汞电极）组成电池，测量电池的电动势即可求得系统（Fe^{3+}/Fe^{2+}-EDTA）的电极电势。与此同时，采用酸度计测出对应条件下的 pH，就可绘制出 Fe^{3+}/Fe^{2+}-EDTA 系统的电势-pH 曲线。

2. 电势-pH 曲线的应用

在实际应用中，Fe^{3+}/Fe^{2+}-EDTA 系统可用于天然气脱硫。天然气中含有 H_2S，它是一种有害物质，大量吸入会损害健康，如当空气中硫化氢浓度达到 20 $mg \cdot m^{-3}$ 时会引起恶心、头晕、头痛、疲倦、胸部压迫及眼、鼻、咽喉黏膜的刺激症状；硫化氢浓度达 60 $mg \cdot m^{-3}$

时，则会出现抽搐、昏迷甚至呼吸中枢麻痹而死亡。利用 Fe^{3+}-EDTA 溶液可将天然气中的 H_2S 氧化为单质 S，然后过滤除去；溶液中的 Fe^{3+}-EDTA 配合物还原为 Fe^{2+}-EDTA 配合物，通入空气又可使 Fe^{2+}-EDTA 迅速氧化为 Fe^{3+}-EDTA，从而使溶液得到再生，循环利用。其反应过程如下

$$2FeY^- + H_2S \xrightarrow{\text{脱硫}} 2FeY^{2-} + 2H^+ + S\downarrow$$

$$2FeY^- + \frac{1}{2}O_2 + H_2O \xrightarrow{\text{再生}} 2FeY^- + 2OH^-$$

我们可根据测定的 Fe^{3+}/Fe^{2+}-EDTA 配合系统的电势-pH 曲线选择较合适的脱硫条件。例如，低含硫天然气 H_2S 含量约为 $1\times10^{-4}\sim6\times10^{-4}$ kg · m^{-3}，在 25 ℃时相应的 H_2S 的分压为 7.29～43.56 Pa。

根据电极反应

$$S + 2H^+ + 2e^- = H_2S(g)$$

在 25 ℃时，其电极电势 $\varphi/V = 0.318-0.0296\lg(p/p^\ominus)-0.0591\text{pH}$。

对于 H_2S 压力确定的 S/H_2S 系统，其 φ 和 pH 的关系，如图 9-1 中的曲线(2)所示。可以看出，对任何具有一定 $\frac{m_{Fe^{2+}}}{m_{Fe^{3+}}}$ 比值的脱硫液而言，此脱硫液的电极电势与反应 $S+2H^++2e^- = H_2S(g)$ 的电极电势之差值在电势平台区的 pH 范围内随着 pH 的增大而增大，到平台区的 pH 上限时，两电极电势的差值最大；超过此 pH，两电极电势差值不再增大而是为定值。这一事实表明，任何具有一定 $\frac{m_{Fe^{2+}}}{m_{Fe^{3+}}}$ 比值的脱硫液在它的电势平台区的 pH 上限时，脱硫的热力学趋势达到最大，超过此 pH 后，脱硫趋势不再随 pH 增大而增加。可见图 9-1 中 A 点以及大于 A 点的 pH 是该系统脱硫的合适条件。

还应指出，脱硫液的 pH 不宜过大。实验表明，如果 pH 大于 12，会有 $Fe(OH)_3$ 沉淀出来，在实验中必须注意。

仪器和试剂

pH-3V 酸度电势测定仪
磁力搅拌器
复合电极
铂电极
150 mL 夹套瓶
$FeCl_3 \cdot 6H_2O$
$FeCl_2 \cdot 4H_2O$
EDTA（四钠盐）
NaOH
HCl
标准缓冲溶液
N_2

实验步骤

1. 仪器的校正

pH−3V 酸度电势测定仪的面板视窗如图 9−2 所示。

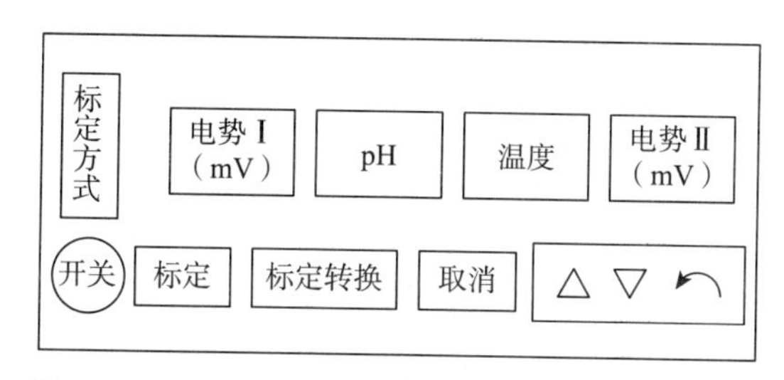

图 9−2　pH−3V 酸度电势测定仪的面板视窗示意图

（1）打开电源开关，仪器预热 15 min。

（2）在仪器测量状态下，按下标定转换键，选择标定方式（1 点法或 2 点法，建议使用 2 点法较准确）。按住标定键 3 s 以上，标定指示灯亮，将复合电极、温度传感器放入装有标准缓冲溶液的小烧杯中，此时 pH 显示窗口的小数点后第三位闪烁，等到电势Ⅰ稳定，根据所显示的温度确定标准溶液的标准 pH，如常用 pH 为 7 的磷酸盐标准缓冲溶液在 25 ℃的 pH 为 6.863，在 pH 小数点后第三位闪烁时，可用增加键△或减小键▽使显示窗口数值为 3；然后按换位键（↶），小数点后第二位数字闪烁，仍用增加△或减小键▽标定，使小数点后第二位数字为 6；以此类推，直至该缓冲溶液标定完毕。注意在标定过程中如果输入失误，可按取消键重新输入。另外，换位键只能从右向左逐位标定。

若设置的是 2 点法标定，则用 pH 为 7 的缓冲溶液标定后，还应将电极、温度传感器清洗干净，继续用 pH 为 4 的标准缓冲溶液重复上述操作，继续标定。

（3）第二次标定后，按换位键（↶）仪器将自动进入测量状态，将复合电极、温度传感器用去离子水清洗干净，待用。

2. 溶液配制

分别配制 4 mol · L^{-1} NaOH、4 mol · L^{-1} HCl 和 0.1 moL · L^{-1} $FeCl_3$ 溶液，并放入相应的瓶中备用。

3. 电池电动势和 pH 的测定

量取 64 mL 0.1 moL · L^{-1} $FeCl_3$ 溶液放入夹套瓶中，加入 7.0 g EDTA、16 mL H_2O，打开磁力搅拌器搅拌，持续通氮气 10 min 后加入 1.0 g $FeCl_2 \cdot 4H_2O$，继续搅拌。

将清洗干净的复合电极、铂电极插入溶液中，按图 9−3 所示，安装仪器和电极。用 4 mol · L^{-1} NaOH 调节溶液的 pH（溶液颜色变为红褐色，pH 大约位于 7.5～8.0），待仪器显示稳定后（约 10 min），可从仪器视窗直接读取溶液的 pH 和电势数据（电势Ⅱ窗口）。滴加 4 mol · L^{-1}的 HCl 溶液调节 pH（每次改变 pH 约 0.3），读取 pH 和相应的电池电动势数据,直到溶液变浑浊为止。

向夹套瓶中再加 1.0 g $FeCl_2 \cdot 4H_2O$,重复前面操作。继续加 1.0 g $FeCl_2 \cdot 4H_2O$,重复前面操作。

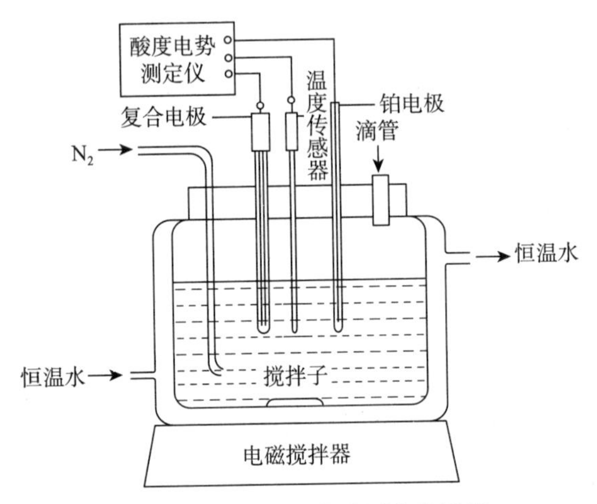

图 9-3 电势-pH 曲线测定装置图

3. S/H_2S 体系电势-pH 曲线的绘制

在含硫低的天然气中,H_2S 含量约为 1×10^{-4} ~ 1×10^{-4} $kg \cdot m^{-3}$,在 25 ℃时相应的 H_2S 的分压为 7.29~43.56 Pa。当 H_2S 的分压分别为 7.29 和 43.56 Pa 时,根据电极电势

$$\varphi/V = -0.072 - 0.0296\ \lg(p/p^{\ominus}) - 0.0591\ \text{pH}$$

可绘制 φ_{H_2S} 和 pH 关系曲线,即 $S+2H^++2e^- \rightleftharpoons H_2S(g)$ 反应的电势-pH 曲线。

根据所得的 Fe^{3+}/Fe^{2+}-EDTA 配合系统及 S/H_2S 系统的电势-pH 曲线可选择适当的脱硫条件。

数据记录与处理

1. 以表格形式正确记录数据,由测定的电池电动势求算出相对标准氢电极的 Fe^{3+}/Fe^{2+}-EDTA 系统的电极电势,绘制电势-pH 曲线,由曲线确定 FeY^- 和 FeY^{2-} 稳定的 pH 范围。

2. 通过所得的 S/H_2S 系统及 Fe^{3+}/Fe^{2+}-EDTA 系统的电势-pH 曲线,确定合适的理论脱硫条件。

实验注意事项

1. 搅拌速度必须加以控制,防止由于搅拌不均匀造成加入 NaOH 时,溶液上部出现少量的 $Fe(OH)_3$ 沉淀。

2. 复合电极不要与强吸水溶剂接触太久,在强碱溶液中使用应尽快操作,用毕立即用水洗净,玻璃电极球泡膜很薄,不能与玻璃杯等硬物相碰。

思考与讨论

1. 写出 Fe^{3+}/Fe^{2+}-EDTA 系统的电势平台区、低 pH 和高 pH 时，系统的基本电极反应及其所对应的电极电势公式的具体表示式，并指出各项的物理意义。

2. 脱硫液的 $m(Fe^{3+})$ / $m(Fe^{2+})$ 比值不同，测得的电势-pH 曲线有什么差异？

参考文献

1. 复旦大学，等. 物理化学实验. 北京：高等教育出版社，2004：73-77.

2. 游效曾. 电势-pH 图及其应用. 化学通报，1975，2：60-65.

3. 四川大学化学系天然气脱硫科研组. Fe(Ⅲ)[Fe(Ⅱ)]-EDTA 配合系统的电位—pH 曲线及用 EDTA 配合铁盐法脱除天然气中 H_2S 时脱硫条件的探讨. 四川大学学报：自然科学版，1976，3：23-31.

Experiment 9

Determination and application of potential-pH curve

Experimental purpose

1. Determine the electrode potential of Fe^{3+}/Fe^{2+} - EDTA system under different pH conditions and plot the potential-pH curve.

2. Design suitable desulfurization conditions based on the measured potential-pH curves and conduct experimental studies.

Basical principles

1. Potential-pH curve of Fe^{3+}/Fe^{2+}-EDTA system

The numerical value of electrode potential reflects the redox ability of substances and can be used to determine whether electrochemical reactions can occur. For electrode reactions involving H^+ or OH^-, their electrode potential values are related to the solution's pH. To gain a comprehensive understanding of the electrode reaction, it is necessary to examine the relationship between electrode potential and pH for systems involving H^+ or OH^-. This relationship is represented graphically by plotting the electrode potential against pH, creating the potentiometric-pH curve. Fig. 9-1 illustrates the relationship between the electrode potential and pH for the Fe^{3+}/Fe^{2+}-EDTA and S/H_2S systems.

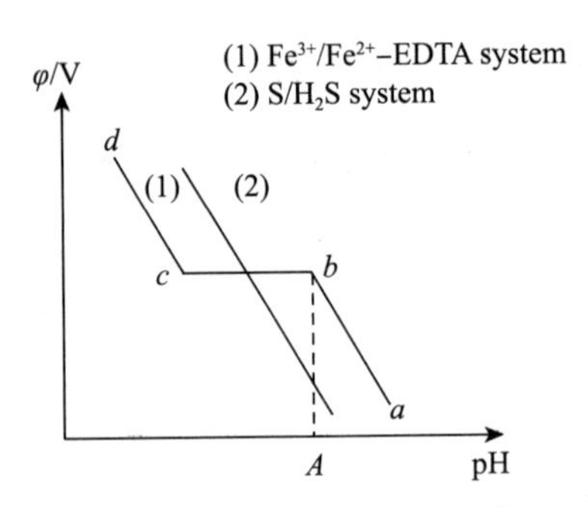

Fig. 9-1 Schematic of potential-pH curve for Fe^{3+}/Fe^{2+}-EDTA and S/H_2S systems

For the Fe^{3+}/Fe^{2+} - EDTA system, the potentiometric behavior varies with different pH values due to the formation of different complexes. Assuming that the acid ion of EDTA is Y^{4-}, the pH values can be divided into three intervals to discuss the variation in the electrode potential of the Fe^{3+}/Fe^{2+}-EDTA system.

(1) At high pH (the interval *ab* in Fig. 9-1), the complexes in the solution are $Fe(OH)Y^{2-}$ and FeY^{2-}, and the electrode reaction is

$$Fe(OH)Y^{2-} + e \longrightarrow FeY^{2-} + OH^-$$

According to Nernst equation, the electrode potential is given by

$$\varphi = \varphi^\ominus - \frac{RT}{F}\ln\frac{a_{FeY^{2-}} \cdot a_{OH^-}}{a_{Fe(OH)Y^{2-}}} \tag{1}$$

where $\varphi^\ominus$ is the standard electrode potential and a is the activity. The relationship between a and the activity coefficient γ, as well as the mass molar concentration (m), is given by

$$a = \gamma \cdot \frac{m}{m^\ominus} \tag{2}$$

Taking into account the activity product K_w of water in dilute solutions, and according to the definition of pH, Equation (1) can be rewritten as

$$\varphi = \varphi^\ominus - \frac{RT}{F}\ln\frac{\gamma_{FeY^{2-}} \cdot K_w}{\gamma_{Fe(OH)Y^{2-}}} - \frac{RT}{F}\ln\frac{m_{FeY^{2-}}}{m_{Fe(OH)Y^{2-}}} - \frac{2.303RT}{F}\text{pH} \tag{3}$$

Let $b_1 = \frac{RT}{F}\ln\frac{\gamma_{FeY^{2-}} \cdot K_w}{\gamma_{Fe(OH)Y^{2-}}}$, with the ionic strength and temperature held constant in the solution, b_1 becomes a constant. Therefore

$$\varphi = (\varphi^\ominus - b_1) - \frac{RT}{F}\ln\frac{m_{FeY^{2-}}}{m_{Fe(OH)Y^{2-}}} - \frac{2.303RT}{F}\text{pH} \tag{4}$$

When EDTA is in excess, the concentration of the complexes can be approximately regarded as the concentration of the corresponding ferrous or ferric ions in the solution, i.e. $m_{FeY^{2-}} \approx m_{Fe^{2+}}$, $m_{Fe(OH)Y^{2-}} \approx m_{Fe^{3+}}$. When the ratio of $m_{Fe^{3+}}$ to $m_{Fe^{2+}}$ is constant, φ exhibits a linear relationship with pH, as seen in the *ab* segment of Fig. 9-1.

(2) Within a specific pH range, stable complexes FeY^{2-} and FeY^{-} are formed by the reaction of Fe^{2+} and Fe^{3+} with EDTA, expressed as

$$FeY^{-} + e \rightleftharpoons FeY^{2-}$$

The expression for the electrode potential is

$$\varphi = \varphi^{\ominus} - \frac{RT}{F}\ln\frac{a_{FeY^{2-}}}{a_{FeY^{-}}} = \varphi^{\ominus} - \frac{RT}{F}\ln\frac{\gamma_{FeY^{2-}}}{\gamma_{FeY^{-}}} - \frac{RT}{F}\ln\frac{m_{FeY^{2-}}}{m_{FeY^{-}}}$$

$$= (\varphi^{\ominus} - b_2) - \frac{RT}{F}\ln\frac{m_{FeY^{2-}}}{m_{FeY^{-}}} \tag{5}$$

in the formula $b_2 = \frac{RT}{F}\ln\frac{\gamma_{FeY^{2-}}}{\gamma_{FeY^{-}}}$, with the ionic strength and temperature held constant in the solution, b_2 becomes a constant. Within this pH range, the electrode potential of the system is only related to the ratio of $\frac{m_{FeY^{2-}}}{m_{FeY^{-}}}$, in other words, it is only related to the ratio of $\frac{m_{Fe^{2+}}}{m_{Fe^{3+}}}$ when the solution is prepared. A plateau area appears in the curve (such as the *bc* segment in Fig. 9-1).

(3) At low pH, the electrode reaction of the system is

$$FeY^{-} + H^{+} + e \rightleftharpoons FeHY^{-}$$

Similarly, we can get

$$\varphi = \varphi^{\ominus} - \frac{RT}{F}\ln\frac{a_{FeHY^{-}}}{a_{FeY^{-}}} = \varphi^{\ominus} - \frac{RT}{F}\ln\frac{\gamma_{FeHY^{-}}}{\gamma_{FeY^{-}}} - \frac{RT}{F}\ln\frac{m_{FeHY^{-}}}{m_{FeY^{-}}}$$

$$= (\varphi^{\ominus} - b_3) - \frac{RT}{F}\ln\frac{m_{FeHY^{-}}}{m_{FeY^{-}}} - \frac{2.303RT}{F}\text{pH} \tag{6}$$

where $b_3 = \frac{RT}{F}\ln\frac{\gamma_{FeHY^{-}}}{\gamma_{FeY^{-}}}$, with the ionic strength and temperature held constant in the solution, b_3 becomes a constant. φ exhibits a linear relationship with pH when $\frac{m_{Fe^{2+}}}{m_{Fe^{3+}}}$ is constant, as seen in the *cd* segment of Fig. 9-1.

Thus, by using an inert metal (such as Pt) as a conductor to form an electrode for the Fe^{3+}/Fe^{2+}-EDTA system, and pairing it with another reference electrode (such as a saturated calomel electrode) to create a cell, measuring the cell's electromotive force (EMF) allows for the determination of the system's (Fe^{3+}/Fe^{2+}-EDTA) electrode potential. Simultaneously, using a pH meter to measure the corresponding pH values under specific conditions enables the plotting of the Fe^{3+}/Fe^{2+}-EDTA system's potential-pH curve.

2. Application of potential-pH curve

In practical applications, the Fe^{3+}/Fe^{2+}-EDTA system can be utilized in the desulfurization of natural gas. Natural gas contains H_2S, a harmful substance that can cause health problems when inhaled in large quantities. For instance, when the concentration of hydrogen sulfide in the air reaches 20 $mg \cdot m^{-3}$, it can lead to symptoms such as nausea, dizziness, headache, fatigue, chest discomfort, and irritation of the eyes, nose, and throat. At concentrations of 60 $mg \cdot m^{-3}$, seizures, unconsciousness, and even respiratory paralysis leading to death may occur.

The Fe^{3+}-EDTA solution can be employed to oxidize H_2S in natural gas to elemental sulfur (S), followed by filtration to remove it. The Fe^{3+}-EDTA complex in the solution will be then reduced to Fe^{2+}-EDTA, and by introducing air, Fe^{2+}-EDTA rapidly oxidizes back to Fe^{3+}-EDTA, allowing the solution to be regenerated and reused. The reactions are as follows

$$2FeY^- + H_2S \xrightarrow{\text{filtration}} 2FeY^{2-} + 2H^+ + S\downarrow$$

$$2FeY^{2-} + \frac{1}{2}O_2 + H_2O \xrightarrow{\text{regeneration}} 2FeY^- + 2OH^-$$

We can select more suitable desulfurization conditions according to the measured potential-pH curve of Fe^{3+}/Fe^{2+}-EDTA complexation system. For example, the H_2S content of low-sulfur natural gas is about 1×10^{-4} to 6×10^{-4} $kg \cdot m^{-3}$, and the corresponding partial pressure of H_2S at 25 ℃ is 7.29 to 43.56 Pa.

According to the electrode reaction

$$S + 2H^+ + 2e^- \xlongequal{\quad} H_2S(g)$$

At 25 ℃, the electrode potential for the above reaction is given by

$$\varphi/V = -0.072 - 0.0296\lg(p/p^\ominus) - 0.0591pH$$

For the S/H_2S system determined by the pressure of H_2S, the relationship between φ and pH is depicted in curve (2) in Fig. 9-1. It can be observed that for any desulfurization solution with a certain $\frac{m_{Fe^{2+}}}{m_{Fe^{3+}}}$ ratio, the difference between the electrode potential of this desulfurization solution and the electrode potential of the reaction $S+2H^+ +2e^- \xlongequal{\quad} H_2S(g)$ increase with pH within the pH range of the potential plateau. When the pH reaches the upper limit of the plateau, the difference in electrode potential is maximized. Beyond this pH value, the difference in electrode potential no longer improving but remains constant. This fact indicates that for any desulfurization solution with a certain S/H_2S ratio, the thermodynamic trend of desulfurization is maximized at the upper limit of the potential plateau pH. After exceeding this pH value, the desulfurization trends no longer improving with increasing pH. Thus, pH values at point A and above A in Fig. 9−1 are suitable conditions for desulfurization in this system.

It should be noted that the pH of the desulfurization solution should not be too high. Experimental evidence suggests that if the pH exceeds 12, $Fe(OH)_3$ precipitates, which must be taken into account during the experiments.

Apparatus and reagents

pH−3V acidity potential meter	Magnetic stirrer
Combination electrode	Platinum electrode
150 mL jacketed bottle	$FeCl_3 \cdot 6H_2O$
$FeCl_2 \cdot 4H_2O$	EDTA (tetrasodium salt)
NaOH	HCl
Standard buffer solutions	N_2

Experimental procedures

1. Calibrate the pH−3V acidity potential meter

The panel window of the pH−3V acidity potential meter is shown in Fig. 9−2.

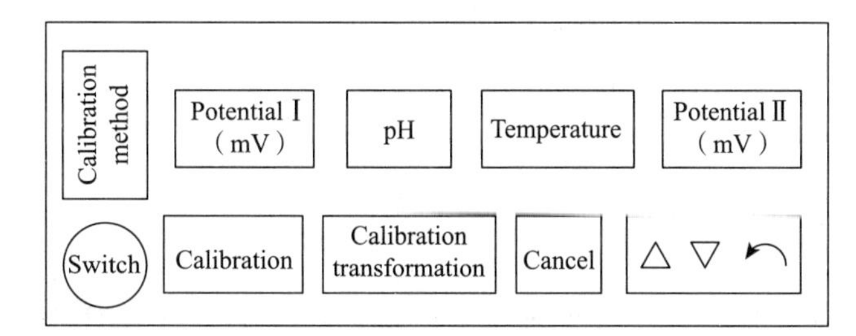

Fig. 9−2 Panel view diagram of pH−3V acidity potential tester

(1) Turn on the power switch, and let the instrument preheat for 15 minutes.

(2) With the instrument in measurement mode, press the calibration switch, and choose the calibration method (1-point or 2-point; it is recommended to use the 2-point method for better accuracy). Hold down the calibration button for more than 3 seconds until the calibration indicator light is on. Place the electrode and temperature sensor into a small beaker containing standard buffer solution. At this point, the third decimal place at the display window of pH value will be flashing. When the value of potential I is stable, determine the standard pH value of the solution based on the displayed temperature. For example, if using a phosphate buffer solution with a common pH of 7 at 25 ℃, and the pH is 6.863 when the third decimal place is flashing, use the increase or decrease key to set the display window value to 3. Then press the shift key (↶) and the second decimal place digit will start flashing. Continue calibration by using the

increase or decrease key to set the second decimal place digit to 6, and so on until the buffer solution is fully calibrated. Note that during the calibration process, if there is an input error, you can press the cancel key to re-enter the data. Additionally, the shift key can only be used to calibrate digits from right to left.

If using the 2-point calibration method, after calibrating with a pH 7 buffer solution, clean the electrode and temperature sensor, and repeat the process with a pH 4 standard buffer solution for the second calibration.

(3) After the second calibration, press the shift key (↶) to automatically enter measurement mode. Clean the combination electrode and temperature sensor with deionized water, and they are ready for use.

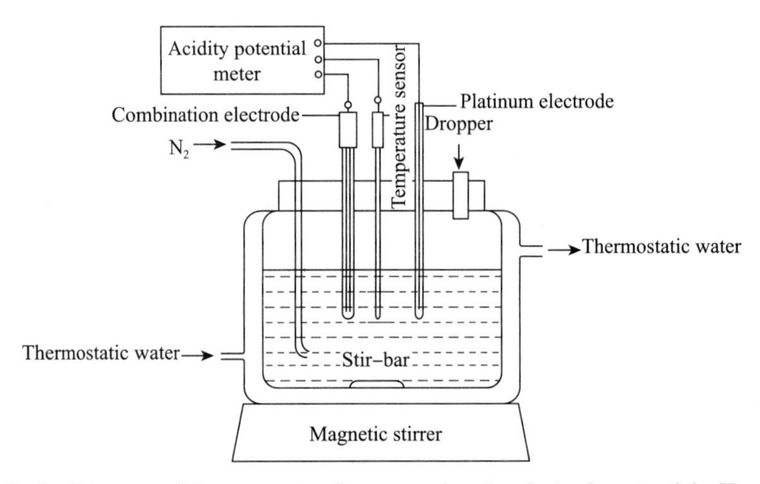

Fig. 9-3 Diagram of the apparatus for measuring the electrode potential-pH curve

2. Solution preparation

Prepare 4 mol · L^{-1} NaOH, 4 mol · L^{-1} HCl and 0.1 moL · L^{-1} $FeCl_3$ solutions separately, and place them in the corresponding bottles for later use.

3. Measurement of cell electric potential and pH

Measure 64 mL 0.1 moL · L^{-1} $FeCl_3$ solutions and place it into a jacketed bottle. Add 7.0 g EDTA and 16 mL H_2O. Turn on the magnetic stirrer and introduce nitrogen gas into the solution. After 10 minutes of continuous nitrogen gas flow, add 1.0 g of $FeCl_2 \cdot 4H_2O$ while maintaining stirring.

Insert the cleaned combination electrode and platinum electrode into the solution. Install the instrument and electrodes, as shown in Fig. 9-3. Adjust the solution's pH using 4 mol · L^{-1} NaOH until the solution turns reddish-brown, with a pH approximately between 7.5 and 8.0. Once the instrument displays stable values (approximately 10 minutes), directly read the solution's pH and EMF data from the instrument window (EMF Ⅱ window). Gradually adjust the pH of the solution by adding 4 mol · L^{-1} HCl solution (changing the pH by approximately 0.3 each time). Record pH values and corresponding cell EMF data until the solution becomes turbid.

Add 1.0 g $FeCl_2 \cdot 4H_2O$ to the jacketed bottle and repeat the above operation. Continue adding 1.0 g $FeCl_2 \cdot 4H_2O$ and repeat the above operation.

4. Potential-pH curve drawing of S/H_2S system

Given that the H_2S content in low-sulfur natural gas is approximately $1\times10^{-4} \sim 6\times10^{-4}$ kg · m^{-3}, with a corresponding H_2S partial pressure of 7.29 ~ 43.56 Pa at 25 ℃, choose appropriate desulfurization conditions based on the measured Fe^{3+}/Fe^{2+}-EDTA complex system's EMF-pH curve. For H_2S partial pressures of 7.29 and 43.56 Pa, based on the electrode potential equation

$$\varphi/V = 0.318 - 0.0296\ \lg(p/p^\ominus) - 0.0591\ \mathrm{pH}$$

Plot the relationship between EMF and pH for the $S + 2H^+ + 2e \Longrightarrow H_2S(g)$ reaction.

Plot the φ-pH curve for the desulfurization solution to determine the optimal desulfurization solutions.

Data recording and processing

1. Record the data correctly in tabular form. Calculate the electrode potential of the Fe^{3+}/Fe^{2+}-EDTA system from the measured cell electric potential. Plot the potential-pH curve and determine from the curve the pH range for FeY^- and FeY^{2-}.

2. Calculate the φ-pH curve for the S/H_2S system. Determine suitable theoretical desulfurization conditions from the results.

Experimental precautions

1. Control the stirring speed to prevent uneven mixing, avoiding the appearance of a small amount of $Fe(OH)_3$ precipitate during NaOH addition due to uneven stirring.

2. Avoid prolonged contact of the combination electrode with strongly absorbent solvents. When used in strong alkaline solutions, operate promptly and wash with water immediately. The glass electrode bulb membrane should be thin and not collide with hard objects such as glass beakers.

Thinking and discussing

1. Write the specific expressions of the basic electrode reactions of the system and their corresponding electrode potential equations in the potential plateau region of the Fe^{3+}/Fe^{2+}-EDTA system, at low pH and high pH, and indicate the physical significance of each.

2. What are the differences in the measured potential-pH curves for different $m(Fe^{3+})$ / $m(Fe^{2+})$ ratios of the desulfurization solution?

Reference materials

1. Fudan University, et al. Physical Chemistry Experiments. Beijing: Higher Education Press, 2004: 73-77.

2. You X. Potentiometric-pH Diagram and Its Applications. Chemical Bulletin, 1975, 2: 60-65.

3. Sichuan University Department of Chemistry Natural Gas Desulfurization Research Group. Discussion on the Potential-pH Curve of Fe(III)[Fe(II)]-EDTA Complex System and the Desulfurization Conditions Using EDTA to Remove H_2S from Natural Gas. Journal of Sichuan University: Natural Science Edition, 1976, 3: 23-31.

实验 10　循环伏安法研究铁氰化钾的电化学行为

实验目的

1. 学习固体电极表面的处理方法,掌握循环伏安法研究电极过程和判断电极过程可逆性的基本原理。

2. 绘制铁氰化钾系统在不同电极材料、扫描速率、电位范围和电解质溶液浓度时的循环伏安图,了解不同电极材料、扫描速率、电位范围和电解质溶液浓度对循环伏安图的影响。

基本原理

1. 循环伏安法

循环伏安法(Cyclic Voltammetry, CV)是一种常用的暂态电化学测量方法,是研究电极反应动力学、反应机理和反应可逆性的重要手段之一。如图 10-1 所示,研究系统通常由工作电极(也称研究电极或待测电极)、参比电极、辅助电极(也称对电极)构成的三电极系统。工作电极和参比电极组成电势测量系统,工作电极和辅助电极组成电流测量系统。常用的工作电极有铂、金、玻璃、石墨及悬汞电极等,根据研究内容的需要,还可用不同材料对工作电极进行修饰。常用的参比电极有饱和甘汞电极、银-氯化银电极等,辅助电极可选用固态的惰性电极,如铂丝或铂片电极、玻璃碳电极等。

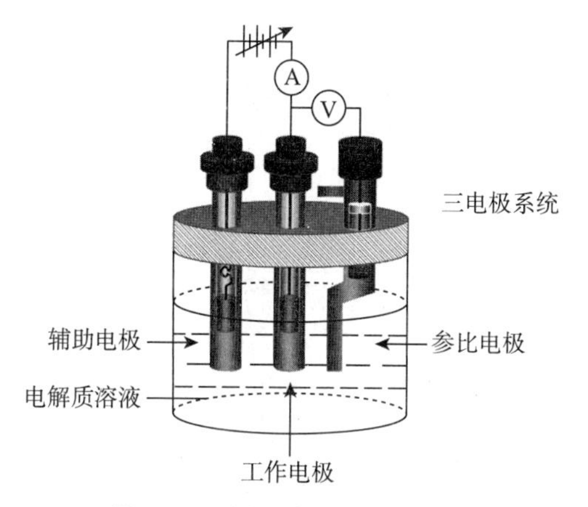

图 10-1 循环伏安法实验装置

循环伏安法的基本原理是：根据研究系统的性质，选择电势扫描范围和扫描速率，从选定的起始电势开始扫描，研究电极的电势按指定的方向和速率随时间线性变化，完成所确定的电势扫描范围，到达终止电势后，自动以同样的扫描速率反向扫描到起始电势，在电势进行扫描的同时，同步测量研究电极的电流响应，所获得的电流电势曲线称为循环伏安曲线或循环伏安扫描图。如图 10-2 所示，控制工作电极的电势以扫速 v 从 E_i 开始向电势负方向扫描，到时间 $t=20$ s（相应电位为 E_b）时，改变电位扫描方向，以相同的扫速回到起始电势 E_i，即完成了一个循环扫描。然后电势再次换向，反复扫描，即采用的电位控制信号为连续三角波信号。

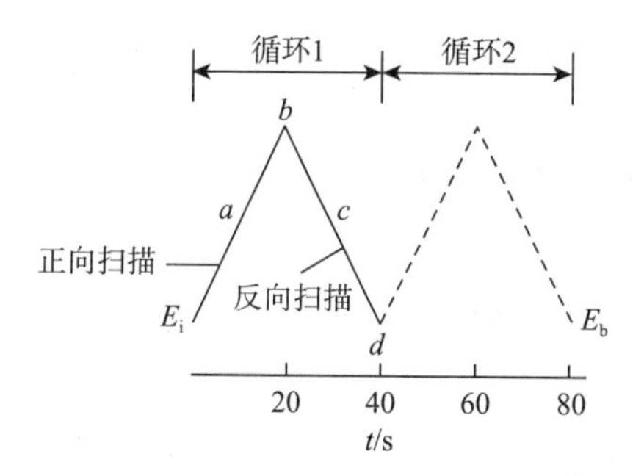

图 10-2 循环伏安法施加的电位示意图

图 10-3 是一张典型的循环伏安图，当电势从 E_i 开始正向扫描到 E_b 时，工作电极上发生还原反应 $Ox+ze^- \longrightarrow Red$，当电势从 E_b 至 E_i 进行反向扫描时，工作电极上发生氧化反应 $Red-ze^- \longrightarrow Ox$。其中 Ox 为电活性物质的氧化态，Red 为还原态。

这样，电势扫描经过了 E_i 至 E_b 再回到 E_i 的 1 次循环，其电流响应如图 10-3 所示，电流随着电势变化而变化，正向扫描时，伏安曲线上出现了 1 个阴极峰，阴极峰电位为 E_{pc}，阴极峰电流为 i_{pc}；反向扫描时，出现了 1 个阳极峰，阳极峰电位为 E_{pa}，阳极峰电流为 i_{pa}。E_{pa}，E_{pc}，i_{pa}，i_{pc}是循环伏安法中的重要参数。

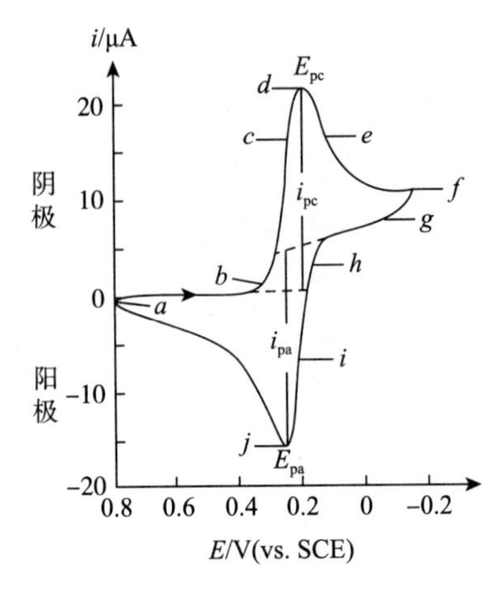

图 10-3　典型的循环伏安曲线

2. 判断电极过程的可逆性

根据 E_{pa},E_{pc},i_{pa},i_{pc}可以判断电化学反应的可逆性。

(1) 可逆电极反应

通常用阳极峰电势(E_{pa})和阴极峰电势(E_{pc})的差值 ΔE_p 来判断电极反应是否为可逆反应,即衡量电极反应的可逆性。对于产物稳定的可逆系统,循环伏安曲线参数具有下述重要特征:

① $|i_{pa}|=|i_{pc}|$,即 $\left|\frac{i_{pa}}{i_{pc}}\right|=1$,并且与扫速、扩散系数等参数无关;

② $|\Delta E_p|=E_{pa}-E_{pc}\approx\frac{2.3RT}{zF}$ 或 $|\Delta E_p|=E_{pa}-E_{pc}\approx\frac{59}{z}$(25 ℃,$z$ 为按所写的电极反应,在反应进度为 1 mol 时,反应式中电子的计量系数)。$|\Delta E_p|$为常数且不随扫速变化。

(2) 准可逆电极反应

大部分电极反应都是介于可逆反应与非可逆反应之间的准可逆反应,准可逆系统循环伏安曲线具有以下特点:

① E_p 随扫速变化而变化;

② 低扫速下,$|\Delta E_p|$随扫速增加而增加;

③ 扫速增加,反应越来越接近不可逆。

(3) 不可逆电极反应

当电极反应不可逆时,氧化峰与还原峰的峰值电位差值较大。

3. 铁氰化钾系统循环伏安曲线的测定

$[Fe(CN)_6]^{3-}$与$[Fe(CN)_6]^{4-}$是典型的可逆氧化还原系统。用玻璃碳、金等电极作为工作电极,进行阴极扫描时,发生还原反应$[Fe(CN)_6]^{3-}+e^- \longrightarrow [Fe(CN)_6]^{4-}$,进行阳极扫描时,发生氧化反应$[Fe(CN)_6]^{4-}-e^- \longrightarrow [Fe(CN)_6]^{3-}$,还原与氧化过程中电荷转移的

速率很快，得到的循环伏安图中阴极波与阳极波基本上是对称的。

仪器和试剂

CHI660E 电化学工作站	铂丝电极
玻璃碳电极	铁氰化钾(分析纯)
金电极	磷酸盐缓冲溶液(pH 7.0 左右即可，含氯化钾 0.1 $mol \cdot L^{-1}$)
饱和甘汞电极	25 mL 烧杯

实验步骤

1. 溶液的配制

以磷酸盐缓冲溶液作为溶剂，准确配制 2×10^{-4}，5×10^{-4}，1×10^{-3}，2.5×10^{-3}，5×10^{-3} $mol \cdot L^{-1}$的铁氰化钾溶液。

2. 固体电极的预处理

将玻璃碳、金电极依次用 1.0 μm 和 0.3 μm 的氧化铝抛光粉抛光，并用蒸馏水冲洗，直至电极表面出现较为平整的镜面状态。

3. 电极材料的选择

在 25 mL 烧杯中放入 5×10^{-3} $mol \cdot L^{-1}$铁氰化钾溶液 10 mL，插入工作电极(玻璃碳或金电极)、参比电极(饱和甘汞电极)和辅助电极(铂丝电极)。将电化学工作站绿色的电极夹夹在工作电极上，白色电极夹夹在参比电极上，红色电极夹夹在辅助电极上。

(1) 依次打开电脑和电化学工作站。电脑屏幕显示清晰后，打开 CHI660E 软件的测量窗口。

(2) 打开 setup 下拉菜单，在"Technique"项选择"Cyclic Voltammetry"方法，在"Parameters"项内设置关键参数：扫描范围-0.3～0.7 V，扫速 0.1 V/s，其他参数可默认。

(3) 完成上述各项设置，再仔细检查一遍电极是否连好、参数是否设置正确后，点击"▶"进行测量。完成后，命名存储。

4. 扫描范围的选择

在步骤 3 后，选择合适的电极作为工作电极，改变扫描范围为-0.6～0.9 V、-0.2～0.5 V，重复上面的操作，考查扫描范围对循环伏安图的影响。

5. 不同铁氰化钾浓度下的循环伏安曲线

在步骤 3 及 4 后，使用合适的电极，选择合适的扫描范围，改变铁氰化钾浓度分别为 2×10^{-4}、5×10^{-4}、1×10^{-3}和 2.5×10^{-3} $mol \cdot L^{-1}$，考查不同铁氰化钾浓度对循环伏安图的影响。

6. 不同扫描速率下的循环伏安曲线

在步骤 3 及 4 后，使用合适的电极及适当的扫描范围，在铁氰化钾浓度为 5×10^{-3} $mol \cdot L^{-1}$时，改变扫描速率，测试扫速为 0.01、0.05、0.1、0.2、0.3、0.4、0.5 $V \cdot s^{-1}$，考查不同扫描速率对循环伏安图的影响。

数据记录与处理

1. 为方便比较，将不同材料、不同扫描范围、不同铁氰化钾浓度及不同扫速下的循环伏安曲线分别叠加，得到4幅循环伏安图，将4幅图打印，并附于实验报告中。

2. 绘制出同一扫描速率下的铁氰化钾浓度（c）与 i_{pa}、i_{pc} 的关系曲线图，讨论电活性物质浓度与循环伏安峰电流的关系。

3. 绘制出同一铁氰化钾浓度下 i_{pa} 和 i_{pc} 与相应的 $v^{1/2}$ 的关系曲线图，讨论循环伏安峰电流与扫速之间的关系。

4. 通过总结各循环伏安曲线的关键测试条件及 E_{pa}，E_{pc}，i_{pa}，i_{pc} 的数据，讨论电极过程可逆性及相关条件对循环伏安结果的影响。

思考与讨论

1. 讨论循环伏安曲线中峰值电流 i_p 的影响因素（可查文献公式）。

2. 参考铁氰化钾溶液的循环伏安图，讨论如何用循环伏安方法来判断电极过程的可逆性。

3. 根据循环伏安图，说明不同电极材料、不同扫描电势范围、不同浓度的铁氰化钾溶液和不同扫描速率对循环伏安曲线的影响，并解释原因。

4. 参比电极应具备什么条件？它有什么作用？

参考文献

1. 朱万春，张国艳，李克昌，等. 基础化学实验（物理化学实验分册）. 高等教育出版社出版，2017.

2. 复旦大学，等. 物理化学实验. 高等教育出版社，2004.

Experiment 10

Study of the electrochemical behavior of potassium ferrocyanide using cyclic voltammetry

Experimental purpose

1. Learn methods for treating the surface of solid electrodes and understand the basic principles of cyclic voltammetry (CV) to study electrode processes and assess their reversibility.

2. Plot the cyclic voltammograms for the potassium ferrocyanide system using different electrode materials, scan rates, potential ranges, and electrolyte concentrations, examine the

effects of these factors on the cyclic voltammograms.

Basical principles

1. Cyclic voltammetry

Cyclic voltammetry (CV) is a commonly used transient electrochemical measurement method, crucial for studying electrode reaction kinetics, mechanisms, and the reversibility of reactions. As illustrated in Fig. 10-1, the system under study typically consists of a three-electrode system comprising a working electrode (also called the research or test electrode), reference electrode, and auxiliary electrode (also called the counter electrode). The working and reference electrodes constitute the potential measurement system, while the working and auxiliary electrodes constitute the current measuring system. Common working electrodes include platinum, gold, glass, graphite, and hanging mercury electrodes. Depending on the research requirements, different materials can be used to modify the working electrode.

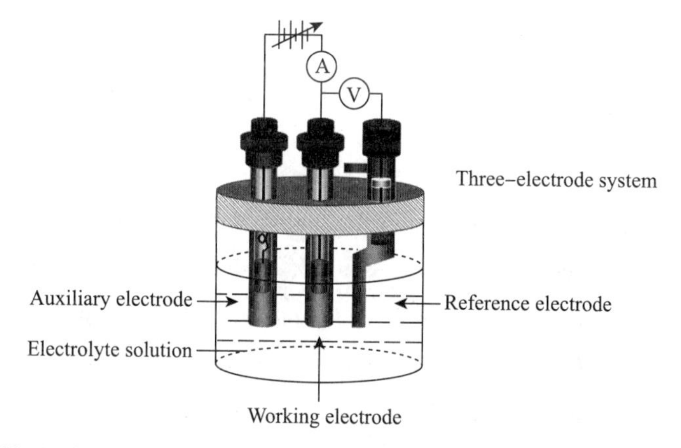

Fig. 10-1 Schematic diagram of the cyclic voltammetry experimental setup

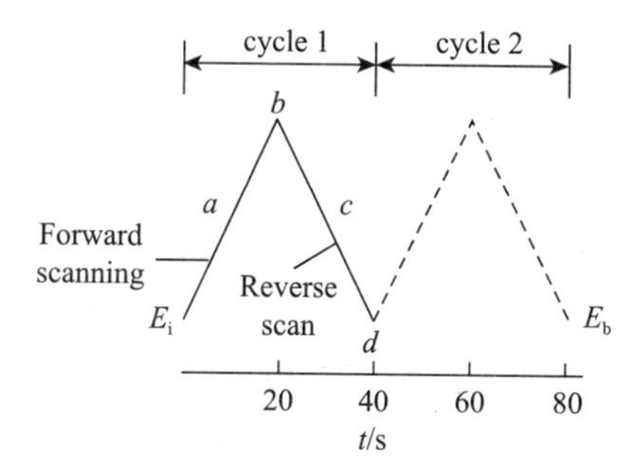

Fig. 10-2 Schematic of the applied potential in cyclic voltammetry

The fundamental principle of cyclic voltammetry involves selecting the potential scan range and scan rate based on the nature of the system under study. This electrochemical measurement method explores the dynamics of electrode reactions, reaction mechanisms, and the reversibility of reactions. As illustrated in Fig. 10-2, the CV process begins by initiating a potential scan from a chosen starting potential. The electrode potential follows a linear variation with time in a specified direction and at a defined rate. Once the predetermined potential scan range is reached, the direction of the scan is automatically reversed at the same scan rate, returning to the starting potential. During the potential scan, the current response of the working electrode is synchronously measured. The resulting plot of current versus potential is termed the cyclic voltammogram or cyclic voltammetric scan.

In the illustrated example, the working electrode potential is controlled to scan from the starting potential (E_i) in the positive direction at a scan rate (v). At $t=20$ s (corresponding to potential E_b), the direction of the potential scan is changed, returning to the starting potential E_i at the same scan rate, completing one cycle. The potential is then reversed, and the process is repeated, utilizing a continuous triangular wave signal as the potential control signal.

Fig. 10-3 is a typical cyclic voltammogram. When the potential is scanned forward from E_i to E_b, the reduction reaction occurs on the working electrode $Ox+ze^- \longrightarrow Red$, and when the potential is scanned backward from E_b to E_i, the oxidation reaction occurs on the working electrode $Red-ze^- \longrightarrow Ox$, where Ox is the oxidation state of the electroactive species and Red is the reduced state.

In the described cyclic voltammetry process, the potential scan completes one cycle from E_i to E_b and back to E_i. The resulting current response, as depicted in Fig. 10-3, exhibits distinct features. During the forward scan, a cathodic peak appears on the voltammogram, denoted as the cathodic peak potential (E_{pc}), accompanied by a cathodic peak current (i_{pc}). Conversely, during the reverse scan, an anodic peak emerges, represented by the anodic peak potential E_{pa} and anodic peak current i_{pa}. These parameters, E_{pa}, E_{pc}, i_{pa} and i_{pc}, hold significant importance in cyclic voltammetry.

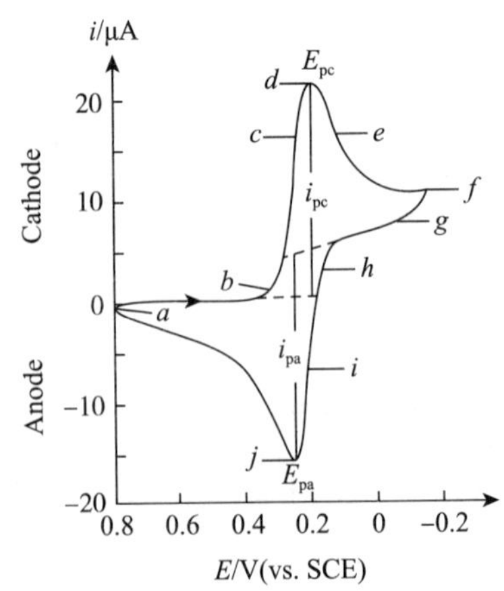

Fig. 10-3 Typical cyclic voltammetry curves

2. Judge the reversibility of electrode process

The reversibility of an electrode process can be evaluated based on the parameters obtained from the cyclic voltammogram.

(1) Reversible electrode reaction

For a system with a reversible electrode reaction, the cathodic peak potential (E_{pc}) and anodic peak potential (E_{pa}) exhibit specific characteristics. For reversible systems with stable products, the cyclic voltammetric curve parameters have the following important characteristics:

① $|i_{pa}| = |i_{pc}|$, in other word, $\left|\frac{i_{pa}}{i_{pc}}\right| = 1$, indicating symmetry between cathodic and anodic peaks, and independence from parameters like scan rate and diffusion coefficients.

② $|\Delta E_p| = E_{pa} - E_{pc} \approx \frac{2.3RT}{zF}$ or $|\Delta E_p| = E_{pa} - E_{pc} \approx \frac{59}{z}$ (25 ℃, n is the number of electrons transferred in the electrode reaction). $|\Delta E_p|$ remains constant and is unaffected by changes in scan rate.

(2) Quasi-reversible electrode reaction

Most electrode reactions fall between reversible and irreversible, representing quasi-reversible systems. The cyclic voltammogram characteristics include:

① E_p varied with scan rate;

② At low scan rates, $|\Delta E_p|$ increases with the scan rate;

③ As the scan rate increases, the reaction becomes increasingly irreversible.

(3) Irreversible electrode reaction

Irreversible electrode reactions exhibit a significant difference in peak potentials between the cathodic and anodic peaks.

3. Cyclic voltammetry of the potassium ferricyanide system

The $[Fe(CN)_6]^{3-}$/ $[Fe(CN)_6]^{4-}$ is typical example of a reversible rexod system. Glass carbon, gold and other electrodes are used as working electrodes. During the cathodic scan, reduction reaction occurs: $[Fe(CN)_6]_3^- + e^- \longrightarrow [Fe(CN)_6]^{4-}$; while during the anodal scan, oxidation reaction occurs: $[Fe(CN)_6]^{4-} - e^- \longrightarrow [Fe(CN)_6]_3^-$. The rate of charge transfer during the reduction and oxidation processes is fast, resulting in a cyclic voltammogram where the cathodic and anodic waves are essentially symmetric.

Apparatus and reagents

CHI660E electrochemical workstation　　Platinum wire electrode

Glassy carbon electrode

Gold electrode

Saturated calomel electrode

Potassium ferriccyanide (analytical pure)

Phosphate buffer solution (pH 7.0 or so, containing potassium chloride 0.1 mol · L^{-1})

25 mL beaker

Experimental procedures

1. Solution preparation

Prepare iron ferrocyanide (potassium ferrocyanide) solutions with concentrations of 2×10^{-4}, 5×10^{-4}, 1×10^{-3}, 2.5×10^{-3} and 5×10^{-3} mol · L^{-1} using phosphate buffer solution as the solvent.

2. Pretreatment of the solid electrode

Polish glassy carbon and gold electrodes successively with 1.0 μm and 0.3 μm alumina polishing powder, followed by rinsing with distilled water until a smooth, mirror-like surface is achieved.

3. Selection of electrode materials

In a 25 mL beaker, place 10 mL of 5×10^{-3} mol · L^{-1} potassium ferrocyanide solution. Insert the working electrode (glassy carbon or gold electrode), reference electrode (saturated calomel electrode), and auxiliary electrode (platinum wire electrode). Secure the electrode clips to the corresponding electrodes on the electrochemical workstation. Attach the green electrode clip of the electrochemical workstation to the working electrode, the white electrode clips to the reference electrode, and the red electrode clip to the auxiliary electrode.

(1) Turn on the computer and electrochemical workstation.

(2) On the CHI660E software, select "Cyclic Voltammetry" under the "Technique" dropdown menu. Set key parameters in the "Parameters" section: scan range −0.3 to 0.7 V, scan rate 0.1 V · s^{-1}, and leave other parameters as default.

(3) After careful checks, initiate the measurement by clicking "▶" and save the data.

4. Selection of the scanning range

After step 3, select an appropriate electrode as the working electrode and change the scan range to −0.6～0.9 V and 0.2～0.5 V. Repeat the measurement to observe the impact of the scan range on the cyclic voltammetry.

5. Cyclic voltammetry curves at different concentrations of potassium ferricyanide

Following steps 3 and 4, use the appropriate electrode, select the appropriate scan range, and vary the potassium ferricyanide concentration to 2×10^{-4}, 5×10^{-4}, 1×10^{-3} and 2.5×10^{-3} mol · L^{-1}. Observe the influence of different concentration on the cyclic voltammogram.

6. Cyclic voltammetry curves at different scanning rates

After steps 3 and 4, use the appropriate electrode and the appropriate scan range. With a potassium ferricyanide concentration of 5×10^{-3} mol · L^{-1}, vary the scan rate to 0.01, 0.05, 0.1, 0.2, 0.3, 0.4 and 0.5 V · s^{-1}. Observe the impact of different scan rates on the cyclic voltammogram.

Data recording and processing

1. Overlay the cyclic voltammograms for different materials, scan ranges, potassium ferrocyanide concentrations, and scan rates to create four plots. Print and include these plots in the experimental report.

2. Plot the relationship between potassium ferrocyanide concentration (c) and i_{pa}, i_{pc} for the same scan rate. Discuss the correlation between the concentration of the electroactive substance and the peak current in cyclic voltammetry.

3. Plot the relationship between i_{pa} and i_{pc} and their corresponding $\upsilon^{1/2}$ for the same potassium ferrocyanide concentration. Discuss the relationship between cyclic voltammetric peak current and scan rate.

4. Summarize the key testing conditions and data (E_{pa}, E_{pc}, i_{pa} and i_{pc}) from various cyclic voltammograms. Discuss the reversibility of electrode processes and the influence of related conditions on cyclic voltammetry results.

Thinking and discussing

1. Discuss the factors affecting the peak current (i_p) in cyclic voltammograms (consult literature for formulas).

2. Using the cyclic voltammogram of potassium ferrocyanide solution as a reference, discuss how cyclic voltammetry can be employed to determine the reversibility of electrode processes.

3. Explain the impact of different electrode materials, scan potential ranges, concentrations of potassium ferrocyanide, and scan rates on cyclic voltammograms, provide reasons for these effects.

4. What conditions should a reference electrode fulfill, and what is its purpose?

Reference materials

1. Zhu W, Zhang G, Li K, et al. Basic Chemistry Experiments (Volume of Physical Chemistry Experiments). Beijing: Higher Education Press, 2017.

2. Fudan University, et al. Physical Chemistry Experiments. Beijing: Higher Education Press, 2004.

实验 11 最大泡压法测定溶液的表面张力

实验目的

1. 了解表面张力、表面自由能的定义,明确表面张力和吸附量的关系。

2. 掌握最大泡压法测定溶液表面张力的原理和技术。

3. 测定不同浓度乙醇水溶液的表面张力,计算表面吸附量和乙醇分子横截面积。

基本原理

1. 表面张力和表面吸附

密切接触的两相之间的过渡区约有几个分子的厚度,称为界面。界面的类型根据物质三态的不同,可以分为气-液、气-固、液-液、液-固和固-固等界面。前两种界面都有气体参加,此类界面习惯上常称为表面。物质表面层的分子与内部分子周围的环境不同,内部分子所受四周邻近相同分子的作用力是对称的,各个方向的力彼此抵消,但表面层的分子一方面受到本相内物质分子的作用,另一方面又受到性质不同的另一相中物质分子的作用,因此表面层的性质与内部不同。

在温度、压力、组成恒定时,每增加单位表面积,系统的 Gibbs 自由能的增值称为表面 Gibbs 自由能($J \cdot m^{-2}$),用 σ 表示。也可以看作是垂直作用在单位长度相界面上的力,即表面张力($N \cdot m^{-1}$)。欲使液体产生新的表面 ΔA_S,就需要对其做表面功,其大小应与 ΔA_S 成正比,系数即为表面张力 σ

$$-W = \sigma \times \Delta A_S \tag{1}$$

当液体中加入某种溶质时,液体的表面张力就会升高或降低,对同一溶质来说,其变化的多少随着溶液浓度不同而异。吉布斯在 1878 年以热力学方法导出溶质的吸附量与溶液的表面张力及溶液浓度之间变化关系的吸附公式。对两组分的稀溶液而言

$$\Gamma = -\frac{c}{RT}\left(\frac{d\sigma}{dc}\right)_T \tag{2}$$

式中,Γ 为溶质在表层的吸附量,单位 $mol \cdot m^{-2}$,σ 为表面张力,c 为溶质的浓度。若 $\left(\frac{d\sigma}{dc}\right)_T < 0$,则 $\Gamma > 0$,称为正吸附,也就是增加浓度时,溶液表面张力降低,表面层的浓度大于溶液内

部的浓度；若 $\left(\frac{d\sigma}{dc}\right)_T > 0$，则 $\Gamma<0$，称为负吸附，也就是增加浓度时，溶液表面张力增加，表面层的浓度小于溶液内部的浓度。

能使表面张力降低的物质称为表面活性物质——表面活性剂。在水溶液中，表面活性物质有显著的不对称结构，它是由极性（亲水）部分和非极性（憎水）部分构成的。在水溶液表面，一般极性部分取向溶液内部，而非极性部分取向空气部分。如图 11-1 所示，在浓度极小情况下，物质分子接近于平躺在溶液表面上；浓度逐渐增加，分子极性部分取向溶液内部，而非极性部分取向空气；当浓度增至一定程度时，溶质分子占据了所有表面，形成饱和吸附层。

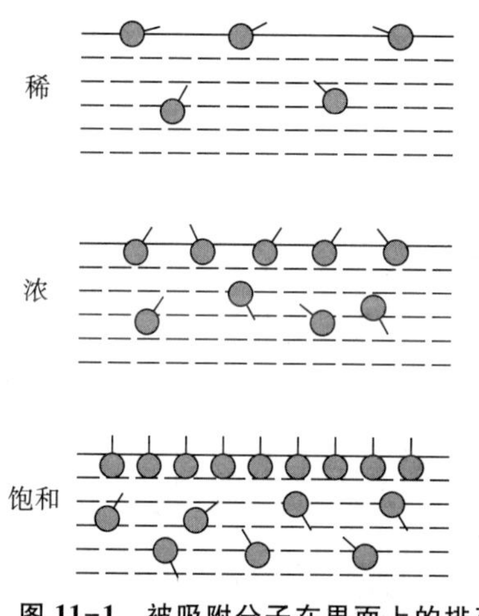

图 11-1　被吸附分子在界面上的排列

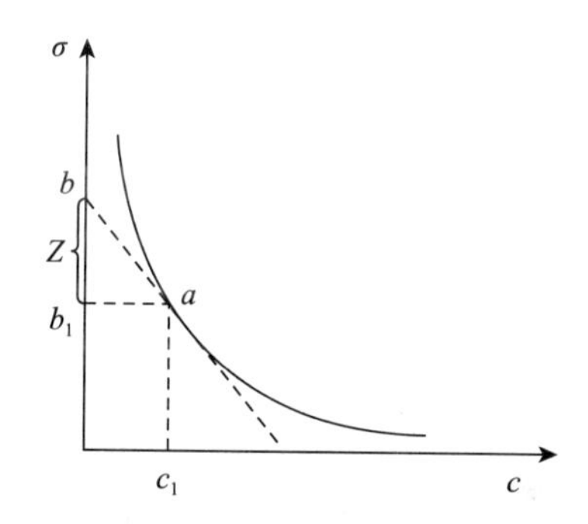

图 11-2　表面张力和浓度的关系

通过实验测得表面张力与溶质浓度的关系，可作 $\sigma-c$ 曲线，如图 11-2 所示。在曲线上任取点 a，过点 a 作曲线的切线以及平行于横坐标的直线，分别交纵轴于 b 和 b_1，令 $bb_1=Z$，得 $\Gamma=\frac{Z}{RT}$，利用此式可求出在该浓度时的溶质吸附量 Γ。

2. 饱和吸附与溶质分子横截面积

在一定温度下，吸附量与溶液浓度之间的关系由 Langmuir 等温式表示为

$$\Gamma=\Gamma_\infty\frac{K_c}{1+K_c}\tag{3}$$

式中 K 为常数，Γ_∞ 为饱和吸附量。

将上式取倒数可得

$$\frac{c}{\Gamma}=\frac{c}{\Gamma_\infty}+\frac{1}{K\Gamma_\infty}\tag{4}$$

由 $\frac{c}{\Gamma}$ 对 c 作图得一直线，由直线的斜率可求得 Γ_∞，如果以 N 代表 1 m^2 表面上溶质

的分子数，则有 $N = \Gamma_\infty N_A$，式中 N_A 为阿伏伽德罗常数，由此可进一步计算，得每个乙醇分子的横截面积

$$S_B = \frac{1}{\Gamma_\infty L} \tag{5}$$

3. 表面张力的测定原理

通常有多种方法来测定表面张力，如毛细管上升法、滴重法、吊环法、最大泡压法、吊片法、静液法等。本实验主要介绍最大泡压法测定表面张力的基本原理，其装置如图 11-3 所示。

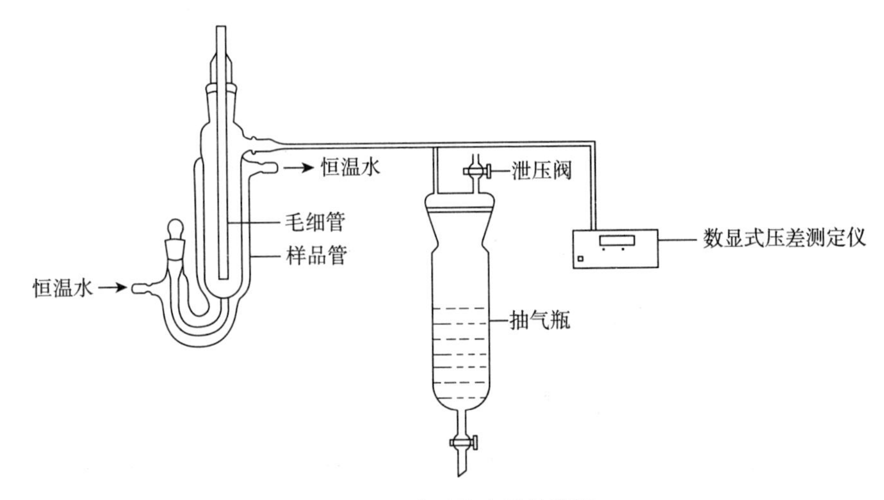

图 11-3　表面张力测量装置

图 11-3 中，毛细管下端与液面相切，毛细管上端的大气压为 p_0。测定体系内毛细管外气压为 p，当打开抽气瓶下端活塞时，抽气瓶中的水流出，系统压力 p 逐渐减小，毛细管上端的大气压力就会把毛细管液面逐渐压至管口，形成气泡，如图 11-4 所示。

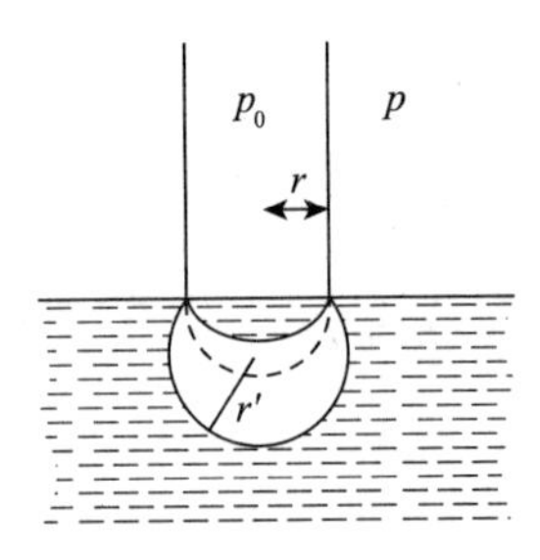

图 11-4　毛细管口出泡示意图

在形成气泡的过程中，气泡曲率半径 r'经历由大→小→大过程，即中间有一极小值等于毛细管半径 r。根据拉普拉斯公式 $\Delta p = p_0 - p = \frac{2\sigma}{r}$，此时气泡承受的压力差最大，此压

力差可由压力计读出,故待测液的表面张力为

$$\sigma = \frac{r}{2} \times \Delta p_{max} \tag{6}$$

若用同一支毛细管测试两种不同液体,其表面张力分别为 σ_1,σ_2,压力计测得最大压力差分别为 Δp_1,Δp_2,则

$$\frac{\sigma_1}{\sigma_2} = \frac{\Delta p_1}{\Delta p_2} \tag{7}$$

若其中一种液体的 σ 已知,例如水,则另一种液体的表面张力可由上式求得。即

$$\sigma_2 = \frac{\sigma_1}{\Delta p_1} \times \Delta p_2 = K \times \Delta p_2 \tag{8}$$

式中 $K = \dfrac{\sigma_1}{\Delta p_1}$ 称为仪器常数,可用某种已知表面张力的液体(常用蒸馏水)测得。如果将已知表面张力的液体作为标准,由实验测得其 Δp 后,就可求出仪器常数 K。然后只要用同一支毛细管测定其他液体的 Δp,通过式(8)计算,即可求得各种液体的表面张力 σ。

仪器和试剂

表面张力测定装置　　恒温水浴

容量瓶(100 mL)　　移液管

乙醇(分析纯)

实验步骤

1. 按表 11-1(100 mL 容量瓶)配制乙醇系列水溶液。

表 11-1　配制 100 mL 溶液所需乙醇的体积

瓶号	1	2	3	4	5	6	7	8
$V_{乙醇}$/mL	5	10	15	20	25	30	35	40

乙醇密度可根据附表 4 数据计算。

2. 仪器检漏

将表面张力仪和毛细管洗净(毛细管可浸泡在 $H_2O_2-H_2SO_4$ 洗液中,使用时用洗耳球反复抽吸毛细管中的洗液,取出后用蒸馏水冲洗干净,再用少量待测液润洗两次)。将毛细管放入表面仪中,打开恒温水,调节恒温槽的温度为 25 ℃。在抽气瓶中放入自来水,打开泄压阀(通大气),打开表面仪左侧小瓶塞,加入蒸馏水至毛细管下端与液面刚好相切,盖紧瓶塞。按压力计上的采零按钮,使压力指示为零,关闭泄压阀。将抽气瓶下端的活塞打开缓慢滴水,使系统内的压力降低,精密数字压力计显示一定数值时,关闭抽气瓶

下端的活塞，若 2～3 min 内精密数字压力计数值基本不变，则说明系统不漏气，可以进行实验。

3. 仪器常数的测定

打开泄压阀，压力计显示为零（若不为零，按一下采零按钮，使压力指示为零），关闭泄压阀，缓慢旋转抽气瓶下端的活塞使抽气瓶中的水慢慢滴出，气泡由毛细管底部冒出，通过调节抽气瓶滴水速度控制毛细管出泡的速度（每分钟 8～12 个泡，整个实验过程的出泡数应尽量相同），当气泡刚脱离管端破裂的一瞬间，记录压力计上显示的压力值，即精密数字压力计瞬间最大值，读 3 次数，取平均值。

4. 不同浓度乙醇水溶液表面张力的测定

按步骤 3 的方法由稀到浓进行测定。注意：每次换溶液时，都要将表面仪和毛细管用少量待测液润洗两次。

5. 实验完毕，使系统与大气相通。洗净玻璃仪器，表面仪中放入蒸馏水，毛细管放到 H_2O_2-H_2SO_4 洗液中。最后结束实验者关闭电源。

数据记录与处理

1. 将所测数据和计算的浓度及表面张力列表表示。

2. 作表面张力-浓度（$\sigma - c$）图（注意曲线须圆滑）。在 $\sigma - c$ 曲线上任取 8 个点，过各点作曲线的切线，求得相应 $Z(bb_1)$ 值（如图 11-2 所示），计算出 Γ，并列表表示所求数据（c, Z, Γ）。

3. 作 Γ-c 图，在曲线上任取 8 个点计算 $\frac{c}{\Gamma}$，并列表表示所求数据 $\left(c, \Gamma, \frac{c}{\Gamma}\right)$。

4. 作 $\frac{c}{\Gamma}$ $-c$ 图，得直线，由直线斜率可求得 Γ_∞，由式(5)计算 S_B 值。

思考与讨论

1. 测定表面张力为什么必须在恒温槽中进行？温度变化对表面张力有何影响？为什么？

2. 毛细管下端为什么要刚好和液面相切？

参考资料

1. 武汉大学化学与分子科学学院实验中心. 物理化学实验.武汉：武汉大学出版社，2012.

2. 复旦大学，等. 物理化学实验. 北京：高等教育出版社，2004.

3. 北京大学化学学院物理化学实验教学组.物理化学实验.北京：北京大学出版社，2002.

4. 玉占君,冯春梁,等. 物理化学实验. 北京:化学工业出版社,2014.

补充资料:镜像法求切线

可使用镜像法作图求曲线上某点的切线,如图 11-5 所示。首先在曲线上任取一点 a,取一面方形小镜子,使镜子某边的直线经过 a 点,以垂直纸面穿过 a 点的直线为轴,转动镜子,同时观察镜子中曲线的镜像,当转动到某一角度发现镜子中曲线的镜像与镜子外实际曲线能连接成一条圆滑曲线时,沿镜子下缘在纸面上画直线。将镜子调转方向(转 180°),镜子的这一边仍应经过 a 点,与上述操作一样,转动镜子,使另半段曲线与其在镜子中的像成圆滑曲线,沿镜子下缘再画直线,作这两条直线的夹角平分线,最后过 a 点作夹角平分线的垂线,该垂线即为切线。

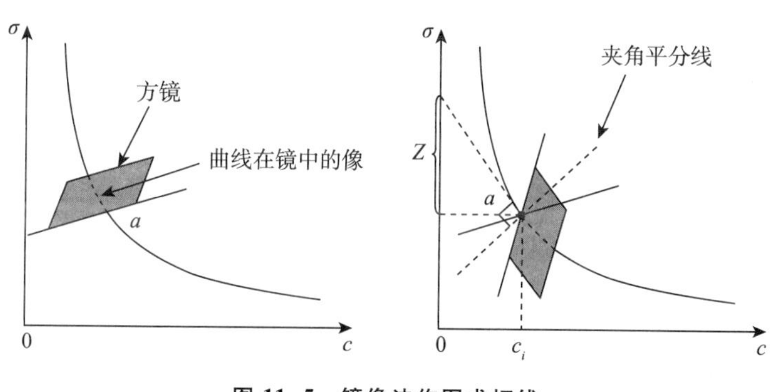

图 11-5 镜像法作图求切线

Experiment 11

Determination of surface tension of solution by maximum bubble pressure method

Experimental purpose

1. Understand the definition of surface tension and surface free energy, and clarifying the relationship between surface tension and adsorption capacity.

2. Master the principle and technology of solution surface tension determined by maximum bubble pressure method.

3. Measure surface tension of ethanol aqueous solution with different concentrations, and calculate surface adsorption capacity and the cross-sectional area of ethanol molecules.

Basical principles

1. Surface tension and surface adsorption

The transition zone between two phases in close contact is about the thickness of several molecules, called the interface. The interface can be divided into gas-liquid, gas-solid, liquid-liquid, liquid-solid and solid-solid according to the three states of matter. The first two interfaces have gas in them, which are customarily called surfaces. The surrounding environment of molecules in the surface layer of matter is different from that of internal molecules, the internal molecules are subjected to symmetrical forces of the same neighboring molecules, and the forces in all directions cancel each other out. However, on one hand, the molecules in the surface layer are affected by the matter molecules in this phase, and on the other hand by the matter molecules in another phase with different properties, so the properties of the surface layer are different from those in the internal layer.

When temperature, pressure and composition are constant, the increment of Gibbs free energy of the system per unit surface area is called surface Gibbs free energy ($J \cdot m^{-2}$), which is expressed by σ. It can also be regarded as the force acting perpendicularly on the phase interface per unit length, that is, the surface tension ($N \cdot m^{-1}$). In order to make a liquid produce a new surface ΔA_S, it is necessary to do surface work on it, the magnitude of which should be proportional to ΔA_S, and the coefficient is the surface tension

$$-W = \sigma \times \Delta A_S \tag{1}$$

When some kind of solute is added to liquid, the surface tension of the liquid increases or decreases, and the amount of variation for the same solute varies with the concentration of the solution. Gibbs derived the adsorption formula for the relationship between the amount of solute adsorbed and the surface tension and concentration of the solution at 1878 by thermodynamic methods. For dilute solutions of two components.

$$\Gamma = -\frac{c}{RT}\left(\frac{d\sigma}{dc}\right)_T \tag{2}$$

In the formula, Γ is the adsorption amount of solute on the surface layer, the unit is $mol \cdot m^{-2}$, σ is the surface tension, and c is the concentration of solute. If $\left(\frac{d\sigma}{dc}\right)_T < 0$, then $\Gamma > 0$, it is called positive adsorption. That is, if the concentration increases, the solution surface tension decreases, the concentration of the surface layer is greater than the concentration of the solution inside. If $\left(\frac{d\sigma}{dc}\right)_T > 0$, then $\Gamma<0$, it is called negative adsorption. That is, the surface tension of the solution increases with the increase of concentration, and the concentration of the surface layer is less than the concentration inside the solution.

Substances that can reduce surface tension are called surfactants. In aqueous solutions, the surfactant has a remarkable asymmetric structure, which consists of polar (hydrophilic) and non-polar (hydrophobic) parts. On the surface of the aqueous solution, the polar parts are generally oriented to the interior of the solution, while the non-polar parts are oriented to the air portion. As shown in Fig. 11-1, at very low concentrations, the molecules of the substance lie flat on the surface of the solution and gradually increase in concentration, with the polar part of the molecule oriented to the interior of the solution and the non-polar part oriented to the air, when the concentration increases to a certain extent, solute molecules occupy all the surface, saturated adsorption layer will form.

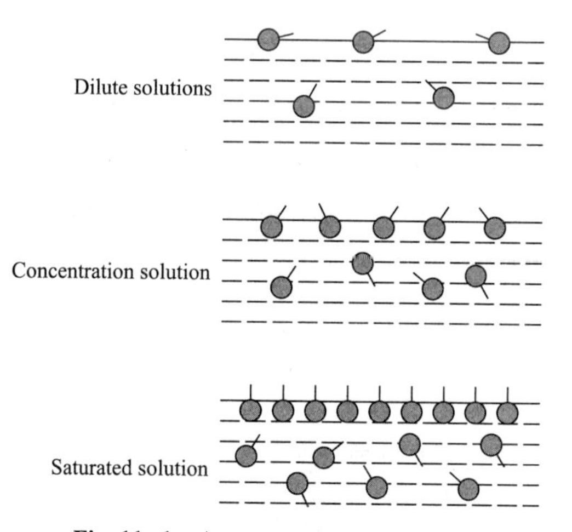

Fig. 11-1 Arrangement of adsorbed molecules at interface

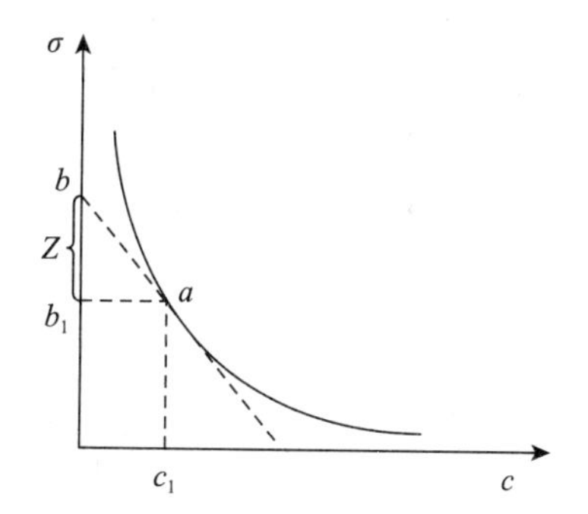

Fig. 11-2 Relationship between surface tension and concentration

The relationship between surface tension and solute concentration is measured experimentally, and the curve $\sigma - c$ is made as shown in Figure 11-2. In the curve, take any point a on the curve, cross point a to make tangent of the curve and parallel to the horizontal coordinate line, respectively, intersection longitudinal axis b and b_1, so that $bb_1 = Z$, then $\Gamma = \frac{Z}{RT}$. The solute adsorption amount Γ at this concentration can be obtained by Formula (2).

2. Saturated adsorption and cross-sectional area of solute molecules

At a certain temperature, the relationship between adsorption capacity and solution concentration is expressed by Langmuir isotherm as follows

$$\Gamma = \Gamma_\infty \frac{K_c}{1 + K_c} \tag{3}$$

where K is a constant, Γ_∞ is the saturated adsorption amount. Take the reciprocal of the above formula to obtain

$$\frac{c}{\Gamma} = \frac{c}{\Gamma_\infty} + \frac{1}{K\Gamma_\infty} \tag{4}$$

A straight line is obtained by drawing $\frac{c}{\Gamma}$ against c, and Γ_∞ can be obtained from the slope of the straight line. If N represents the number of solute molecules on the surface of 1 m^2, then $N = \Gamma_\infty N_A$, where N_A is the Avogadro constant, from which the cross-sectional area of each ethanol molecule can be further calculated

$$S_B = \frac{1}{\Gamma_\infty N_A} \tag{5}$$

3. Measuring mechanism of surface tension

There are many methods to determine the surface tension, such as capillary rise method, drop weight method, hanging ring method, maximum bubble pressure method, hanging sheet method, hydrostatic method, etc. This experiment mainly introduces the basic principle of measuring surface tension by the maximum bubble pressure method, and the device is shown in Fig. 11-3.

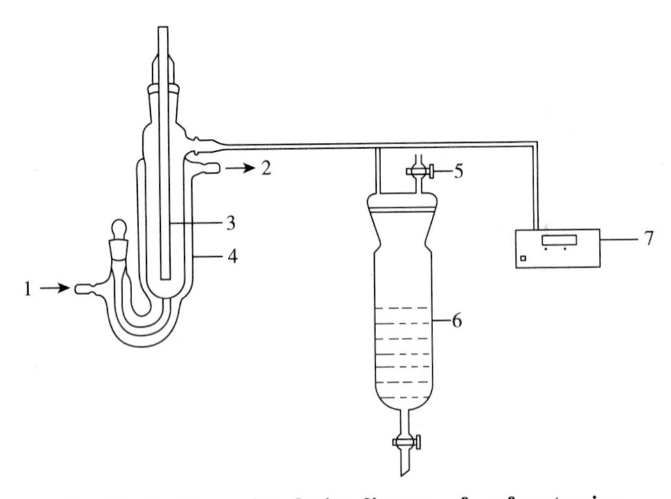

Fig. 11-3 Measuring device diagram of surface tension

1,2-Constant temperature water; 3-Capillary; 4-Sample tube; 5-Pressure relief valve; 6-Suction cylinder; 7-Measuring instrument of digital pressure difference

In Fig. 11 − 3, the lower end of the capillary is tangent to the liquid surface, and the atmospheric pressure at the upper end of the capillary is p_0. The outside air pressure of capillary in the system is determined as p. When the piston at the lower end of the pump cylinder is opened, the water in the pump cylinder flows out, the system pressure p gradually decreases, and the atmospheric pressure at the upper end of the capillary will gradually press the capillary liquid level to the nozzle, forming bubbles, as shown in Fig. 11−4.

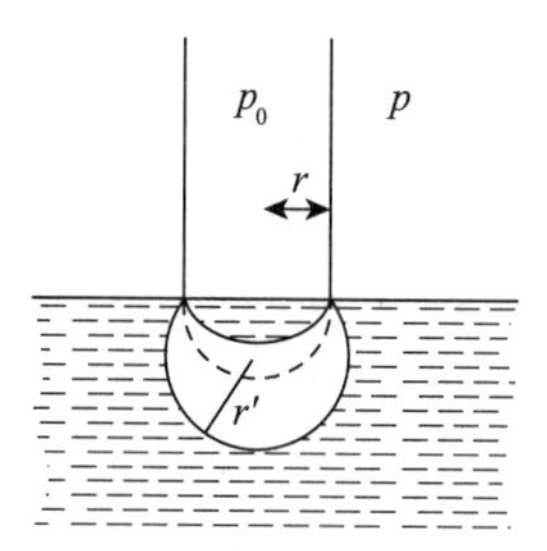

Fig. 11−4 Schematic diagram of capillary outlet bubble

In the process of forming bubbles, the radius of curvature r' of bubbles undergoes a process from large to small to large, that is, there is a minimum value equal to the radius r of capillary tube. According to Laplace formula $\Delta p = p_0 - p = \frac{2\sigma}{r}$, the pressure difference borne by bubbles is the largest at this time, and this pressure difference can be red out by pressure gauge, so the surface tension of the liquid to be measured is

$$\sigma = \frac{r}{2} \times \Delta p_{max} \tag{6}$$

If the same capillary tube is used to test two different liquids, the surface tension is σ_1 and σ_2, respectively, and the maximum pressure difference measured by the pressure gauge is Δp_1 and Δp_2 respectively

$$\frac{\sigma_1}{\sigma_2} = \frac{\Delta p_1}{\Delta p_2} \tag{7}$$

If one of σ is known, such as water, the surface tension of the other can be obtained from the above formula. Namely

$$\sigma_2 = \frac{\sigma_1}{\Delta p_1} \times \Delta p_2 = K \times \Delta p_2 \tag{8}$$

The formula $K = \frac{\sigma_1}{\Delta p_1}$ is called instrument constant, which can be measured by a liquid with known surface tension (usually distilled water). If the liquid with known surface tension is taken as the standard, the instrument constant K can be obtained after Δp being measured experimentally. Then, as long as the Δp values of other liquids are measured by the same capillary tube, the surface tension of various liquids can be obtained by calculation of Formula (8).

Apparatus and reagents

Surface tension measuring device
Constant temperature water bath
Volumetric flask (100 mL)
Pipette
Ethanol (AR)

Experimental procedures

1. Prepare ethanol series solution according to Table 11-1 (100 mL volumetric flask)

Table 11-1 Volume of ethanol required to prepare 100 mL solution

Bottle number	1	2	3	4	5	6	7	8
$V_{ethanol}$/mL	5	10	15	20	25	30	35	40

The ethanol density can be calculated according to the data in appendix Table 4.

2. Instrument leak detection

The surface tension meter and the capillary are cleaned (the capillary can be soaked in the H_2O_2-H_2SO_4 lotion, and the lotion in the capillary is repeatedly aspirated with the washing ear ball when using, and then washed with distilled water after being taken out, and then washed twice with a small amount of the solution to be measured). Put the capillary into the surface instrument, turn on the constant temperature water and adjust the temperature of the constant temperature tank to 25 ℃. Fill the suction cylinder with tap water, open the pressure relief valve (vent air), open the small stopper on the left side of the surface tension meter, add distilled water to the lower end of the capillary tube tangent to the liquid level, and close the stopper. Press the zero button on the manometer to set the pressure indicator to zero and close the relief valve. Open the piston at the bottom of the pump cylinder and slowly drip water, so as to reduce the pressure in the system. When the precise digital pressure gauge shows a certain value, close the piston at the bottom of the pump cylinder. If the numerical value of the manometer is basically unchanged within 2～3 minutes, the system is not leaking and the experiment can be carried out.

3. Determination of instrument constants

Open the pressure relief valve, the pressure gauge shows zero (if it is not zero, press the zero-picking button, make the pressure indication zero). Close the pressure relief valve, slowly rotate the piston at the lower end of the pump cylinder to make the water in the pump cylinder drip out slowly, bubbles come out from the bottom of capillary tube, and the bubble speed of capillary tube is controlled by adjusting the dripping speed of extraction cylinder (8 ~ 12 bubbles per minute, and the number of bubbles in the whole experimental process should be the same as far as possible). When the bubble just breaks away from the tube end, record the pressure value displayed on the pressure gauge, that is the instantaneous maximum value of a precision digital manometer, read it for 3 times, and take the average value.

4. Surface tension determination of ethanol aqueous solution with different concentrations

Measure from dilute to concentrated solution according to the method in step 3. Note that when changing the solution every time, the surface meter and capillary should be moistened and washed twice with a small amount of liquid to be measured.

5. After the experiment, the system communicates with the atmosphere. Wash the glass instrument, put distilled water into the surface tension meter, and put the capillary into H_2O_2-H_2SO_4 washing solution. Finally, the experimenter turns off the power supply.

Data recording and processing

1. Tabulate measured data, calculated concentrations and surface tension.

2. Make a surface tension-concentration ($\sigma - c$) diagram (note that the curve should be smooth). Take any eight points in the $\sigma - c$ curve, make the tangent line of the curve through each point, get the corresponding Z (bb_1) value (as shown in Fig. 8-3), calculate Γ, and list the data (c, Z, Γ).

3. Make a Γ-c diagram, calculate $\frac{c}{\Gamma}$ by taking any eight points in the curve, and list the data $\left(c, \Gamma, \frac{c}{\Gamma}\right)$.

4. Make a $\frac{c}{\Gamma}-c$ diagram to get a straight line, Γ_∞ can be obtained from the slope of the line, and the S_B value is calculated by Equation (5).

Thinking and discussing

1. Why must surface tension be measured in a constant temperature tank? What is the effect of temperature change on surface tension? Why?

2. Why should the lower end of capillary just be tangent to the liquid surface?

Reference materials

1. Experimental Center, School of Wuhan University Chemistry and Molecular Sciences.

The Physical Chemistry Experiment. Wuhan: Wuhan University Press, 2012.

2. Fudan University, et al. The Physical Chemistry Experiment. Beijing: Higher Education Press, 2004.

3. Physical chemistry experimental teaching group, School of Chemistry, Peking University. The Physical Chemistry Experiment. Beijing: Peking University Press, 2002.

4. Yu Z, Feng C. The Physical Chemistry Experiment. Beijing: Chemical Industry Press, 2014.

Supplementary Information: Tangent lines drawn by the image method

The tangent of a point on the curve can be obtained by using the mirror method, as shown in Figure 11-5. Firstly, take any point a on the curve, take a small square mirror, and make one side of the mirror pass through the point a. Make a straight line as the axis to rotate the mirror. The straight line runs perpendicular to the paper and through point a. When it is found that the image of the curve in the mirror and the actual curve outside the mirror can be connected to form a smooth curve, draw a line along the bottom edge of the mirror on the paper. Turn the mirror (180°) so that this side of the mirror still passes through point a. As above, turn the mirror so that the other half of the curve is a smooth curve with its image in the mirror, and then draw a straight line along the lower edge of the mirror, do the angle bisector of these two lines. Finally make vertical line of the angle bisector through point a, the vertical line is tangent.

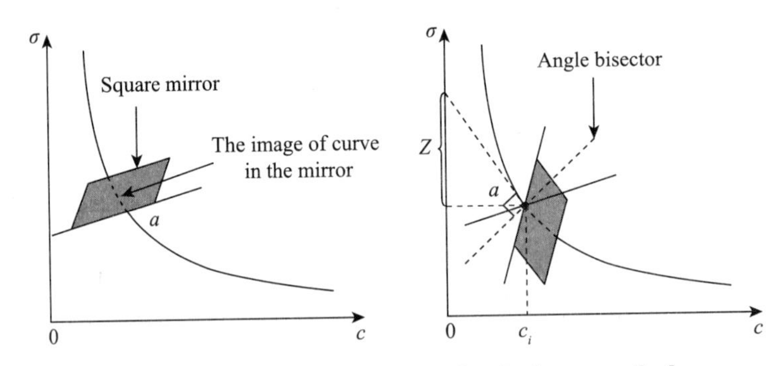

Fig. 11-5 Tangent lines drawn by the image method

实验 12 溶液吸附法测量固体物质的比表面

实验目的

1. 了解溶液吸附法测定固体比表面的原理和方法。

2. 用溶液吸附法测定活性炭的比表面。

基本原理

比表面,又叫比表面积,即单位物质所有能接触空气的表面积的总和,是固体物质在工业吸附剂、催化剂、颜料、水泥和聚合物等多孔结构应用中最有意义的形态特征。孔直径小于 2 nm 的可被归类为小孔,位于 2～50 nm 的是中孔,直径超过 50 nm 可称为大孔。微孔可进一步细分为超微孔(直径<0.7 nm),中等大小的微孔(0.7 nm<直径<0.9 nm)和直径大于 0.9 nm 的微孔。B.E.T 法(B.E.T 是三位科学家 Brunauer、Emmett 和 Teller 的首字母缩写)、色谱法等是目前广泛采用的测定比表面的方法。溶液吸附法测定固体物质的比表面,虽不如上述方法准确,但设备简单,操作、计算简便,是了解固体吸附剂性能的一种简单途径。

在一定温度下,固体在某些溶液中吸附溶质的情况与固体对气体的吸附很相似,可用 Langmuir 单分子层吸附方程来处理。即

$$\Gamma = \Gamma_m K_c/(1 + K_c) \tag{1}$$

式中 Γ 为平衡吸附量,即单位质量吸附剂达吸附平衡时,吸附溶质的物质的量($mg \cdot g^{-1}$);Γ_m 为饱和吸附量,即单位质量吸附剂的表面上吸满一层吸附质分子时所能吸附的最大量($mol \cdot g^{-1}$);c 为达到吸附平衡时,吸附质在溶液本体中的平衡浓度($mol \cdot L^{-1}$);K_c 为经验常数,与溶质(吸附质)、吸附剂性质有关。若能求得 Γ_m,则可由下式求得吸附剂比表面 $S_{比}$

$$S_{比} = \Gamma_m N_A A \tag{2}$$

式中,N_A 为阿伏伽德罗常数;A 为每个吸附质分子在吸附剂表面占据的面积。

将式(1)写成

$$c/\Gamma = c/\Gamma_m + 1/\Gamma_m K_c \tag{3}$$

配制不同吸附质浓度的样品溶液,测量达吸附平衡后吸附质的浓度 c,可用式(4)计算各份样品中吸附剂的吸附量

$$\Gamma = (c_0 - c)V/m \tag{4}$$

式中,c_0 为吸附前吸附质浓度($mol \cdot L^{-1}$),c 为达吸附平衡时吸附质的浓度($mol \cdot L^{-1}$),V 为溶液体积(L),m 为吸附剂质量(g),根据式(3)作图得直线斜率,可求得 Γ_m。

研究表明,在一定浓度范围内,大多数固体对亚甲基蓝的吸附是单分子层吸附,即 Langmuir 型吸附。本实验选用活性炭为吸附剂,亚甲基蓝为吸附质,溶剂为水。如果溶液浓度过高时,可能出现多分子层吸附,实验中要选择合适的吸附剂用量及吸附质原始浓度。亚甲基蓝水溶液为蓝色,可用分光光度法,在波长 665 nm 处测定其浓度。亚甲基蓝的分子式为 $C_{16}H_{18}ClN_3S \cdot 3H_2O$,

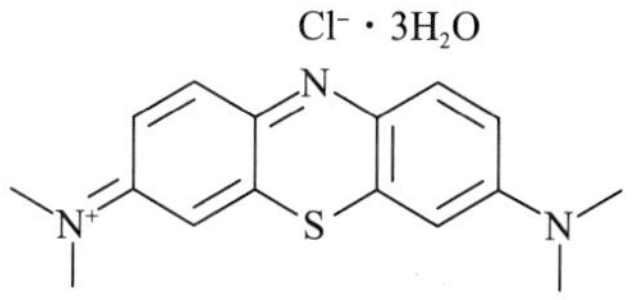

$Cl^- \cdot 3H_2O$

其摩尔质量为 373.9 $g \cdot mol^{-1}$，假设吸附质分子在吸附剂表面是直立的，S 值（每个吸附质分子占据的面积）取 1.52×10^{-18} m^2。这样根据前面算式，可求活性炭的比表面积。

仪器和试剂

721 型分光光度计　　恒温振荡器

干燥器　　锥形瓶（磨口 100 mL）

容量瓶（50 mL，100 mL）　　移液管（20 mL，25 mL，50 mL）

活性炭　　亚甲基蓝（分析纯）

实验步骤

1. 配制 1.000×10^{-3} $mol \cdot L^{-1}$ 亚甲基蓝水溶液。

2. 用电子天平称取 100.00 mg 左右活性炭六份，分别放入六只洗净干燥的 100 mL 磨口锥形瓶中，用移液管在六只锥形瓶中分别加入亚甲基蓝水溶液（1.000×10^{-3} $mol \cdot L^{-1}$）及去离子水，加入的量如表 12-1 所示。将六只锥形瓶的瓶盖塞好，放在恒温振荡器内，在恒温下振荡 1～3 天（此步骤应提前做好）。

表 12-1　加入的亚甲基蓝及水的体积

锥形瓶号	1	2	3	4	5	6
亚甲基蓝/mL	20.0	25.0	30.0	35.0	40.0	50.0
水/mL	30.0	25.0	20.0	15.0	10.0	0.0

3. 配制亚甲基蓝标准溶液

用 50 mL 容量瓶配制浓度为 1.000×10^{-5} $mol \cdot L^{-1}$ 的亚甲基蓝标准溶液，根据实验情况可以改变标准溶液的浓度。

4. 取步骤 2 中振荡平衡后的上清液，注意不要有活性炭微粒，用分光光度计在 665 nm 处分别测其浓度。如果溶液浓度过大（$A>0.8$），用去离子水稀释一定倍数后测定。

5. 实验完毕，对比色皿和盛过亚甲基蓝溶液的玻璃器皿，先用酸洗，再用自来水清洗，最后用去离子水涮洗。

数据记录与处理

1. 列表记录各样品吸附前及达吸附平衡后的浓度、活性炭质量，并记录实验温度。

2. 计算各份样品的吸附量。

3. 作 $c/\Gamma - c$ 图，通过线性拟合，求得直线斜率，根据斜率求得 Γ_m。

4. 由式（2）计算活性炭的 $S_{比}$。

思考与讨论

1. 如何确定吸附质浓度 c 是已达吸附平衡的浓度？

2. 本实验中溶液浓度太浓时，为什么要稀释后再测量？

参考资料

1. 复旦大学，等.物理化学实验.北京：高等教育出版社，2004.

2. Bardestani R, Patience G S, Kaliaguine S. Experimental methods in chemical engineering: specific surface area and pore size distribution measurements—BET, BJH, and DFT. The Canadian Journal of Chemical Engineering, 2019, 97(11): 2781-2791.

Experiment 12

The specific surface area determination of solid materials by solution adsorption methods

Experimental purpose

1. Understand the principle and method of solid specific surface measurement by solution adsorption.

2. Determinate specific surface area of activated carbon by solution adsorption.

Basical principles

Specific surface, also known as specific surface area (SSA), is the sum of all the surface areas per unit of matter that can come into contact with air. SSA is the most meaningful morphological characteristic of solid substances in applications related to porous structures, such as industrial adsorbents, catalysts, pigments, cement, and polymers. Pores for which d <2 nm are classified as micropores, while mesopores lie between 2 to 50 nm, and macropores exceed 50 nm in diameter. Micropores are further subdivided into ultra micropores (d<0.7 nm), medium-sized micropores (0.7 nm<d<0.9 nm), and super-micropores (d>0.9 nm). B.E.T method (the acronym of three scientists Brunauer, Emmett and Teller) and chromatography are widely used to determine SSA. Although the solution adsorption method is not as accurate as the method mentioned above, its equipment is handy, the operation and the calculation are simple, so it is the simple way to understand the solid adsorbent performance.

At a certain temperature, the adsorption of solute by solid in some solutions is similar to the adsorption of gas by solid, which can be treated by Langmuir monolayer adsorption equation. Namely

$$\Gamma = \Gamma_m K_c / (1 + K_c) \tag{1}$$

In the equation, Γ is the equilibrium adsorption amount, that is, the amount of substance that adsorbs solute when the unit mass of adsorbent reaches adsorption equilibrium, its unit is $mg \cdot g^{-1}$. Γ_m is the saturated adsorption amount, that is, the maximum amount of adsorbent that can be adsorbed when a layer of adsorbent molecules is adsorbed on the surface of the adsorbent per unit mass. c is the equilibrium concentration ($mol \cdot L^{-1}$) of the adsorbent in the solution when the adsorption equilibrium is reached. K_c is the empirical constant, which has relation with solute (adsorbent) and adsorbent properties. If Γ_m is calculated, the specific surface of the adsorbent can be obtained from the following equation

$$S_{SSA} = \Gamma_m N_A A \tag{2}$$

In the equation, N_A is the Avogadro constant and A is the area occupied by each adsorbent molecule on the surface of the adsorbent. Write Equation (1) as follows

$$c/\Gamma = c/\Gamma_m + 1/\Gamma_m K_c \tag{3}$$

Sample solutions with different adsorbent concentrations are prepared, and the concentration of adsorbent after reaching adsorption equilibrium is measured by the following equation

$$\Gamma = (c_0 - c)V/m \tag{4}$$

In the equation, c_0 is the concentration of adsorbent before adsorption ($mol \cdot L^{-1}$), c is the concentration of adsorbent at adsorption equilibrium ($mol \cdot L^{-1}$), V is the volume of solution (L), m is the mass of adsorbent (g). Drawing according to Equation (3), the slope of a straight line can be obtained to calculate Γ_m.

The researches show that the adsorption of methylene blue on most solids is monolayer adsorption (Langmuir adsorption). Activated carbon is used as adsorbent, and methylene blue as adsorbent, water as solvent. If the concentration of the solution is too high, multi-molecular layer adsorption may occur. In the experiment, the appropriate amount of adsorbent and the original concentration of the adsorbent should be selected. The methylene blue solution is blue in colour and the concentration can be determined using spectrophotometer at a wavelength of 665 nm. The formula of methylene blue is expressed as $C_{16}H_{18}ClN_3S \cdot 3H_2O$, that is

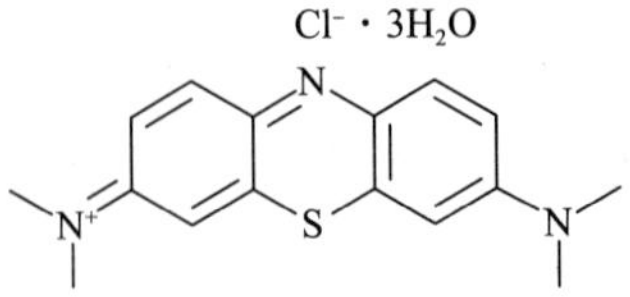

$Cl^- \cdot 3H_2O$

Its molar mass is 373.9 $g \cdot mol^{-1}$, assuming that the adsorbent molecules are upright on the surface, the S value (the area occupied by each adsorbent molecule) is 1.52×10^{-18} m^2. Thus, the specific surface area of the activated carbon can be obtained according to the preceding equation.

Apparatus and reagents

Type 721 photometer

Desiccator

Capacity bottles(50 mL,100 mL)

Activated carbon

Thermostatic oscillator (20 mL, 25 mL, 50 mL)

Conical bottle (grinding mouth 100 mL)

Pipette(20 mL, 25 mL, 50 mL)

Methylene blue (AR)

Experimental procedures

1. Prepare 1.000×10^{-3} mol · L^{-1} methylene blue solution.

2. Take about 100.00 mg of activated carbon in six parts by electronic balance, and put them into six 100 mL conical bottles with grinding mouth, respectively. Add methylene blue solution (1.000×10^{-3} mol · L^{-1}) and deionized water to the six conical bottle by pipette, the volume is shown in Table 12-1. Put the caps on the six conical bottles and put them into a thermostatic oscillator to oscillate 1～3 days at a constant temperature (this step should be done in advance).

Table 12-1 The volume of methylene blue solution and deionized water

Conical bottle number	1	2	3	4	5	6
Methylene blue solution/mL	20.0	25.0	30.0	35.0	40.0	50.0
Deionized water/mL	30.0	25.0	20.0	15.0	10.0	0.0

3. Preparation of methylene blue standard solution.

Methylene blue standard solution of 1.000×10^{-5} mol · L^{-1} is prepared in a 50 mL capacity bottles, the concentration of which can be changed according to the experimental conditions.

4. Take the supernatant from step 2 after the balance of the oscillations, be careful not to have activated carbon particles, and measure its concentration with the photometer at 665 nm. If the concentration of the solution is too high ($A > 0.8$), diluted with deionized water and determine after a certain multiple.

5. At the end of the experiment, the color dishes and the glassware containing the methylene blue solution are washed with acid, then with tap water and finally with deionized water.

Data recording and processing

1. Record the concentration before adsorption and after adsorption equilibrium, the quality

of activated carbon and the experimental temperature.

2. Calculate the adsorption capacity of each sample.

3. Draw $c/\Gamma - c$ curve, according to the slope of straight line, get Γ_m.

4. Calculate the S_{SSA} from the equation (2).

Thinking and discussing

1. How to determine whether the adsorbent concentration c is at the adsorption equilibrium concentration?

2. In this experiment, when the solution concentration is too thick, why should it be measured after dilution?

Reference materials

1. Fudan University, et al. The Physical Chemistry Experiment. Beijing: Higher Education Press, 2004:131-1353.

2. Bardestani R, Patience G S, Kaliaguine S. Experimental methods in chemical engineering: specific surface area and pore size distribution measurements—BET, BJH, and DFT. The Canadian Journal of Chemical Engineering, 2019, 97(11): 2781-2791.

实验 13 溶胶的制备和性质研究

实验目的

1. 利用不同的方法制备胶体溶液,并利用热渗析法进行纯化。

2. 了解胶体的光学性质,探讨不同电解质对所制备溶胶的聚沉值,掌握通过聚沉值判断溶胶荷电性质的方法。

基本原理

1. 溶胶的制备

把一种或几种物质分散在另一种物质中,就构成分散系统。在分散系统中被分散的物质叫作分散相,另一种物质叫作分散介质。按分散相粒子的大小,通常把分散系统区分为分子(或离子)分散系统、胶体分散系统和粗分散系统(r>100 nm)。胶体分散系统在生物界和非生物界都普遍存在,在实际生活和生产中也占有重要地位,如在石油、冶金、造纸、橡胶、塑料、纤维、肥皂等领域可广泛接触到与胶体分散系统有关的问题。固体以胶体分散程度分散在液体介质中即组成溶胶。溶胶具有如下的基本特征:

(1) 多相体系,相界面很大;

（2）胶粒大小在1～100 nm；

（3）热力学不稳定体系（要依靠稳定剂使其形成离子或分子吸附层，才能得到暂时的稳定）。

溶胶的制备方法可分为两类。

（1）分散法

这种方法是用适当方式将大块物质在有稳定剂存在的情况下分散成胶体粒子的大小。常用的分散法有：①机械作用法，这种方法通常适用于脆而易碎的物质，如用胶体磨或其他研磨方法把物质分散。②胶溶法，它不是使粗粒分散成溶胶，而只是使暂时凝聚起来的分散相又重新分散。许多新的沉淀，经洗涤除去过多的电解质，再加少量的稳定剂后，又可以制成溶胶，这种作用称为胶溶作用。③超声波法，利用超声波场的空化作用，将物质撕碎成细小的质点，它适用于分散硬度低的物质或制备乳状液。④电弧法，此法是以金属为电极，通电产生电弧，金属受高热变成蒸气，并在液体中凝聚成胶体质点。⑤气相沉积法，在惰性气体中，用电加热、高频感应、电子束或激光等热源，将要制备成纳米级粒子的材料气化，处于气态的分子或原子，按照一定规律共聚或发生化学反应，形成纳米级粒子，再将它用稳定剂保护（此法是先分散再聚合，故也可归入凝聚法）。

（2）凝聚法

这个方法的一般特点是先制成难溶物的分子（或离子）的过饱和溶液，再使之互相结合成胶体粒子而得到溶胶。通常可以分成三种方法：①化学凝聚法，通过化学反应，如复分解反应、水解反应、氧化或还原反应等，使生成物呈过饱和状态，然后再结合成胶粒。例如，铁、铝、铬、铜等金属的氢氧化物溶胶可以通过其盐类的水解而制得。②物理凝聚法，利用适当的物理过程，如蒸气骤冷，可以使某些物质凝聚成胶体粒子的大小。例如将汞的蒸气通入冷水中，就可以得到汞溶胶。此时，高温下的汞蒸气与水接触时生成的少量氧化物起稳定剂的作用。③更换溶剂法，更换溶剂也可以制得溶胶，例如将松香的酒精溶液滴入水中，由于松香在水中的溶解度很低，使溶质以胶粒的大小析出，形成松香的水溶胶。

2. 溶胶的纯化

在制得的溶胶中常含有一些电解质，通常除了形成胶团所需要的电解质以外，过多的电解质存在反而会破坏溶胶的稳定性，因此必须将溶胶净化。常用方法有以下几种：

（1）渗析法

由于溶胶粒子不能通过半透膜，而分子离子能通过，故可把溶胶放在装有半透膜的容器内，常见的半透膜如羊皮纸、动物膀胱膜、硝酸纤维、醋酸纤维等。膜外放纯溶剂。由于膜内外杂质的浓度有差别，膜内的离子或其他能透过半透膜的杂质小分子向半透膜外迁移，不断更换膜外的溶剂，则可逐渐降低溶胶中的电解质或杂质的浓度，而达到净化的目的，这种方法叫渗析法。

(2) 超过滤法

用孔径细小的半透膜(约 10～300 nm),在加压吸滤的情况下,使胶粒与介质分开,这种方法称为超过滤法。可溶性杂质能透过滤板而被除去,有时可将第一次超过滤得到的胶粒再加到纯的分散介质中,再加压过滤,如此反复进行,达到净化的目的。

3. 溶胶聚沉值的测定

带电质点对电解质十分敏感,在电解质作用下溶胶质点因聚结而下沉的现象称为聚沉。在指定条件下使某溶胶聚沉时,电解质的最低浓度称为聚沉值,常用 $mmol \cdot L^{-1}$表示。影响聚沉的主要因素是与胶粒电荷相反的离子的价数、离子的大小及同号离子的作用等。一般来说,反号离子价数越高,聚沉效率越高,聚沉值越小,聚沉值大致与反离子价数的 6 次方成反比。同价无机小离子的聚沉能力常随其水合半径增大而减小,这一顺序称为感胶离子序。与胶粒带有同号电荷的二价或高价离子对胶体体系常有稳定作用,使该体系的聚沉值有所增加。此外,当使用高价或大离子聚沉时,少量的电解质可使溶胶聚沉:电解质浓度大时,聚沉形成的沉淀物又重新分散,浓度再提高时,又可使溶胶聚沉,这种现象称为不规则聚沉。不规则聚沉的原因是,低浓度的高价反离子使溶胶聚沉后,增大反离子浓度,它们在质点上强烈吸附使其带有反离子符号的电荷而重新稳定;继续增大电解质浓度,重新稳定的胶体质点的反离子又可使其聚沉。

仪器和试剂

试管
锥形瓶
滴定管
移液管
烧杯
量筒
观察丁达尔现象的暗箱
NaCl 溶液($5\ mol \cdot L^{-1}$)
Na_2SO_3 溶液($0.1\ mol \cdot L^{-1}$)
H_2SO_4($0.1\ mol \cdot L^{-1}$)
$AlCl_3$ 溶液($0.001\ mol \cdot L^{-1}$)
$FeCl_3$ 溶液(10%,20%)
K_2SO_4 溶液($0.01\ mol \cdot L^{-1}$)
KI 溶液($0.01mol \cdot L^{-1}$)
$K_3Fe(CN)_6$ 溶液($0.001\ mol \cdot L^{-1}$)
$AgNO_3$ 溶液($0.01\ mmol \cdot L^{-1}$)
10% $NH_3 \cdot H_2O$
松香
硫黄
酒精

实验步骤

1. 胶体溶液的制备

(1) 化学反应法

① $Fe(OH)_3$ 溶胶(水解法)

在 250 mL 烧杯中加入 95 mL 蒸馏水,加热至沸腾,慢慢地滴入 5 mL 10% $FeCl_3$ 溶液,

并不断搅拌，加完后继续沸腾几分钟水解后，得红棕色的氢氧化铁溶胶，其结构可用下式表示

$$\{m[Fe(OH)_3]\cdot nFeO^+\cdot(n-x)Cl^-\}^{x+}\cdot xCl^-$$

② 硫溶胶

取 Na_2SO_3 溶液（0.1 mol · L^{-1}）5 mL 放入试管中，再取 H_2SO_4（0.1 mol · L^{-1}）5 mL，将两液体混合，观察丁达尔现象。同法配制混合液，在亮处仔细观察透射光和散射光颜色的变化；当混浊度增加到盖住颜色时（约经 5 min），把溶胶稀释 1 倍，继续观察颜色，记下透射光和散射光颜色随时间变化的情形。

③ AgI 溶胶

AgI 溶胶微溶于水（9.7×10^{-7} mol · L^{-1}），当硝酸银溶液与易溶于水的碘化物混合时，应析出沉淀。但是如果混合稀溶液并且取其中之一过剩，则不产生沉淀，而形成胶体溶液，胶体溶液的性质与过剩离子种类有关。在此，胶粒的电荷是由过剩的离子被 AgI 所吸附，在 $AgNO_3$ 过剩时，得正电性的胶团，其结构为

$$\{m[AgI]\cdot nAg^+\cdot(n-x)NO_3^-\}^{x+}\cdot xNO_3^-$$

在 KI 过剩时，得负电性的胶团，其结构为

$$\{m[AgI]\cdot nI^-\cdot(n-x)K^+\}^{x-}\cdot xK^+$$

取 30 mL KI 溶液（0.01 mol · L^{-1}）注入 100 mL 的锥形瓶中，然后用滴定管把 20 mL $AgNO_3$ 溶液（0.01 mmol · L^{-1}）慢慢地滴入，制得负电性的 AgI 溶胶（A）。

按此法取 30 mL $AgNO_3$ 溶液（0.01 mmol · L^{-1}），慢慢加入 20 mL KI 溶液（0.01 mol · L^{-1}），制得正电性溶胶（B）。

（2）改变分散介质和实验条件（硫溶胶）

取少量硫黄置于试管中，注入 2 mL 酒精，加热到沸腾（重复数次，使硫得到充分溶解），在未冷却前把上部清液倒入盛有 20 mL 水的烧杯中，搅匀，观察变化情况及丁达尔现象。

（3）胶溶法

取 1 mL 20% $FeCl_3$ 溶液放在小烧杯中，加水稀释到 10 mL。用滴管逐渐加入 10% $NH_3\cdot H_2O$ 到稍微过量时为止。过滤，用水洗涤数次。取下沉淀放在另一烧杯中，加水 20 mL，再加入约 1 mL 20% $FeCl_3$ 溶液，用玻璃棒搅动，并用小火加热，沉淀消失，形成透明的胶体溶液，利用溶胶的光性加以鉴定。

2. 溶胶的纯化

把制得的 $Fe(OH)_3$ 溶胶置于半透膜袋内，用线拴住袋口，置于 400 mL 烧杯内，用蒸馏水渗析，保持温度在 60～70 ℃。半小时换一次水，并取 1 mL 检验其 Cl^- 及 Fe^{3+}（分别用 $AgNO_3$ 及 KSCN 溶液检验），直至不能检查出 Cl^- 及 Fe^{3+} 为止。也可通过测溶胶的电导率，来判断溶胶纯化的程度。

3. 溶胶的聚沉作用

用10 mL移液管在3个干净的50 mL锥形瓶中各注入10 mL前面用水解法制备的$Fe(OH)_3$溶胶(若条件许可,应使用经渗析纯化过的溶胶),然后在每个瓶中分别用滴定管逐滴慢慢加入0.5 $mol \cdot L^{-1}$的KCl溶液、0.01 $mol \cdot L^{-1}$ K_2SO_4溶液、0.001 $mol \cdot L^{-1}$ $K_3Fe(CN)_6$溶液,不断摇动。在开始有明显聚沉物出现时,停止加入电解质。若加电解质的量达到10 mL后仍无聚沉物出现,则不再继续加入该电解质。

数据记录与处理

1. 将前面制得的溶胶A和溶胶B按表13-1的量混合。观察混合后的现象、溶胶颜色的变化、透过光颜色的变化,说明其稳定性的程度和原因。

表13-1 溶胶A和溶胶B的体积

试管编号	1	2	3	4	5	6	7
V_A/mL							
V_B/mL							

2. 记下溶胶聚沉实验中每次所用溶液的体积,计算聚沉值大小,说明溶胶带何种电荷,并与理论值比较。

思考与讨论

1. 试比较不同溶胶的制备方法有什么共同点和不同点?
2. 不同电解质对同一溶胶的聚沉值是否一样? 为什么?

参考资料

1. 北京大学化学学院物理化学实验教学组.物理化学实验.北京:北京大学出版社,2002.
2. 普季洛娃.胶体化学实验作业指南. 北京:高等教育出版社,1955.
3. Crockford H D, et al. Laboratory Manual of Physical Chemistry. New York: John Wiley, 1975.
4. 周祖康,顾惕人,马季铭.胶体化学基础. 北京大学出版社,1991.
5. 傅献彩,沈文霞,姚天杨,等.物理化学(第五版)下册. 北京:高等教育出版社,1990.

Experiment 13

Preparation and properties of sol

Experimental purpose

1. Prepare colloidal solutions by different methods and purify solutions by thermodialysis.

2. Understand the optical properties of colloid, investigate the precipitation of sol prepared by different electrolytes, and master the method of judging the sol charge by the precipitation value.

Basical principles

1. Preparation of sol

Dispersion system is formed by dispersing one or more substances into another. The matter that is dispersed in a dispersion system is called dispersion phase, and the other matter is called dispersion medium. According to the size of dispersed phase particles, the dispersed system is usually divided into molecular (or ion) dispersed system, colloidal dispersed system and coarse dispersed system. Colloidal dispersion systems are ubiquitous in both the biological and non-biological worlds, and play an important role in real life and production, such as in petroleum, metallurgy, paper, rubber, plastic, fiber, soap and other fields in which colloidal dispersion system-related problems can be widely exposed. The solid disperses in the liquid medium by the colloid dispersion degree, that is, formation of the sol. The basic characteristics of sol are as follows:

(1) It is a multiphase system with a large phase interface.

(2) The size of colloidal particles is 1～100 nm.

(3) It is a thermodynamic unstable system, which can be temporarily stabilized only by the formation of an ion or a molecular adsorbent by a stabilizer.

The preparation method of sol can be divided into two types:

(1) Dispersion method

The method is to disperse the bulk materials in a suitable manner to the size of colloidal particles in the presence of a stabilizer. The commonly used dispersion methods are: ① mechanical method, which is usually suitable for brittle and fragile substances, such as colloid grinding or other grinding methods to disperse the material. ② The gelation method. It does not disperse the coarse particles into a sol, but simply re-disperses the dispersed phase that has temporarily coalesced. Many of the new precipitates are washed away to remove the excess electrolytes, and a small amount of a stabilizer is added, which can also be made into a sol, and this process is called gelatinization.③ The ultrasonic method, which uses the cavitation effect of ultrasonic field to tear the material into small particles. It is suitable for dispersing the low hardness material or preparing emulsion. ④ Arc method. This method is to take the metal as the electrode. The electric current produces arc, and the metal becomes the steam by the high heat, it condenses in the liquid to form the colloid particle.⑤ Meteorological deposition. Materials to be prepared into nano-particles are vaporized by electric heating, high-frequency induction, electron beams or lasers in an inert atmosphere. Molecules or atoms in the gaseous state form nano-particles according to certain laws of copolymerization or chemical reaction.Then it will be protected with a stabilizer (in this method, dispersion occurs firstly and then polymerization, it can also be classified as condensation method) .

(2) Condensation method

The general feature of this method is that the supersaturation of insoluble molecules (or ions) are first made and then bound to each other to form colloidal particles to produce sol. There are usually three methods: ① Chemical agglomerations. In which the products are supersaturated by chemical reactions, such as metathesis, hydrolysis, oxidation or reduction, and then bound together to form colloidal particles. For example, the hydroxide sol of metals such as iron, aluminium, chromium and copper can be made by hydrolysis of their salts. ② Physical condensation method. Certain substances can be condensed to the size of colloidal particles by appropriate physical processes, such as steam cooling. For example, mercury sol can be obtained by passing mercury vapor into cold water. At this point, mercury vapor at high temperature in contact with water when the formation of a small amount of oxide acts as a stabilizer.③ Replacement solvent method. For example, the alcohol solution of rosin is dropped into water, because the solubility of rosin in water is very low, so that the size of the solute into colloidal particles precipitated, the formation of rosin hydrosol.

2. Purification of sol

The prepared sol often contains some electrolytes. In addition to the electrolyte needed to form micelles, too many electrolytes will destroy the stability of the sol, so the sol must be purified. There are several common methods.

(1) Dialysis

As the sol particles can not pass through the semi-permeable membrane while the molecular ions can pass through, the sol can be placed in a container containing semi-permeable membrane. Commonly used semi-permeable membrane are parchment, animal bladder membrane, nitrocellulose, cellulose acetate and so on. Put pure solvent outside the membrane. Because of the difference between concentration of impurities inside and outside the membrane, ions inside the membrane or other small molecules of impurities that can pass through the semi-permeable membrane migrate to the outside of membrane, constantly changing the solvent outside the membrane can gradually reduce the concentration of electrolyte or impurities in the sol, and achieve the goal of purification. This method is called dialysis.

(2) Ultrafiltration method

Use a small pore size semi-permeable membrane (about 10 to 300 nanometers) to separate the colloidal particles from the medium under the condition of pressure suction filtration, which is called the ultrafiltration method. Soluble impurities can be removed through the filter plate. Sometimes the first ultra-filtered particles can be added to the pure dispersion medium, and then pressure filtration, repeat to achieve the goal of purification.

3. Measurement of sol precipitation value

Charged particles are very sensitive to electrolytes, and under the action of electrolytes, sol particles sink due to coalescence, which is called coalescence. When a sol is precipitated under specified conditions, the lowest concentration of the electrolyte is called the precipitating value, which is usually expressed in $mmol \cdot L^{-1}$. The main factors affecting the precipitation are the valence number of the ion opposite to the colloidal charge, the size of the ion and the action of the same ion. In general, the higher number of anti-sign ions have, the higher efficiency of precipitation will be, and the smaller value the precipitation has. The deposition value is approximately inversely proportional to the 6th power of the counter-ion valence.The precipitation ability of isovalent inorganic small ions often improve with the increase of hydration radius, which is called colloidal ion sequence. The divalent or high valent ions with the same charge as colloidal particles often have a stabilizing effect on the colloidal system, which increases the precipitation value of the system. In addition, when high valence or large ion is used, a small amount of electrolyte can make the sol precipitate: when the concentration of electrolyte is high, the precipitate formed by the precipitate can be dispersed again; when the concentration is increased again, the sol can be precipitated again. This phenomenon is called irregular agglomeration. The reason of irregular agglomeration is that the high valence counter-ions of low concentration make the sol agglomerate, then increase the concentration of counter-ions, and they adsorb strongly on the particle to make the charge with the sign of counter-ions stable again continuing to increase the concentration of electrolyte, re-stable colloidal particles of the counter-ion and make its precipitation.

Apparatus and reagents

Test tube
Titrator
Beaker
Obscura for observing Tyndall effect
Na_2SO_3 solution(0.1 mol · L^{-1})
$AlCl_3$ solution (0.001 mol · L^{-1})
K_2SO_4 solution (0.01 mol · L^{-1})
$K_3Fe(CN)_6$ solution (0.001 mol · L^{-1})
10% $NH_3 \cdot H_2O$
Sulfur
Conical bottle
Pipette
Measuring tube
NaCl solution(5 mol · L^{-1})
H_2SO_4 solution(0.1 mol · L^{-1})
$FeCl_3$ solution(10%,20%)
KI solution(0.01 mol · L^{-1})
$AgNO_3$ solution(0.01 mmol · L^{-1})
Rosin
Alcohol

Experimental procedures

1. Preparation of colloid solution

(1) Chemical reaction method

① $Fe(OH)_3$ sol (hydrolysis)

Add 95 mL distilled water to a 250 mL beaker, heat it to boil. Slowly drop in 5 mL 10% $FeCl_3$ solution and stir it constantly. Continue to boil it for a few minutes. After hydrolysis, the red-brown iron(Ⅲ) oxide-hydroxide sol is produced, its structure can be represented by the following formula

$$\{m[Fe(OH)_3] \cdot n\,FeO^+ \cdot (n-x)\,Cl^-\}^{x+} \cdot x\,Cl^-$$

② Sulfur sol

5 mL of Na_2SO_3 solution (0.1 mol · L^{-1}) is put into the test tube and 5 mL H_2SO_4(0.1 mol · L^{-1}) is taken. The two liquids are mixed and the Tyndall effect is observed. Mixing solution is prepared by the same method, and the color changes of transmitted light and scattered light are observed carefully in the bright place. When the turbidity increases to cover the color (about 5 min), the sol is diluted 1 times and the color is observed again, note how the color of transmitted and scattered light changes over time.

③ AgI sol

AgI sol is slightly soluble in water (9.7×10^{-7} mol · L^{-1}). When silver nitrate solution is mixed with water-soluble iodide, precipitation should occur. But if there is an excess of one of these two solutions, no precipitation occurs and a colloid solution is formed. The properties of colloid solutions are related to the type of excess ions. For example, when $AgNO_3$ is in excess, a positively charged micelle form with a structure of

$$\{m[AgI] \cdot nAg^+ \cdot (n-x)NO_3^-\}^{x+} \cdot xNO_3^-$$

If KI is excess, negative micelles is get

$$\{m[AgI] \cdot nI^- \cdot (n-x)K^+\}^{x-} \cdot xK^+$$

30 mL KI solution (0.01 mol · L^{-1}) is injected into a 100 mL conical flask, and then 20 mL $AgNO_3$ solution (0.01 mmol · L^{-1}) is slowly dripped into the flask with a titration tube to prepare a negatively charged AgI sol (A). According to this method, 30 mL $AgNO_3$ solution (0.01 mmol · L^{-1}) is slowly added into 20 mL KI solution (0.01 mol · L^{-1}) to prepare positively charged sol (B).

(2) Changing the dispersion medium and experimental conditions (sulfur sol)

Take a small amount of sulfur into the test tube, inject 2 mL of alcohol, heat it to boil (repeated several times, so that the sulfur is fully dissolved), pour the liquid into a beaker with 20 mL water before cooling. Stir well and observe the changes and Tyndall effect.

(3) The gelation method.

Place 1 mL 20% $FeCl_3$ solution in a beaker and dilute to 10 mL with water. Gradually add 10% $NH_3 \cdot H_2O$ until slightly excessive. Filter and wash several times with water. The precipitate is put into another beaker, add 20 mL water, add 20% $FeCl_3$ solution about 1 mL, stir with glass rod, and heat with small fire. The precipitate disappear, forming a transparent colloid solution, which is identified by the light property of the sol.

2. Purification of sol

The prepared $Fe(OH)_3$ sol is placed in a semi-permeable membrane bag, the mouth of the bag is bolted with thread, and placed in a 400 mL beaker. The dialysis is conducted by distilled water. The water is changed at 60～70 ℃ for half an hour. 1 mL water is taken to test for Cl^- and Fe^{3+} (using $AgNO_3$ and KSCN solutions, respectively) until Cl^- and Fe^{3+} could not be detected. The degree of sol purification can also be judged by measuring the conductivity of sol.

3. Aggregation and sedimentation of sol

Three clean 50 mL conical flasks are filled with 10 mL of $Fe(OH)_3$ sol prepared by hydrolysis with 10 mL pipette (the sol should be purified by dialysis if the conditions permit). Then slowly add 0.5 mol · L^{-1} KCl solution, 0.01 mol · L^{-1} of K_2SO_4 solution, and 0.001 mol · L^{-1} of $K_3Fe(CN)_6$ solution, drop by drop, respectively, in each bottle, with constant shaking. When obvious sediment appears, stop adding electrolyte. If the amount of electrolyte reaches 10 mL and no sediment appears, the electrolyte will not be added.

Data recording and processing

1. Mix sol A and sol B previously prepared as shown in Table 13-1. Observe the mixing phenomenon, the change of the sol color, the change of the transmission light color, and explain the degree and reason of its stability.

Table 13-1 Volume of sol A and sol B

Tube number	1	2	3	4	5	6	7
V_A/mL							
V_B/mL							

2. Record the solution volume used each time in the sol precipitation experiment, calculate the value of precipitation, explain what kind of charge the sol take. Compared with the theoretical value.

Thinking and discussing

1. What are the similarities and differences between different sol preparation methods?

2. Do different electrolytes have the same deposition value for the same sol? Why?

Reference materials

1. Physical chemistry experimental teaching group, School of Chemistry, Peking University. The Physical Chemistry Experiment. Beijing: Peking University Press, 2002.

2. Protilova. Guide to Colloid Chemistry Experiment, Beijing: Higher Education Press, 1955.

3. Crockford H D, et al. Laboratory Manual of Physical Chemistry. New York: John Wiley, 1975.

4. Zhou Z, Gu T, Ma J. The Basis of Colloid Chemistry. Peking University Press, 1991.

5. Fu X, Shen W, Yao T, et al. Physical Chemistry (2nd ed. Volume Ⅱ). Beijing: Higher Education Press, 1990.

实验 14 黏度法测定水溶性高聚物的平均摩尔质量

实验目的

1. 掌握使用三管黏度计测定黏度的方法。

2. 掌握测定黏均摩尔质量的原理和方法。

3. 测定聚乙二醇的黏均摩尔质量。

基本原理

在高聚物的研究中，相对分子质量是一个不可缺少的重要数据，它不仅反映了高聚物分子的大小，而且关系到高聚物的物理性能。例如，纤维素若是短链分子多，就不适宜做纺织材料；又如天然橡胶，若含低摩尔质量的物质多，生胶的硫化效果也就不好。准确测定高聚物摩尔质量的分布是一件极其复杂的工作，因此常采用高聚物的平均摩尔质量来反映高聚物的某些特性。通常平均摩尔质量随所用测定方法的不同而不同，所得到平均值的含义也有所差异，如数均摩尔质量 $\overline{M}_n$ 、质均摩尔质量 $\overline{M}_m$ 、Z 均摩尔质量 $\overline{M}_Z$ 、黏均摩尔质量 $\overline{M}_\eta$ 等。

溶液的黏度随着聚合物分子的大小及性质、温度、溶剂的性质、浓度的不同而不同。在温度、聚合物溶剂系统选定后，溶液黏度仅与浓度和聚合物分子的大小有关。黏度法测聚合物的摩尔质量是目前最常用的方法，原因在于设备简单、操作便利、耗时较少，且精确度较高。此外，黏度法与其他方法配合，还可以研究聚合物分子在溶液中的形态、尺寸及大分子与溶剂分子的相互作用等。

黏度是指液体对流动所表现的阻力，这种力反抗液体中邻接部分的相对移动，因此可看作是一种内摩擦。纯溶剂黏度用 η_0 表示，η_0 是由溶剂分子之间的内摩擦所表现出来的黏度；溶液的黏度用 η 表示，η 是由溶剂分子之间的内摩擦、高聚物分子相互之间的内摩擦以及高分子与溶剂分子之间的内摩擦所表现的黏度总和。在同一温度下，一般来说，$\eta > \eta_0$。相对于溶剂，其溶液黏度增加的分数，称为增比黏度，记作 η_{sp}，即

$$\eta_{sp} = \frac{\eta - \eta_0}{\eta_0} = \frac{\eta}{\eta_0} - 1 = \eta_r - 1 \tag{1}$$

式中的 η_r 称为相对黏度，η_r 反映的是整个溶液的黏度行为；增比黏度 η_{sp} 则反映的是扣除了溶剂分子之间的内摩擦效应后，高聚物分子之间以及高聚物分子与纯溶剂分子之间的内摩擦效应。

对于高分子溶液，增比黏度 η_{sp} 往往随溶液浓度 c 的增加而增加。为了便于比较，将单位浓度下所显示出的增比黏度 $\frac{\eta_{sp}}{c}$ 称为比浓黏度；而 $\frac{\ln\eta_r}{c}$ 称为比浓对数黏度。η_r 和 η_{sp} 都是无因次的量。

为了进一步消除高聚物分子之间的内摩擦效应，必须将溶液无限稀释，使得每个高聚物分子彼此相隔极远，其相互干扰可以忽略不计。这时溶液所呈现出的黏度行为基本上反映了高分子与溶剂分子之间的内摩擦。这一黏度的极限值记为 $[\eta]$

$$[\eta] = \lim_{c \to 0} \frac{\eta_{sp}}{c} = \lim_{c \to 0} \frac{\ln \eta_r}{c} \tag{2}$$

$[\eta]$ 称为特性黏度,其值与浓度无关。实验表明,当聚合物、溶剂和温度确定以后,高分子溶液的特性黏度与高聚物分子黏均摩尔质量的关系可用 Mark Houwink 经验方程式表示

$$[\eta] = K\overline{M}_{\eta}^{\alpha} \tag{3}$$

式中 K 为比例常数,α 是与分子形状有关的经验常数。它们都与温度、聚合物和溶剂性质有关,在一定的相对分子质量范围内与相对分子质量无关。

K 和 α 的数值,只能通过其他方法确定,例如渗透压法、光散射法等等。黏度法只能测定 $[\eta]$,利用式(3)可求算出 $\overline{M}_{\eta}$。

表 14-1　溶液黏度的命名

名称	符号和定义
黏度(系数)	η
相对黏度	$\eta_r = \frac{\eta}{\eta_0}$ (η_0 为溶剂的黏度)
增比黏度	$\eta_{sp} = \frac{\eta - \eta_0}{\eta_0} = \eta_r - 1$
比浓黏度	$\frac{\eta_{sp}}{c}$
比浓对数黏度	$\frac{\ln \eta_r}{c}$
特性黏度	$[\eta] = \left(\frac{\eta_{sp}}{c}\right)_{c=0} = \left(\frac{\ln \eta_r}{c}\right)_{c=0}$

测定液体黏度的方法主要有三类:(1)用毛细管黏度计测定液体在毛细管里的流出时间;(2)用落球式黏度计测定圆球在液体里的下落速率;(3)用旋转式黏度计测定液体与同心轴圆柱体相对转动的情况。

测定高分子的 $[\eta]$ 时,用毛细管黏度计最为方便。液体在毛细管黏度计内因重力作用而流出时,遵守泊肃叶(Poiseuille)定律

$$\frac{\eta}{\rho} = \frac{\pi h g r^4 t}{8lV} - m\frac{V}{8\pi l t} \tag{4}$$

式中 ρ 为液体的密度;l 是毛细管长度;r 是毛细管半径;t 是流出时间;h 是流经毛细管液体的平均液柱高度;g 为重力加速度;V 是流经毛细管液体的体积;m 是与仪器的几何形状有关的常数,在 $\frac{r}{l} \ll 1$ 时,可取 $m = 1$。

对某一支指定的黏度计而言，令 $\alpha=\frac{\pi hgr^4}{8lV}$，则式(4)可改写为

$$\frac{\eta}{\rho}=\alpha t-\frac{\beta}{t} \tag{5}$$

式中 $\beta<1$，当 $t>100$ s 时，等式右边第二项可以忽略。设溶液的密度 ρ 与溶剂密度 ρ_0 近似相等。这样，通过分别测定溶液和溶剂的流出时间 t 和 t_0，就可求算相对黏度 η_r

$$\eta_r=\frac{\eta}{\eta_0}=\frac{t}{t_0} \tag{6}$$

进而可分别计算得到 η_{sp}、$\frac{\eta_{sp}}{c}$ 和 $\frac{\ln\eta_r}{c}$ 的值。配制一系列不同浓度的溶液分别进行测定，以 $\frac{\eta_{sp}}{c}$ 和 $\frac{\ln\eta_r}{c}$ 为同一纵坐标，c 为横坐标作图，得两条直线，分别外推到 $c=0$ 处（如图 14-1 所示），其截距即为特性黏度 $[\eta]$，代入(3)式（K、α 已知），即可得到 $\overline{M}_\eta$。

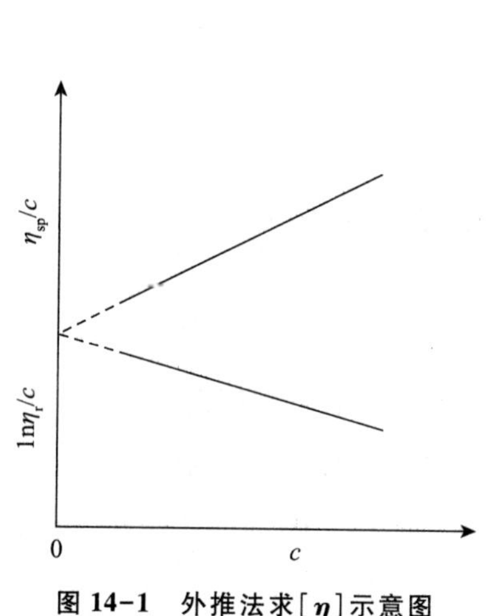

图 14-1 外推法求[η]示意图

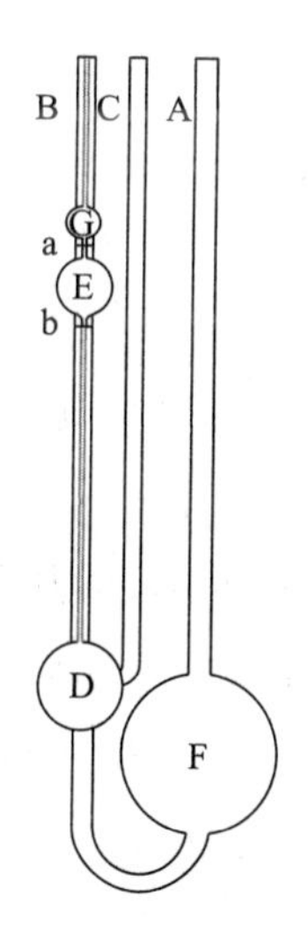

图 14-2 三管黏度计示意图

表 14-2 不同温度下聚乙二醇水溶液的 K、α 值[4]

t/℃	$K\times10^3$/ $dm^3\cdot kg^{-1}$	α	$\overline{M}_\eta\times10^{-4}$
25	156	0.50	0.019～0.1
30	12.6	0.78	2～500
35	6.4	0.82	3～700
40	16.6	0.82	0.04～0.4
45	6.9	0.81	3～700

仪器和试剂

三管黏度计
移液管(2 mL,5 mL,10 mL)
聚乙二醇
恒温水浴
秒表

实验步骤

1. 溶液流出时间(t)的测定

将三管黏度计放入恒温槽中(注意:黏度计要垂直),用移液管移取 10 mL 10 $g \cdot L^{-1}$的聚乙二醇溶液,通过 A 管放入黏度计中,恒温 10 min 开始测流出时间。B、C 管可套上乳胶管,先夹住 C 管使其与空气隔绝,用洗耳球从 B 管往上吸溶液使溶液超过 a 线至 G 球,松开 C 管使其通大气,用洗耳球挤压 B 管中的溶液流出,重复操作 5 次,使毛细管得到充分润洗。

再用洗耳球从 B 管往上吸溶液使溶液超过 a 线至 G 球,松开 C 管,使 B 管中的溶液自然向下流,当液面降至 a 线时启动秒表开始计时,液面降至 b 线时停表,记下溶液流经 ab 段所需要的时间,即溶液流出时间 t。再重复测定流出时间两次,将三次流出时间取平均值(注意:每次误差不能超过 0.2 s)。

将溶液进行稀释,依次往黏度计中加入二次蒸馏水 2,3,5,10 mL,重复上述操作。每次加完水后,可用洗耳球通过 C 管向溶液中吹气,将溶液混合均匀,混合时应注意不要把溶液吸到洗耳球中。

2. 溶剂流出时间(t_0)的测定

测完全部溶液流出时间后,将黏度计从恒温槽中取出倒掉溶液,反复用自来水、5%醋酸清洗黏度计,最后再用二次蒸馏水润洗黏度计 3 次(注意:一定要将毛细管部分洗涤干净)。将黏度计注入 10 mL 蒸馏水,放入恒温槽中恒温 10 min,测水的流出时间,方法同步骤 1。

3. 黏度计的洗涤

实验完毕后,将黏度计中水倒掉,加少量乙醇润洗,回收乙醇,再将黏度计放在干燥箱中烘干,以便下组同学使用。

数据记录与处理

1. 记录实验所得数据,并计算 η_r,η_{sp},$\frac{\eta_{sp}}{c}$,$\frac{\ln\eta_r}{c}$。

2. 绘图求 $[\eta]$ 值。

以 $\frac{\eta_{sp}}{c}$ 和 $\frac{\ln\eta_r}{c}$ 分别对 c 作图,得两条直线,将直线外推至 $c=0$,求出截距即为 $[\eta]$ 值。

3. 由式(3)求出聚乙二醇的黏均摩尔质量。

思考与讨论

1. 三管黏度计中的支管 C 有什么作用？除去支管 C 是否仍可以测黏度？

2. 黏度计毛细管的粗细对实验有何影响？

3. 请评价黏度法测定高聚物平均摩尔质量的优缺点，指出影响测定结果准确性的主要因素。

参考资料

1. 钱人元.高聚物分子量的测定.北京：科学出版社，1958.

2. 何曼君，陈维孝，董西侠.高分子物理.上海：复旦大学出版社，1982：127-132.

3. Daniels F, Alberty R A, Williams J W, et al. Experimental Physical Chemistry. 7th ed. New York: McGraw-Hill, Inc, 1975: 329.

4. 淮阴师范专科学校化学科.物理化学实验.北京：高等教育出版社，2003.

Experiment 14

Determination of average molar mass for water-soluble macromolecules by viscosity method

Experimental purpose

1. Master the method of measuring viscosity with three-tube viscometer.

2. Master the principle and method of measuring the viscosity average molar mass.

3. Determinate the viscosity-average molar mass for the polyethylene glycol.

Basical principles

In the research of macromolecules, the relative molecular weight is an indispensable important data, which not only reflects the size, but also be related to the physical properties of the macromolecules. For example, if cellulose has many short-chain molecules, it is not suitable for textile materials; if natural rubber contain many materials of low mole mass, the vulcanization effect of raw rubber is not good. The accurate determination of molar mass distribution for macromolecules is a very complex work, so the average molar mass is often used to reflect some characteristics of macromolecules. In general, the average molar mass varies with the method used, and the meaning of the average values obtained also varies, such as number-average molar mass $\overline{M}_n$, mass-average molar mass $\overline{M}_m$, Z-average molar mass $\overline{M}_Z$, viscosity-average molar mass $\overline{M}_\eta$, and so on.

The viscosity of solution varies with the size and property of macromolecules, temperature, solvent properties and concentration. The viscosity of the solution depends only on the concentration and the size of the macromolecules after the temperature and the polymer solvent system are selected. The viscosity method is the most commonly used method to measure the molar mass of polymer, because of its simple equipment, convenient operation, less time-consuming and high accuracy. In addition, the viscosity method, together with other methods, can also be used to study the morphology and size of polymer molecules in solution and the interaction between macromolecules and solvent molecules.

Viscosity refers to the resistance of a liquid to flow, which resists the relative movement of adjacent parts of the liquid and can therefore be regarded as a kind of internal friction. The viscosity of a pure solvent is represented by η_0, which is the internal friction between the solvent molecules. The viscosity of a solution is represented by η, which is the sum of the internal friction between solvent molecules, internal friction between macromolecules and internal friction between macromolecules and solvent molecules.

Generally, at the same temperature, $\eta > \eta_0$. The fraction by which the viscosity of the solution increases relative to that of the solvent is called the increasing viscosity, which is denoted as η_{sp}

$$\eta_{sp} = \frac{\eta - \eta_0}{\eta_0} = \frac{\eta}{\eta_0} - 1 = \eta_r - 1 \tag{1}$$

For macromolecules solutions, the increased viscosity η_{sp} often increases with the improvement of the solution concentration c. For comparison, the increased specific viscosity shown at the unit concentration is called the specific viscosity $\frac{\eta_{sp}}{c}$; $\frac{\ln\eta_r}{c}$ is called the logarithmic specific concentration viscosity. η_r and η_{sp} are both causeless quantities.

In order to further eliminate the internal friction effect between macromolecules, the solution must be diluted infinitely, so that each macromolecule is extremely far apart from each other, and its mutual interference is negligible. At this time, the viscosity behavior of the solution basically reflects the internal friction between the macromolecules and the solvent molecule. The limit value of this viscosity is denoted as $[\eta]$

$$[\eta] = \lim_{c\to 0}\frac{\eta_{sp}}{c} = \lim_{c\to 0}\frac{\ln\eta_r}{c} \tag{2}$$

$[\eta]$ is called intrinsic viscosity, and its value is independent of concentration. Experiments show that when the polymer, solvent and temperature are determined, the relationship between the intrinsic viscosity of the macromolecules solution and the average molecular weight of the macromolecules can be expressed by Mark-Houwink equation

$$[\eta] = K\overline{M}_{\eta}^{\alpha} \tag{3}$$

where K is the constant of proportionality, α is an empirical constant related to the shape of the molecule. They are all related to temperature, polymer and solvent properties, and are independent of relative molecular mass within a certain range.

The value of K and α can only be determined by other methods, such as osmotic pressure, light scattering, and so on. The viscosity method can only determine $[\eta]$, and $\overline{M}_{\eta}$ is calculated by using Equation (3).

Table 14-1 Nomenclature of solution viscosity

Name	Symbols and definitions
Viscosity (coefficient)	η
Relative viscosity	$\eta_r = \dfrac{\eta}{\eta_0}$ (η_0 is the viscosity of the solvent)
Increased viscosity	$\eta_{sp} = \dfrac{\eta - \eta_0}{\eta_0} = \eta_r - 1$
Specific viscosity	$\dfrac{\eta_{sp}}{c}$
Logarithmic specific concentration viscosity	$\dfrac{\ln\eta_r}{c}$
Intrinsic viscosity	$[\eta] = \left(\dfrac{\eta_{sp}}{c}\right)_{c=0} = \left(\dfrac{\ln\eta_r}{c}\right)_{c=0}$

There are three main types of methods for determining the viscosity of liquids: (1) using a capillary viscometer to determine the outflow time of liquid in the capillary; (2) using a falling ball viscometer to determine the falling rate of the ball in the liquid; (3) using a rotary viscometer to determine the relative rotation of liquid and concentric shaft cylinder.

When measuring the viscosity of the macromolecules solution, it is most convenient to use a capillary viscometer. Poiseuille's law is obeyed when a liquid flows by gravity in a capillary viscometer

$$\frac{\eta}{\rho} = \frac{\pi h g r^4 t}{8lV} - m\frac{V}{8\pi l t} \tag{4}$$

where ρ is the density of the liquid; l is the length of the capillary; r is the capillary radius; t is the outflow time; h is the average column height of the liquid flowing through the capillary; g is the acceleration of gravity; V is the volume of liquid flowing through the capillary; m is a constant related to the geometry of the instrument, $m=1$ is desirable when $\frac{r}{l} \ll 1$.

For a given viscometer, let $\alpha = \frac{\pi hgr^4}{8lV}$, then Equation (4) can be rewritten a

$$\frac{\eta}{\rho} = \alpha t - \frac{\beta}{t} \tag{5}$$

where $\beta < 1$, the second term on the right side of the equation can be ignored when $t > 100$ s. Let the density of the solution ρ be approximately equal to that of the solvent ρ_0. In this way, by measuring the outflow time t and t_0 of the solution and solvent respectively, the relative viscosity η_r can be calculated

$$\eta_r = \frac{\eta}{\eta_0} = \frac{t}{t_0} \tag{6}$$

Further, the values of η_{sp}, $\frac{\eta_{sp}}{c}$ and $\frac{\ln\eta_r}{c}$ can be calculated respectively. Prepare a series of solutions with different concentrations for determination, plot $\frac{\eta_{sp}}{c}$ and $\frac{\ln\eta_r}{c}$ as the same ordinate and c as the abscissa, and get two straight lines, which are extrapolated to $c=0$ respectively (as shown in Fig. 14-1), and the intercept is the intrinsic viscosity $[\eta]$, and $\overline{M}_\eta$ can be obtained by substituting into Equation (3) (K、α have been known).

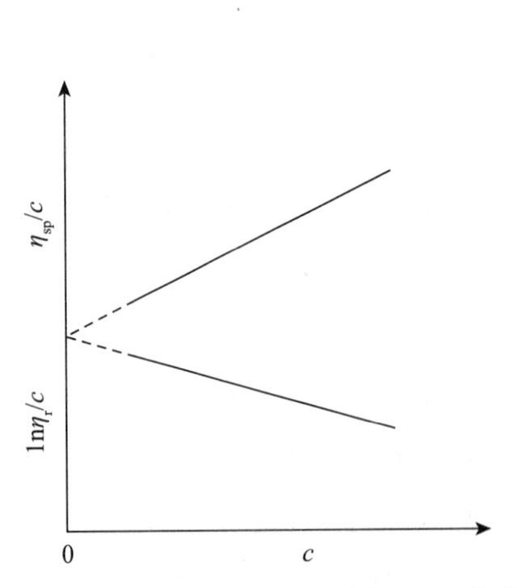

Fig. 14-1 Schematic diagram of finding $[\eta]$ by extrapolation method

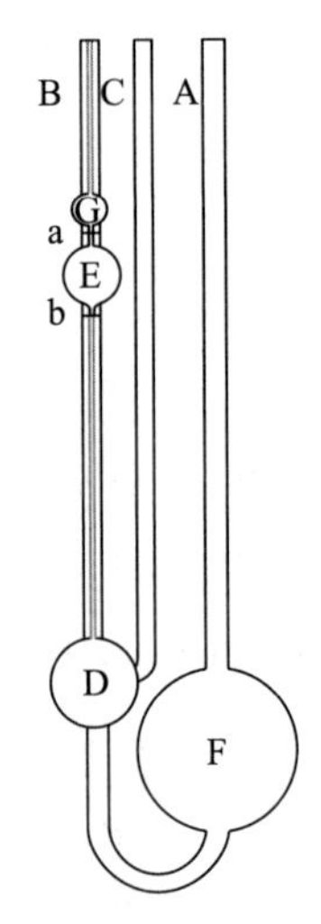

Fig. 14-2 Schematic diagram of three tube viscometer

Table 14-2 K、α value of polyethylene glycol aqueous solution at different temperatures [4]

t/℃	$K\times10^3$/ $dm^3\cdot kg^{-1}$	α	$\overline{M}_\eta\times10^{-4}$
25	156	0.50	0.019～0.1
30	12.6	0.78	2～500
35	6.4	0.82	3～700
40	16.6	0.82	0.04～0.4
45	6.9	0.81	3～700

Apparatus and reagents

Three tube viscometer
Thermostatic water bath
Pipettes (2 mL, 5 mL, 10 mL)
Stopwatch
Polyethylene glycol

Experimental procedures

1. Determination of solution outflow time (t).

Place the three-tube viscometer in a thermostatic bath (note: the viscometer should be vertical). Remove 10 mL of 10 $g\cdot L^{-1}$ polyethylene glycol solution by pipette and place it into the viscometer from tube A. Maintain the viscometer in the thermostatic bath for 10 min, then start to measure the flow time. Tube B and C can be covered with latex tube, clamp the C tube to isolate it from the air firstly, use the ear wash ball from the tube B to suck the solution up to exceed the line a to the ball G, loosen the tube C to make it ventilated, use the ear wash ball to squeeze the solution in the tube B to flow out, repeat the operation 5 times, so that the capillary tube is fully rinsed.

Use the ear wash ball again to suck the solution up from the tube B to make the solution exceed the line a to the ball G, loosen the tube C, and make the solution in the tube B flow down naturally. When the liquid level drops to the line a, start the stopwatch to record the time, stop the watch when the liquid level drops to the line b, and note the time required for the solution to flow through the ab segment, that is, the solution outflow time t. Repeat the operation of outflow time twice, averaging the three t (note: the error cannot exceed 0.2 s each time).

Dilute the solution, add 2, 3, 5, 10 mL of secondary distilled water to the viscometer in turn, and repeat the operation. After each addition of water, ear wash ball can be used to blow into the solution through the tube C. Be sure to mix the solution well after each addition of water, and be careful not to suck the solution into the ear wash bull when mixing.

2. Determination of solvent outflow time (t_0).

After measuring the outflow time of all the solution, remove the viscometer from the thermostatic bath and pour out the solution. Wash the viscometer repeatedly with tap water and 5% acetic acid, and finally rinse the viscometer 3 times with secondary distilled water (note: be sure to wash the capillary part clean). The viscometer is added 10 mL distilled water, put into a thermostatic bath for 10 min, and the outflow time of water is measured. The measurement method is the same as step 1.

3. Washing of the viscometer

After the experiment, the water of viscometer is poured out, a small amount of ethanol is added to rinse the viscometer. Recovery the ethanol, and then dry the viscometer in a drying oven for the next group of students.

Data recording and processing

1. Record the experimental data and calculate $\eta_r, \eta_{sp}, \frac{\eta_{sp}}{c}, \frac{\ln\eta_r}{c}$。

2. Draw a plot to calculate the value of $[\eta]$.

Plotting c with $\frac{\eta_{sp}}{c}$ and $\frac{\ln\eta_r}{c}$ respectively, two straight lines are obtained, extrapolated to $c=0$, and the intercept is found as the $[\eta]$ value.

3. The viscosity-average molar mass of polyethylene glycol is calculate from Equation (3).

Thinking and discussing

1. What is the role of tube C in a three-tube viscometer? Can viscosity be measured without tube C?

2. How does the thickness of the viscosimeter capillary affect the experiment?

3. Could you evaluate the advantages and disadvantages of measuring average molar mass of macromolecules by viscosity method, and point out the main factors affecting the accuracy of determination results?

Reference materials

1. Qian R. Determination of Molecular Weight of Polymers. Beijing: Science Press, 1958.

2. He M, Chen W, Dong X. Polymer Physics. Shanghai: Fudan University Press, 1982: 127-1323.

3. Daniels F, Alberty R A, Williams J W, et al. Experimental Physical Chemistry.7th ed. New York: McGraw-Hill, Inc, 1975: 329.

4. Department of Chemistry, Huaiyin Normal College. The Physical Chemistry Experiment. Beijing: Higher Education Press, 2003.

实验 15 配合物的磁化率测定

实验目的

1. 掌握古埃(Gouy)法磁天平测定物质磁化率的基本原理和实验方法。

2. 测定一些化合物的磁化率,了解磁化率数据对推断未成对电子数和分子配键类型的作用。

基本原理

1. 分子磁矩与磁化率

物质在外磁场的作用下会被磁化,被磁化的程度可用磁化强度 $\boldsymbol{M}$ 来描述。$\boldsymbol{M}$ 是一个矢量,它与磁场强度 $\boldsymbol{H}$ 成正比

$$\boldsymbol{M} = \chi \boldsymbol{H} \tag{1}$$

χ 称为物质的体积磁化率,是物质的一种宏观性质。在化学上常用质量磁化率 χ_m 或摩尔磁化率 χ_M 来表示物质的磁性质

$$\chi_m = \frac{\chi}{\rho} \tag{2}$$

$$\chi_M = M \cdot \chi_m = \frac{\chi M}{\rho} \tag{3}$$

式中 ρ、M 分别为物质的密度和摩尔质量。χ_m 的单位为 $m^3 \cdot kg^{-1}$,χ_M 的单位为 $m^3 \cdot mol^{-1}$。

物质的原子、分子或离子在外磁场作用下的磁化现象存在三种情况。

(1) 逆磁性物质

物质本身并不呈现磁性,但由于它内部的电子轨道运动,在外磁场作用下会产生拉摩进动,感应出一个诱导磁矩来,表现为一个附加磁场,磁矩的方向与外磁场相反,其磁化强度与外磁场强度成正比,并随着外磁场的消失而消失。这类物质称为逆磁性物质,其相对磁导率 $\mu < 1$,$\chi_M<0$。

(2) 顺磁性物质

物质的原子、分子或离子本身具有永久磁矩 μ_m,由于热运动,永久磁矩指向各个方向的机会相同,所以该磁矩的统计值等于零。但它在外磁场作用下,一方面永久磁矩会顺着外磁场方向排列,其磁化方向与外磁场相同,而磁化强度与外磁场强度成正比;另一方面物质内部的电子轨道运动会产生拉摩进动,其磁化方向与外磁场相反,因此这类物质在外磁场下表现的附加磁场是上述两者作用的总结果。通常称具有永久磁矩的物质为顺磁性物质。显然,此类物质的摩尔磁化率 χ_M 是摩尔顺磁化率 χ_μ 和摩尔逆磁化率 χ_0 两部分之和

$$\chi_M = \chi_\mu + \chi_0 \tag{4}$$

但由于$\chi_\mu \gg |\chi_0|$,故顺磁性物质的$\mu > 1, \chi_M > 0$。可以近似地把χ_μ当作χ_M,即

$$\chi_M \approx \chi_\mu \tag{5}$$

(3)铁磁性物质

物质被磁化的强度与外磁场强度之间不存在正比关系,而是随着外磁场强度的增加而剧烈增强。当外磁场消失后,这种物质的磁性并不消失,而是呈现出滞后现象,这种物质称为铁磁性物质。

2. 磁化率与未成对电子数

磁化率是物质的宏观性质,分子磁矩是物质的微观性质。假定分子间无相互作用,应用统计力学的方法,可以导出摩尔顺磁磁化率χ_μ和永久磁矩μ_m之间的定量关系

$$\chi_\mu = \frac{N_A \mu_m^2 \mu_0}{3kT} = \frac{C}{T} \tag{6}$$

式中N_A为阿伏伽德罗常数,k为玻尔兹曼常数,T为热力学温度。物质的摩尔顺磁磁化率与热力学温度成反比这一关系,是居里(Curie P.)在实验中首先发现的,所以该式称为居里定理,C称为居里常数。

分子的摩尔逆磁磁化率χ_0是由诱导磁矩产生的,它与温度的依赖关系很小。因此具有永久磁矩的物质的摩尔磁化率χ_M与永久磁矩μ_m间的关系为

$$\chi_M = \chi_0 + \frac{N_A \mu_m^2 \mu_0}{3kT} \approx \frac{N_A \mu_m^2 \mu_0}{3kT} \tag{7}$$

该式将物质的宏观物理性质(χ_M)和其微观性质(μ_m)联系起来了,因此只要实验测得χ_M,代入式(7)就可算出永久磁矩μ_m。

物质的顺磁性来自与电子的自旋相联系的磁矩。电子有两个自旋状态。如果原子、分子或离子中两个自旋状态的电子数不相等,则该物质在外磁场中就呈现顺磁性。这是由于每一轨道上不能存在两个自旋状态相同的电子(泡利原理),因而各个轨道上成对电子自旋所产生的磁矩是相互抵消的,所以只有存在未成对电子的物质才具有永久磁矩,它在外磁场中表现出顺磁性。

物质的永久磁矩μ_m和它所包含的未成对电子数n的关系可用下式表示

$$\mu_m = \mu_B \sqrt{n(n+2)} \tag{8}$$

μ_B称为玻尔(Bohr)磁子,其物理意义是单个自由电子自旋所产生的磁矩

$$\mu_B = \frac{eh}{4\pi m_e} = 9.274\ 078 \times 10^{-24}\ \mathrm{A \cdot m^2} \tag{9}$$

式中h为普朗克常数,m_e为电子质量。因此,对于顺磁性物质只要实验测得χ_M,即可求出μ_m,进而算得未成对电子数n。

由(7)至(9)各式，以及各物理常量的数值，可得未成对电子数 n 的表达式

$$n = \sqrt{797.7^2 \chi_M T + 1} - 1 \tag{10}$$

这对于研究某些原子或离子的电子组态，以及判断配合物分子的配键类型是很有意义的。

3. 物质配键类型

通常认为配合物可分为电价配合物和共价配合物两种。配合物的中央离子与配位体之间依靠静电库仑力结合起来的化学键叫电价配键，这时中央离子的电子结构不受配位体的影响，基本上保持自由离子的电子结构。共价配合物则是以中央离子的空的价电子轨道接受配位体的孤对电子形成共价配键，这时中央离子为了尽可能多地成键，往往会发生电子重排，以腾出更多空的价电子轨道来容纳配位体的电子对。例如 Fe^{2+} 在自由离子状态下的外层电子组态(如图 15-1 所示)。

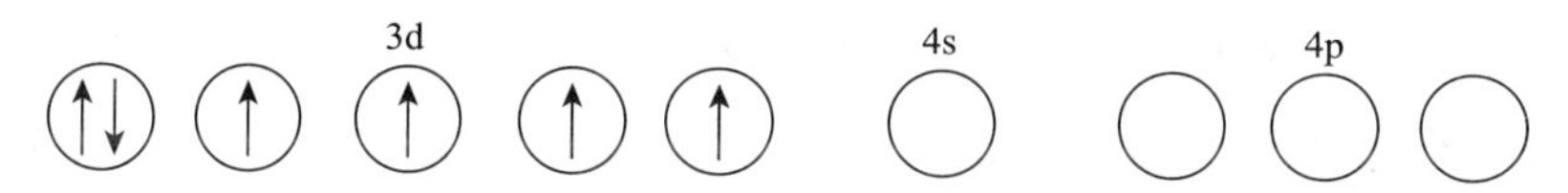

图 15-1　Fe^{2+} 在自由离子状态下的外层电子组态示意图

当它与 6 个 H_2O 配位形成络离子 $[Fe(H_2O)_6]^{2+}$ 时，中央离子 Fe^{2+} 仍保持着上述自由离子状态下的电子组态，故此配合物是电价配合物。当 Fe^{2+} 与 6 个 CN^- 配位体形成络离子时，Fe^{2+} 的电子组态发生重排(如图 15-2 所示)。

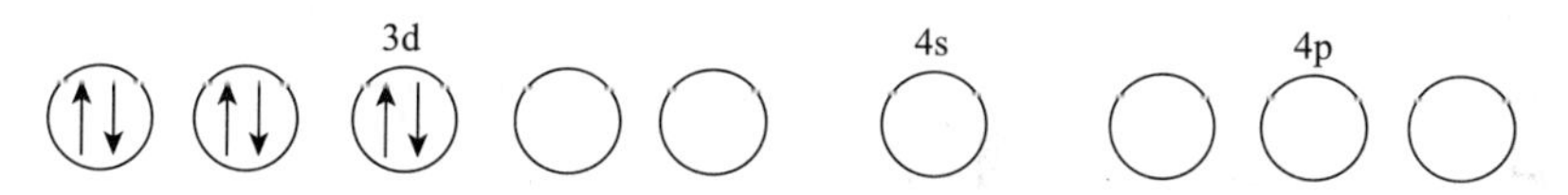

图 15-2　Fe^{2+} 外层电子组态重排示意图

Fe^{2+} 的 3d 轨道上原来未成对电子重新配对，腾出两个 3d 空轨道来，再与 4s 和 4p 轨道进行 d^2sp^3 杂化，构成以 Fe^{2+} 为中心的指向正八面体各个顶角的 6 个空轨道，以此来容纳 6 个 CN^- 中 C 原子上的孤对电子，形成 6 个共价配键。如图 15-3 所示。

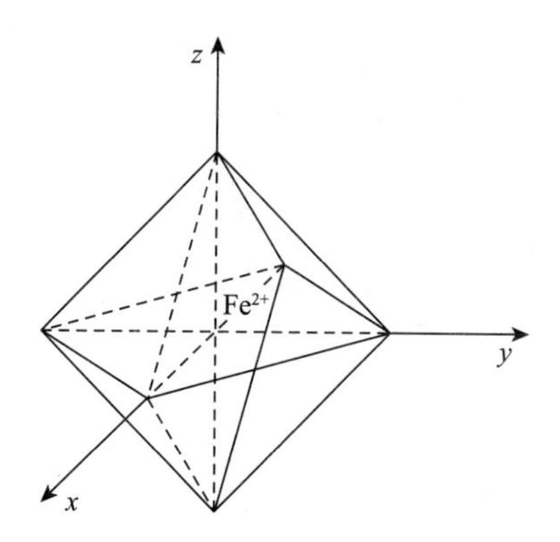

图 15-3　$[Fe(CN)_6]^{4-}$ 离子中共价键的相对位置

一般认为中央离子与配位原子之间的电负性相差很大时，容易生成电价配键；而电负性相差很小时，则生成共价配键。

4. 古埃磁天平

本实验用古埃磁天平测定物质的摩尔磁化率，其工作原理如图 15-4 所示。将圆柱形玻璃样品管（内装粉末状或液体样品）悬挂在分析天平的一个臂上，使样品管底部处于电磁铁两极的中心（即处于均匀磁场区域），样品管的顶端离磁场中心较远，磁场强度很弱，整个样品处于一个非均匀的磁场中。但由于沿样品轴心方向，即图示 z 方向，存在一磁场强度梯度 $\partial \boldsymbol{H}/\partial z$，故样品沿 z 方向受到磁力的作用，它的大小为

$$f_z = \int_{H}^{H_0} (\chi - \chi_{空}) \mu_0 S \boldsymbol{H} \frac{\partial \boldsymbol{H}}{\partial z} \mathrm{d}z \tag{11}$$

式中 $\boldsymbol{H}$ 为磁场中心磁场强度，$\boldsymbol{H}_0$ 为样品顶端处的磁场强度，χ 为样品的体积磁化率，$\chi_{空}$ 为空气的体积磁化率，S 为样品的截面积（位于 x、y 平面），μ_0 为真空磁导率。

$\boldsymbol{H}_0$ 为当地的地磁场强度，约为 40 A · m^{-1}，一般可略去不计，则作用于样品的力为

$$f_z = \frac{1}{2}(\chi - \chi_{空}) \mu_0 \boldsymbol{H}^2 S \tag{12}$$

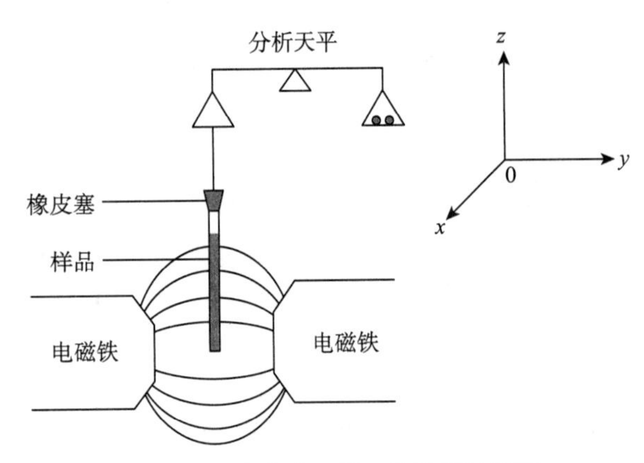

图 15-4　古埃磁天平工作原理示意图

由天平分别称得装有被测样品的样品管和不装样品的空样品管在有外加磁场和无外加磁场时的质量变化，则有

$$\Delta m = m_{磁场} - m_{无磁场} \tag{13}$$

显然，某一不均匀磁场作用于样品的力可由下式计算

$$f_z = (\Delta m_{样品+空管} - \Delta m_{空管}) g \tag{14}$$

于是有

$$\frac{1}{2}(\chi - \chi_{空}) \mu_0 \boldsymbol{H}^2 S = (\Delta m_{样品+空管} - \Delta m_{空管}) g \tag{15}$$

整理后得

$$\chi = \frac{2(\Delta m_{样品+空管} - \Delta m_{空管})g}{\mu_0 \boldsymbol{H}^2 S} + \chi_{空} \tag{16}$$

物质的摩尔磁化率$\chi_M = \frac{M\chi}{\rho}$，而$\rho = \frac{m}{hS}$

故
$$\chi_M = \frac{M}{\rho}\chi = \frac{2(\Delta m_{样品+空管} - \Delta m_{空管})ghM}{\mu_0 m \boldsymbol{H}^2} + \frac{M}{\rho}\chi_{空} \tag{17}$$

式中 h 为样品的实际高度，m 为无外加磁场时样品的质量，M 为样品的摩尔质量，ρ 为样品的密度（固体样品则指装填密度），μ_0 为真空磁导率；$\chi_{空}$为空气的体积磁化率，其值为 3.64×10^{-7}（SI 单位），但因样品管体积很小，故常予以忽略。

所以
$$\chi_M = \frac{2(\Delta m_{样品+空管} - \Delta m_{空管})ghM}{\mu_0 m \boldsymbol{H}^2} \tag{18}$$

对于莫尔氏盐，则有

$$\chi_{M_{莫}} = \frac{2(\Delta m_{莫+空管} - \Delta m_{空管})ghM_{莫}}{\mu_0 m_{莫} \boldsymbol{H}^2} \tag{19}$$

若每次装样高度 h 相同，同时控制每次磁场强度 $\boldsymbol{H}$ 相同，将（18）和（19）两式相除，可得

$$\frac{\chi_M}{\chi_{M_{莫}}} = \frac{M \cdot m_{莫}}{M_{莫} \cdot m} \times \frac{(\Delta m_{样+管} - \Delta m_{空管})}{(\Delta m_{莫+管} - \Delta m_{空管})} \tag{20}$$

将 $\frac{\chi_{M_{莫}}}{M_{莫}} = \chi_{m_{莫}}$ 代入上式，并将上式变形得

$$\chi_{M_{样}} = \frac{\chi_{m_{莫}} \cdot M_{样} \cdot m_{莫}}{m_{样}} \times \frac{(\Delta m_{样+管} - \Delta m_{空管})}{(\Delta m_{莫+管} - \Delta m_{空管})} \tag{21}$$

该式乘号左边各项可通过计算或直接测试得到，右边的各项为空管、空管中装样品或莫尔氏盐时，有磁和无磁情况下的质量差，可通过实验测得。因此样品的摩尔磁化率可由式（21）算得。莫尔氏盐的质量磁化率计算公式为

$$\chi_{m_{莫}} = \frac{9\ 500}{T+1} \times 4\pi \times 10^{-9}\ \mathrm{m^3 \cdot kg^{-1}}$$

仪器和试剂

古埃磁天平

装样品工具（包括研钵、角匙、小漏斗、玻棒）

$FeSO_4 \cdot 7H_2O$（分析纯）

软质玻璃样品管

莫尔氏盐$(NH_4)_2SO_4 \cdot FeSO_4 \cdot 6H_2O$（分析纯）

$K_4Fe(CN)_6 \cdot 3H_2O$（分析纯）

实验步骤

1. 打开仪器预热 15 min，同时将莫尔氏盐及其他固体样品在研钵中研细，装在小广口

瓶中备用。

2. 取一支洁净干燥的空样品管,在台秤上粗称其质量,然后挂在磁天平的挂钩上,使样品管底部与磁极中心线平齐,准确称量空样品管的质量 m_0^1;缓慢增加电流调节磁场强度至 300 mT,待天平读数稳定后,记录空管的质量 m_{300}^1;继续缓慢增加电流调节磁场强度至 500 mT,待天平读数稳定后,记录空管的质量 m_{500}^1;再缓慢增加电流调节磁场强度至 600 mT,不记录数据;稍停片刻后,缓慢减小电流将磁场强度降至 500 mT,记录此时空管的质量 m_{500}^2;继续缓慢减小电流将磁场强度降至 300 mT,读取空管的质量 m_{300}^2;将电流缓降至 0 mT,再次记录空管的质量 m_0^2。

3. 取下样品管,将事先研细的莫尔氏盐通过小漏斗装进样品管,在填装时每装入 1 cm 样品需要将样品管在木垫上敲击 20 次,务必使粉末样品装填的均匀紧实。用直尺准确量取样品高度,使其大于 20 cm。重复步骤 2 的操作,测定样品管装莫尔氏盐时的相关数据,测量完毕后将莫尔氏盐回收,样品管用蘸有少量乙醇的棉球擦净。

4. 用同一样品管分别对 $FeSO_4 \cdot 7H_2O$ 和 $K_4Fe(CN)_6 \cdot 3H_2O$ 进行测试,方法同步骤 2、3,注意每次装填的样品高度应相等。

数据记录与处理

1. 计算物质有磁和无磁情况下的质量差

以空管为例,将 0 mT 时测得的质量值平均,得空管在无磁场时的质量。即 $m_{空管}^{无磁} = \dfrac{m_0^1 + m_0^2}{2}$,将 300 mT 时测得的空管质量平均,得该磁场下空管的质量,即 $m_{空管}^{有磁} = \dfrac{m_{300}^1 + m_{300}^2}{2}$,则 $\Delta m_{空管} = m_{空管}^{有磁} - m_{空管}^{无磁}$,类似方法可计算样品管分别装有莫尔氏盐、$FeSO_4 \cdot 7H_2O$、$K_4Fe(CN)_6 \cdot 3H_2O$ 的质量差。

2. 由标准物(莫尔氏盐)的质量磁化率 $\chi_{m莫}$,计算所测样品的 $\chi_{M样}$

将前面所得 $\Delta m_{空管}$,样品管分别装有 $FeSO_4 \cdot 7H_2O$ 和莫尔氏盐时在有磁和无磁情况下的质量差、莫尔氏盐的质量、莫尔氏盐质量磁化率以及 $FeSO_4 \cdot 7H_2O$ 的质量、摩尔质量代入式(21),可计算 $FeSO_4 \cdot 7H_2O$ 在 300 mT 时的摩尔磁化率。

类似方法可计算 $K_4Fe(CN)_6 \cdot 3H_2O$ 在 300 mT 时的摩尔磁化率,以及 $FeSO_4 \cdot 7H_2O$ 和 $K_4Fe(CN)_6 \cdot 3H_2O$ 在 500 mT 时的摩尔磁化率。

3. 计算所测样品的未成对电子数

由所得摩尔磁化率数据、式(7)和式(8)算出所测样品的 μ_m,由式(10)求出未成对电子数 n。

4. 根据未成对电子数,讨论 $FeSO_4 \cdot 7H_2O$ 和 $K_4Fe(CN)_6 \cdot 3H_2O$ 中 Fe^{2+} 的最外层电子结构,并推测分子的配键类型。

思考与讨论

1. 用古埃磁天平测定样品磁化率的精度与哪些因素有关？
2. 不同磁场强度下测得的样品摩尔磁化率是否相同？为什么？

参考资料

1. 徐光宪，王祥云. 物质结构(第二版). 北京:高等教育出版社,1987: 457.
2. 项一非,李树家. 中级物理化学实验. 北京:高等教育出版社,1988: 146.
3. 游效曾. 结构分析导论. 北京:科学出版社,1980.
4. Selwood P W. Magnetochemistry. 2nd ed. New York: Interscience Publishers, Inc, 1956.
5. 欧晓波,王清叶,王成瑞. 用 HNMR 法测量过渡元素离子的磁矩. 大学化学,1990,(5): 46.

Experiment 15

Determination of the magnetic susceptibility for the complex

Experimental purpose

1. Master the basic principle and experimental method of measuring magnetic susceptibility for materials by Gouy magnetic balance.

2. Determine magnetic susceptibility for some compounds, understand the role of magnetic susceptibility data in inferring the number of unpaired electrons and the type of molecular bonding.

Basical principles

1. The molecular magnetic moment and magnetic susceptibility

Substances will be magnetized under the action of an external magnetic field. The degree of magnetization can be described by the magnetic intensity M, which is a vector and proportional to the magnetic field intensity H.

$$\boldsymbol{M} = \chi \boldsymbol{H} \tag{1}$$

χ is the volume magnetic susceptibility of substance and is a macroscopic property of a substance.

In chemistry, mass susceptibility χ_m (or molar susceptibility χ_M) is commonly used to express the magnetic properties of substances

$$\chi_m = \frac{\chi}{\rho} \tag{2}$$

$$\chi_M = M \cdot \chi_m = \frac{\chi M}{\rho} \tag{3}$$

where ρ and M are the density and molar mass of the substance, respectively. The unit of χ_m is $m^3 \cdot kg^{-1}$, and the unit of χ_M is $m^3 \cdot mol^{-1}$.

There are three situations in which atoms, molecules or ions of substances are magnetized under the action of an external magnetic field.

(1) Inverse magnetic substance

The substance itself does not show magnetism.However, because of the electrons motion in the internal orbit of substance, under the action of an external magnetic field, there will be a Larmor precession, which will induce an induced magnetic moment, and behave as an additional magnetic field. The magnetic moment is opposite to the external magnetic field, and its magnetization is proportional to the external magnetic field and disappears with the disappearance of the external magnetic field. This kind of substance is called inverse magnetic substance, its relative permeability $\mu < 1$, $\chi_M < 0$.

(2) Paramagnetic substance

The atoms, molecules or ions of the substance themselves have a permanent magnetic moment μ_m, and due to thermal motion, the chance of the permanent magnetic moment pointing in all directions is the same, so the statistical value of the magnetic moment is equal to zero. However, under the action of the external magnetic field, on one hand, the permanent magnetic moment will be arranged in the direction of the external magnetic field, and its magnetization direction is the same as the external magnetic field, the magnetization intensity is proportional to the strength of the external magnetic field.

On the other hand, the electron orbital motion inside the substances will produce Larmor precession, and its magnetization direction is opposite to the external magnetic field, so the additional magnetic field that such substances exhibited under the external magnetic field is the total result of the above two. The substance with a permanent magnetic moment is usually called a paramagnetic substance. Obviously, the molar magnetic susceptibility χ_M of such substances is the sum of the two parts of the molar paramagnetic susceptibility χ_μ and the molar inverse magnetic susceptibility χ_0

$$\chi_M = \chi_\mu + \chi_0 \tag{4}$$

However, due to $\chi_\mu \gg |\chi_0|$, so the paramagnetic substance $\mu > 1$, $\chi_M > 0$. χ_μ can be approximated as χ_M, i.e.

$$\chi_M \approx \chi_\mu \tag{5}$$

(3) Ferromagnetic substance

The magnetized strength of the substance has no proportional relationship with the strength of the external magnetic field, but sharply increases with the increasing of the external magnetic field strength. When the external magnetic field disappears, the magnetism of this substance does not disappear, showing a lag phenomenon, this substance is called ferromagnetic substance.

2. Susceptibility and number of unpaired electrons

The magnetic susceptibility is the macroscopic property of a substance, and the molecular magnetic moment is the microscopic property of a substance. Assuming that there is no interaction between molecules, statistical mechanics can be applied to derive a quantitative relationship between molar paramagnetic susceptibility χ_μ and μ_m permanent magnetic moment

$$\chi_\mu = \frac{N_A \mu_m^2 \mu_0}{3kT} = \frac{C}{T} \tag{6}$$

where N_A is the Avogadro constant, k is the Boltzmann constant, and T is the thermodynamic temperature. The inverse relationship between the molar paramagnetic susceptibility of a substance and the thermodynamic temperature was firstly discovered by Curie P. in experiments, so the formula is called Curie's theorem, and C is called Curie's constant.

The molar inverse magnetic susceptibility of the molecule, χ_0, is generated by the induced magnetic moment, which has little dependence on temperature. Therefore, the relationship between the molar magnetic susceptibility χ_M of a substance with a permanent magnetic moment and the permanent magnetic moment μ_m is

$$\chi_M = \chi_0 + \frac{N_A \mu_m^2 \mu_0}{3kT} \approx \frac{N_A \mu_m^2 \mu_0}{3kT} \tag{7}$$

This equation relates the macroscopic physical properties (χ_M) of a substance to its microscopic properties (μ_m), so as long as χ_M is measured by the experiment, the permanent magnetic moment μ_m can be calculated by the substitution equation (7).

This relates the macroscopic physical properties of substance (χ_M) to its microscopic properties (μ_m), so that the substitution equation (7) can be used to calculate μ_m as long as the permanent magnetic moment χ_M is experimentally measured.

The paramagnetism of substance comes from the magnetic moment associated with the electron spin. The electron has two spin states. If the electron number in the two spin states of an atom, molecule, or ion is not equal, the substance will show paramagnetic in an external magnetic field. This is because there can not be two electrons in the same spin state in each orbit

(Pauli principle), so the magnetic moments produced by the spins of electron pairs in each orbit cancel each other out, therefore, only substance with unpaired electron has a permanent magnetic moment, which exhibits paramagnetism in an external magnetic field.

The relationship between the permanent magnetic moment of a substance μ_m and the number n of its unpaired electrons can be expressed by the following equation

$$\mu_m = \mu_B \sqrt{n(n+2)} \tag{8}$$

μ_B is called Bohr magneton, and its physical meaning is the magnetic moment produced by the spin of a single free electron

$$\mu_B = \frac{eh}{4\pi m_e} = 9.274\ 078 \times 10^{-24}\ \mathrm{A \cdot m^2} \tag{9}$$

where h is Planck constant and m_e is electron mass. Therefore, the number n of unpaired electrons can be calculated for paramagnetic materials as long as its χ_M is measured experimentally.

From (7) to (9) and the values of each physical constant, the expression of the unpaired electron number n can be obtained

$$n = \sqrt{797.7^2 \chi_M T + 1} - 1 \tag{10}$$

It is useful for studying the electronic configuration of certain atoms or ions, and for determining the ligand type of complex molecules.

3. Ligand type of substance

It is generally considered that the complexes can be divided into two kinds: valence complex and covalent complex. The chemical bond between the central ion of the complex and the ligand, which is bound by an electrostatic Coulomb force, is called a valence bond, in which the electronic structure of the central ion is not affected by the ligand, the electronic structure of free ions is maintained substantially.

Covalent complex is one in which the empty valence electron orbitals of the central ion receive the lone pair electrons of the ligand to form a covalent bond. At this time, in order to form as many bonds as possible, the central ions often rearrange their electrons to free up more empty valence orbitals to accommodate the electron pairs of the ligands.

For example, the outer electron configuration of Fe^{2+} in the free ion state (as shown in Fig. 15-1).

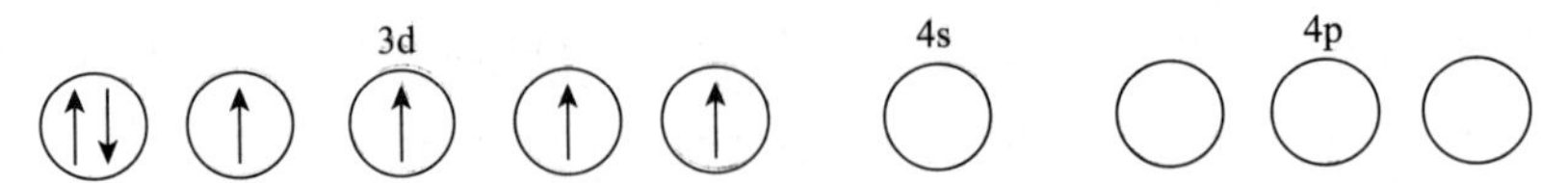

Fig. 15-1 Schematic diagram of the outer electron configuration of Fe^{2+} in the free ion state

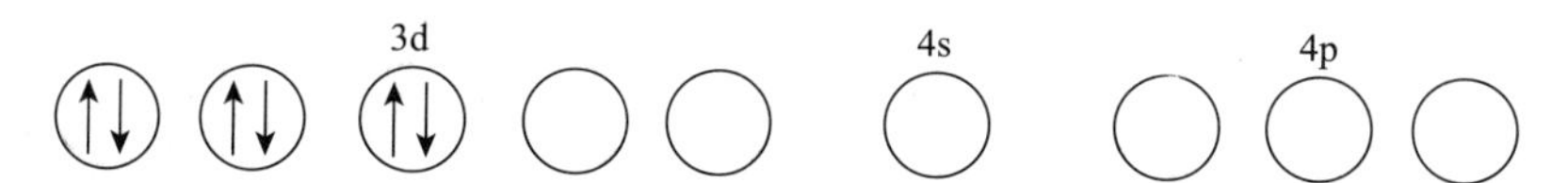

Fig. 15-2 Schematic diagram of outer electron configuration rearrangement of Fe^{2+}

When it coordinates with six H_2O to form a complex ion $[Fe(H_2O)_6]^{2+}$, the central ion Fe^{2+} still maintains the electronic configuration in the free ion state mentioned above, so the complex is a valence complex. When Fe^{2+} forms a complex ion with six CN^- ligands, the electronic configuration of Fe^{2+} rearranges (as shown in Fig. 15-2).

The original unpaired electrons in the 3d orbitals of Fe^{2+} are re-paired to free up two 3d empty orbitals, and then hybridized with the 4s and 4p orbitals for d^2sp^3 to form 6 empty orbitals centered on Fe^{2+} pointing to each apex corner of the regular octahedron, so as to accommodate the solitary pair of electrons on the C atom in 6 CN^- and form 6 covalent bonds. This is shown in Fig. 15-3.

It is generally believed that covalent bonds are formed when the electronegativity difference between the central ion and the ligand atom is very large, and covalent bonds are formed when the electronegativity difference is very small.

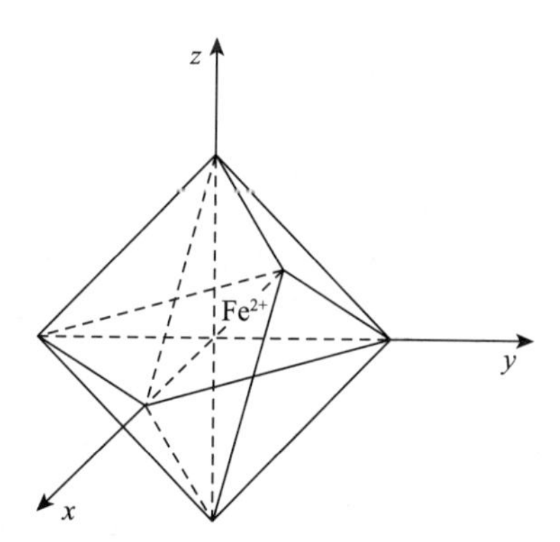

Fig. 15-3 Schematic diagram of the relative position of $[Fe(CN)_6]^{4-}$ ion covalent bonds

4. Gouy magnetic balance

In this experiment, the molar magnetic susceptibility of substance is measured by the Gouy magnetic balance, and its working principle is shown in Fig. 15-4. A cylindrical glass tube (containing a powder or liquid sample) is suspended from an arm of the analytical balance so that the bottom of the tube is in the center of the electromagnet poles (i.e. , in the region of a uniform magnetic field).

The top of the sample tube is far away from the center of the magnetic field, the magnetic field intensity is very weak, and the whole sample is in a non-uniform magnetic field. But because there is a magnetic field intensity gradient $\partial \boldsymbol{H}/\partial z$ along the axis of the sample in the z direction, the sample is affected by magnetic force along the z direction. The magnetic force can be expressed as

$$f_z = \int_{H}^{H_0} (\chi - \chi_{\text{air}}) \mu_0 S \boldsymbol{H} \frac{\partial \boldsymbol{H}}{\partial z} \mathrm{d}z \tag{11}$$

where $\boldsymbol{H}$ is the strength of the magnetic field center, $\boldsymbol{H}_0$ is the magnetic field strength at the top of the sample, χ is the volume magnetic susceptibility of the sample, χ_{air} is the volume magnetic susceptibility of air, S is the cross-sectional area of the sample (located in the x and y planes), and μ_0 is the vacuum magnetic permeability.

$\boldsymbol{H}_0$ is the local geomagnetic field strength, about 40 $\text{A} \cdot \text{m}^{-1}$, generally omitted, so the force acting on the sample is

$$f_z = \frac{1}{2}(\chi - \chi_{\text{air}}) \mu_0 H^2 S \tag{12}$$

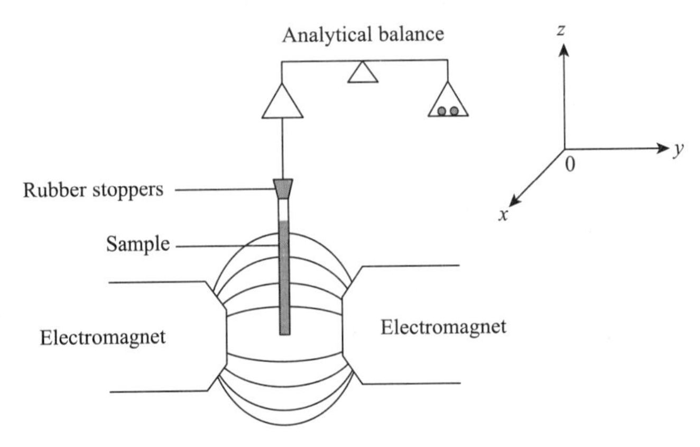

Fig. 15-4　Working principle scheme of the Gouy magnetic balance

The mass change of a sample tube with and without a sample in the presence and absence of an applied magnetic field, respectively, is determined by the balance. There is

$$\Delta m = m_{\text{magnetic field}} - m_{\text{non-magnetic field}} \tag{13}$$

Obviously, the force of an in homogeneous magnetic field acting on the sample can be calculated by

$$f_z = (\Delta m_{\text{sample+empty tube}} - \Delta m_{\text{empty tube}}) g \tag{14}$$

So, there is

$$\frac{1}{2}(\chi - \chi_{\text{air}}) \mu_0 H^2 S = (\Delta m_{\text{sample+empty tube}} - \Delta m_{\text{empty tube}}) g \tag{15}$$

after finishing it

$$\chi = \frac{2(\Delta m_{\text{sample+empty tube}} - \Delta m_{\text{empty tube}})g}{\mu_0 H^2 S} + \chi_{\text{air}} \tag{16}$$

The molar magnetic susceptibility of substance $\chi_M = \frac{M\chi}{\rho}$, whereas $\rho = \frac{m}{hS}$

$$\chi_M = \frac{M}{\rho}\chi = \frac{2(\Delta m_{\text{sample+empty tube}} - \Delta m_{\text{empty tube}})ghM}{\mu_0 m H^2} + \frac{M}{\rho}\chi_{\text{air}} \tag{17}$$

where h is the actual height of the sample, m is the mass of the sample without an applied magnetic field, M is the molar mass of the sample, ρ is the density of the sample (solid samples refer to the packing density), μ_0 is the vacuum permeability; χ_{air} is the volumetric magnetic susceptibility of air, which is 3.64×10^{-7} (SI units), but it is often ignored due to the small volume of the sample tube

thus
$$\chi_M = \frac{2(\Delta m_{\text{sample+empty tube}} - \Delta m_{\text{empty tube}})ghM}{\mu_0 m H^2} \tag{18}$$

For Moir's salt, there is

$$\chi_{\text{M,Moir}} = \frac{2(\Delta m_{\text{Moir+empty tube}} - \Delta m_{\text{empty tube}})ghM_{\text{Moir}}}{\mu_0 m_{\text{Moir}} H^2} \tag{19}$$

If the height h of each sample is the same, at the same time, the magnetic field intensity H is controlled to be the same. Divide the Equation (18) by (19) to obtain

$$\frac{\chi_M}{\chi_{\text{M,Moir}}} = \frac{M \cdot m_{\text{Moir}}}{M_{\text{Moir}} \cdot m} \times \frac{(\Delta m_{\text{sample+empty tube}} - \Delta m_{\text{empty tube}})}{(\Delta m_{\text{Moir+tube}} - \Delta m_{\text{empty tube}})} \tag{20}$$

Bring $\frac{\chi_{\text{M,Moir}}}{M_{\text{Moir}}} = \chi_{\text{m,Moir}}$ into the upper equation and deform the above equation to get

$$\chi_{\text{M,sample}} = \frac{\chi_{\text{m,Moir}} \cdot M_{\text{sample}} \cdot m_{\text{Moir}}}{m_{\text{sample}}} \times \frac{(\Delta m_{\text{sample+tube}} - \Delta m_{\text{empty tube}})}{(\Delta m_{\text{Moir+tube}} - \Delta m_{\text{empty tube}})} \tag{21}$$

The left terms of the multiplication sign can be calculated or directly tested. The right items are empty tubes, empty tubes filled with samples or Moir's salts, the mass difference in magnetic and non-magnetic conditions, which can be measured experimentally, so the molar magnetic susceptibility of the sample can be calculated by Formula (21). The formula for calculating the mass susceptibility of Moir's salt is

$$\chi_{m,\text{Moir}} = \frac{9500}{T+1} \times 4\pi \times 10^{-9}\ m^3 \cdot \text{kg}^{-1}$$

Apparatus and reagents

Gouy magnetic balance

Flexible glass tubes

Sample loading tools (including mortar, angle spoon, small funnel, glass rod)

$FeSO_4 \cdot 7H_2O$ (AR)

Moir's salt[$(NH_4)_2SO_4 \cdot FeSO_4 \cdot 6H_2O$] (AR)

$K_4Fe(CN)_6 \cdot 3H_2O$ (AR)

Experimental procedures

1. Turn on the instrument to preheat for 15 min, and at the same time grind Moir's salt and other solid samples in a mortar and store in a small jar for later use.

2. Take a clean and dry empty sample tube, weigh its mass on the platform scales, then hang it on the hook of the magnetic balance, make the bottom of the sample tube flush with the center line of the magnetic pole, accurately weigh the quality of the empty tube m_0^1. Slowly increase the current to adjust the magnetic field intensity to 300 mT, after the balance reading stability, record the quality of empty tube m_{300}^1 . Continue to slowly increase the current to adjust the magnetic field strength to 500 mT. After the balance is stabilized, record the mass of the empty tube m_{500}^1, and then the magnetic field is adjusted to 600 mT by increasing the current slowly, without recording the data. After a short pause, slowly reducing the current will reduce the magnetic field strength to 500 mT, record the mass of the empty tube m_{500}^2, continue to slowly reduce the current will reduce the magnetic field strength to 300 mT. Read the mass of the empty tube m_{300}^2 . Slowly reduce the current to 0 mT, and record the mass of the empty tube m_0^2.

3. Take the sample tube and fill the pre-refined Moir's salt into the tube through a small funnel. The tube should be knocked 20 times on the mat for every 1 cm of the sample filled. Make sure that the powder sample is filled evenly and tightly. Use a ruler to accurately measure the height of the sample, make it higher than 20 cm. Repeat step 2 to determine the relative data when the sample tube is filled with Moir's salt. After the measurement, the sample tube is cleaned with a cotton ball dipped in a small amount of ethanol.

4. $FeSO_4 \cdot 7H_2O$ and $K_4Fe(CN)_6 \cdot 3H_2O$ are tested with the same sample tube, and the methods are the same as steps 2 and 3. Note that the sample height should be equal for each filling.

Data recording and processing

1. Calculate the mass difference of the substance between the magnetic and non-magnetic conditions

Taking the empty tube as an example, the mass value measured at 0 mT is averaged to obtain the mass of the empty tube in the non-magnetic field. That is $m_{\text{empty tube}}^{\text{non-magnetic field}} = \frac{m_0^1 + m_0^2}{2}$. By averaging the mass of the empty tube measured at 300 mT, the mass of the empty tube under the magnetic field is obtained, that is, $m_{\text{empty tube}}^{\text{magnetic field}} = \frac{m_{300}^1 + m_{300}^2}{2}$, thus $\Delta m^{\text{empty tube}} = m_{\text{empty tube}}^{\text{magnetic field}} - m_{\text{empty tube}}^{\text{non-magnetic field}}$. The similar method can be used to calculate the mass difference of sample tubes loaded with Moir's salt, $FeSO_4 \cdot 7H_2O$, $K_4Fe(CN)_6 \cdot 3H_2O$, respectively.

2. Calculate the $\chi_{M,\text{sample}}$ of the sample based on the mass magnetic susceptibility of the reference material (Moir's salt)

When the sample tube is filled with $FeSO_4 \cdot 7H_2O$ and Moir's salt, respectively, substitute the mass difference between magnetic and non-magnetic conditions, the mass of Moir's salt, the mass magnetic susceptibility of Moir's salt, the $FeSO_4 \cdot 7H_2O$ mass, molar mass of $FeSO_4 \cdot 7H_2O$, and $\Delta m_{\text{empty tube}}$ to Equation (21), thus the molar susceptibility of $FeSO_4 \cdot 7H_2O$ at 300 mT can be calculated.

The molar susceptibility of $K_4Fe(CN)_6 \cdot 3H_2O$ at 300 mT and that of $FeSO_4 \cdot 7H_2O$ and $K_4Fe(CN)_6 \cdot 3H_2O$ at 500 mT can be calculated by similar methods.

3. Calculate unpaired electrons number of the sample measured

μ_m of the sample can be calculated from the molar susceptibility data obtained, Equation (7) and Equation (8). The unpaired electrons number n can be calculated from Equation (10).

4. The outermost electronic structure of Fe^{2+} in $FeSO_4 \cdot 7H_2O$ and $K_4Fe(CN)_6 \cdot 3H_2O$ are discussed according to the unpaired electron number, and the bonding type of Fe^{2+} can be presumed.

Thinking and discussing

1. What are the factors related to the magnetic susceptibility accuracy of samples determined by the Gouy magnetic balance?

2. Is the molar susceptibility of the sample the same under different magnetic field intensities? Why?

Reference materials

1. Xu G, Wang X. The Structure of Matter (2nd ed). Beijing: Higher Education Press, 1987: 457.

2. Xiang Y, Li S. Intermediate Physical Chemistry. Beijing: Higher Education Press, 1988: 146.

3. You X. Introduction to Structural Analysis. Beijing: Science Press, 1980.

4. Selwood P W. Magnetochemistry (2nd ed). New York: Interscience Publishers, Inc, 1956.

5. Ou X, Wang Q, Wang C. The magnetic moment of transition element ions was measured by HNMR method. University Chemistry, 1990, (5): 46.

实验 16　甲醛分子的结构和性质的计算化学研究

实验目的

1. 了解量子化学入门软件 Gaussian 和 GaussView 的使用方法。

2. 通过极性分子甲醛的相关计算，掌握用量子化学方法研究分子结构和性质的基本方法。

基本原理

随着化学科学研究水平的日益提高，人们对物质世界的分子水平的理解也在逐步加深。化学传统上以实验为主，而进入现代化学之后，研究者对理论展现出了强烈的依赖性。借助计算机科学技术的飞速进步和理论化学计算方法的迅速发展，以量子化学为主体的计算化学，已成为化学领域科学研究不可缺少的得力工具。化学学科不再是传统意义上的纯实验学科，理论计算的结果可以与实验结果互相比较和印证，前者还能在原子、分子的水平上对化学性质、化学反应及其规律进行合理解释，更提供对实验观察的预言和对材料药物等的合理设计。迄今，实验与理论计算已成为化学学科的两个重要支柱。因此，学习和掌握计算化学的基本知识和方法对于化学及有关专业的学生十分必要。

Gaussian 程序简便易学，易于操作，功能强大，是计算化学领域内最流行、应用范围最广的综合性量子化学计算程序包。它包含了目前主流的量子力学计算方法，如从头算、半经验方法、密度泛函理论等。Gaussian 程序能够计算分子的能量、几何结构、电子结构、化学热力学和动力学参数、光谱性质、激发态性质以及晶体体系等，适合于从入门到一般专业级的化学工作者使用。此外，GaussView 是与 Gaussian 程序配套的直观且强有力的图形界面软件，用于观察、构建、修改分子模型，设置和提交计算任务，显示计算结果。

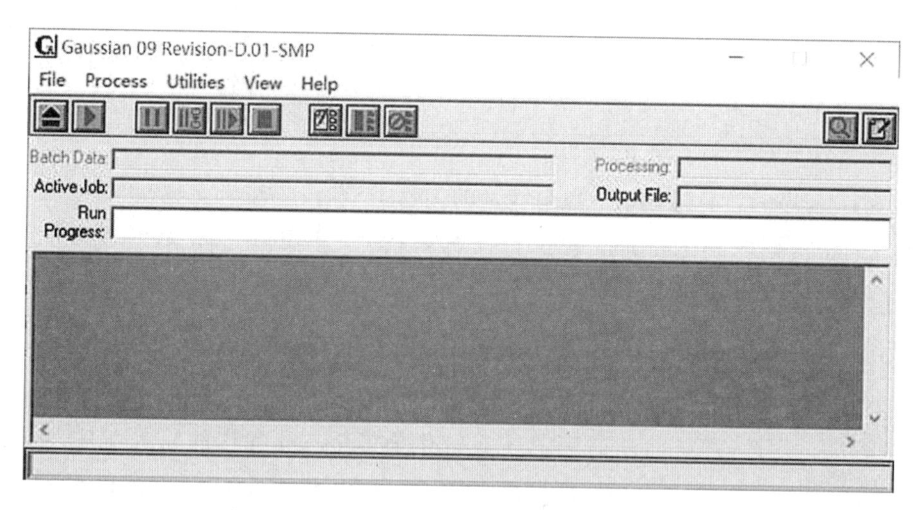

图 16-1　Gaussian 09(G09W)程序操作界面

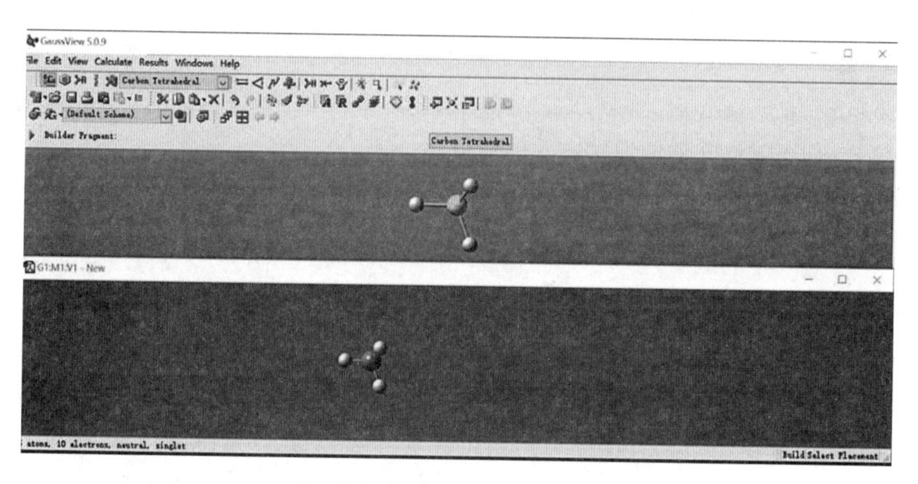

图 16-2　GaussView 程序操作界面

对于一般的体系,进行其分子几何结构、电子结构和性质计算的基本步骤为:

(1) 构建初始构型。依据化学理论或文献数据构建合理的分子初始构型。

(2) 选择计算方法和基组。查阅文献,从研究体系的性质、计算机性能等方面,选定量子力学计算方法和基组,试做单点能计算。若分子构型采用了准确的结构参数,可直接用单点能的计算结果研究分子的电子结构和性质。

(3) 几何构型优化计算。在每次单点计算中,按照分子总能量对坐标的一阶导数确定每个原子的坐标移动方向及大小,通过此步骤的多次重复,找到分子势能面上的极小点,即为分子的平衡几何构型,也称为优化构型。

(4) 振动频率计算。在优化构型的基础上,计算分子总能量对坐标的二阶导数,并得到分子各简振模式的振动力常数和频率。N 原子分子在稳定的能态下应有 $3N-5$(线性分子)或 $3N-6$(非线性分子)个振动实频(力常数为正)而无虚频(力常数为负);化学反应过渡态应有且仅有一个虚频。

(5) 化学性质计算。在优化构型下,计算分子的电荷分布、轨道成分、键级等电子结构信息,以及分子的若干物理、化学性质。

Gaussian 程序的计算过程中一般会涉及的文件类型有:①输入文件,扩展名为“.gjf”、“.in”和“.com”,用于向 Gaussian 程序传递各种信息;②输出文件,扩展名为“.out”和“.log”,包含计算所得结果,可用 GaussView 或文件编辑器进行结果查看;③检查点文件,扩展名为“.chk”和“.fchk”,包含计算过程相关信息;④一些临时文件。

Gaussian 程序是将所输入的分子转化为薛定谔方程,并进行求解,最终得到需要的有关分子的性质。因此,需要将所研究的体系转化为 Gaussian 程序的输入文件,才能调用 Gaussian 程序的运行。输入文件内容包含分子的初始构型、计算的方法和基组、计算任务的类型等,输入文件内容的合理性将直接决定据此得到的薛定谔方程的合理性,也影响求解的准确性。因此,创建一个合理的输入文件是 Gaussian 程序进行计算模拟的一个重要环节。

Gaussian 程序输入文件的通常由 5 部分组成:

1. Link 0 命令段

包括对内存、CPU、chk 文件、Linda 并行等进行设置,可包含多个输入行,每行行首均用%号开始。其中常用的“%chk=”命令用于对过程和结果信息的检查点文件“*.chk”进行命名,同时给出保存路径,该路径必须真实存在,否则在后续计算过程中会报错。

2. Route 部分

用于制定计算类型(需要以#开头),制定计算方法、基组以及所需计算项目。可包含多个输入行,每行的行首均用#号开始。Route 部分结束时以一个空行结尾。Route 部分语法为:

```
# Method/Basis Job Type/Keyword1=(option1,…) Keyword2=(option2,…)
```

“#”及其后面的字母用于打印输出的控制。其中,“#N(缺省)”指以默认形式打印输出;“#P”表示以更详细的形式输出结果;“#T”表示输出最基本的信息和结果。

“Method/Basis”为计算所选择的泛函和基组。

“Job Type/Keyword”指计算采用的任务类型/关键词。这些任务类型用各自的关键词来代表和执行。常用的任务类型有:“SP”表示单点能计算,可不指定,默认任务类型;“opt”表示几何结构优化;“freq”表示频率与热化学分析。一般而言,只能指定一种任务类型,也有例外,比如 opt 和 freq 可以放在一起。

3. Title 部分

标题行,用于书写计算的简要说明、作业内容与目的的简要描述,便于输出文件的阅读。可包含多个输入行，在 Title 部分结束时以一个空行结尾。标题部分应避免使用以下字符：@ #! -\控制字符（尤其是 Ctrl-G）。

4. 电荷与自旋多重度

给定电荷与自旋态，定义分子体系所带的净电荷数（正负整数）及自旋多重度（正整数）。两个整数以空格或逗号分隔。

5. 分子坐标

分子性质的任何计算都必须输入分子几何构型，可用笛卡儿坐标或分子内坐标，给定分子中各原子的坐标，程序能自动识别。亦可从检查点文件“. chk”文件中读取，在输入文件的 Route 部分加上关键词“Geom = Check”。分子坐标结束时以一个空行结尾。

6. 可附加部分

通常用于特殊作业类型的输入，如自定义基组/赝势基组，多步任务，计算控制语句等。

输入文件内容的最末尾务必保证存在空行，否则在计算过程中会报错。

仪器和程序

计算机　　　　　　　　　　Gaussian 程序

GaussView 程序

实验步骤

1. 构建分子初始构型

（1）在 GaussView 5.0 的图形界面上绘制甲醛（CH_2O）分子，点击左上角 element fragment 图标进行绘制，如图 16-3 所示。

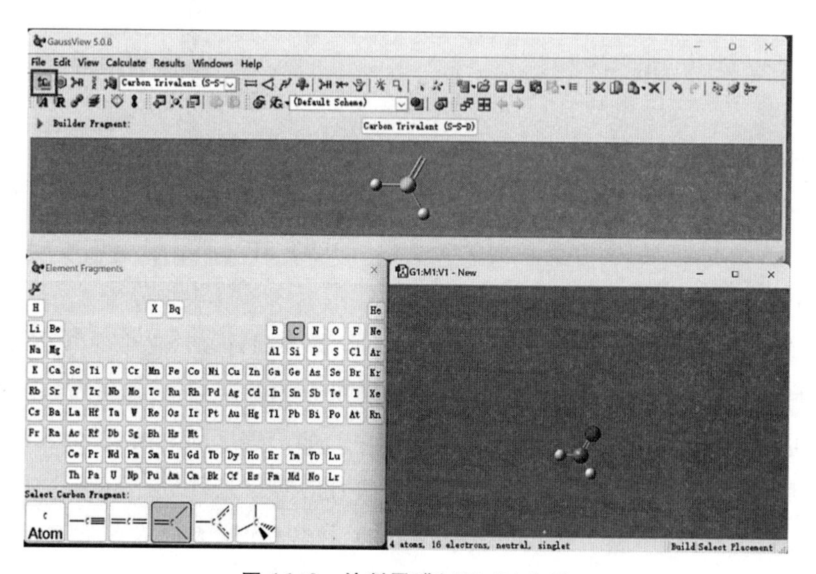

图 16-3　绘制甲醛（CH_2O）分子

（2）点击 Edit 下拉菜单中点选“Point Group”功能，选择弹出的窗口中的“Enable Point Group Symmetry”，点击“Symmetrize”命令，强制其对称性为“$C2_v$”。如图 16-4 所示。

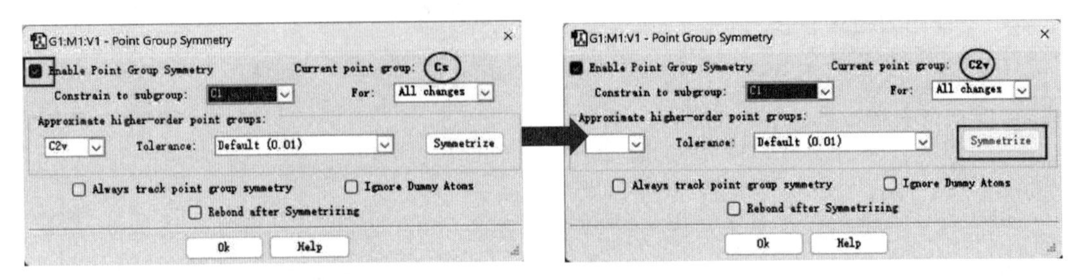

图 16-4　强制甲醛对称性为“$C2_v$”

2. 保存输入文件

将文件保存为“. gjf”格式。点击菜单栏中“File—Save”，将文件命名为“CH2O. gjf”，并保存至指定的文件夹中。需要注意的是文件名和路径名（文件夹名）中不要出现中文字符和空格。得到的初步的输入文件如图 16-5(A)所示。

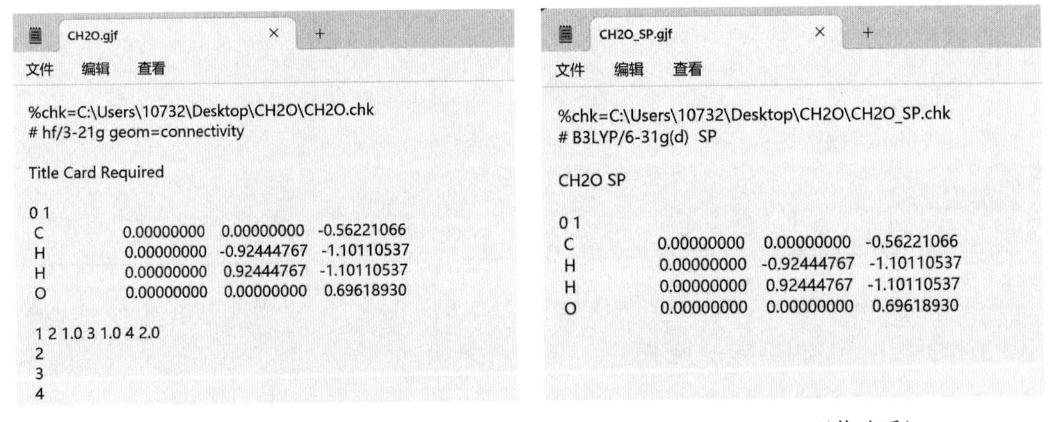

（A）CH2O. gjf（修改前）　　　　（B）CH2O_SP. gjf（修改后）

图 16-5　“. gjf”格式的输入文件格式及内容

3. 单点能试算

（1）对上一步的输入文件进行编辑，采用 B3LYP/6-31G(d)命令试算分子的单点能（SP），如图 16-5(B)所示，另存为新的输入文件“CH2O_SP. gjf”。注意需要对“%chk =”路径及“. chk”文件命名进行检查和修改。

（2）启动 G09W 程序，点击菜单栏“File—open”，选择目标文件夹中的“CH2O_SP. gjf”。在弹出的窗口中（图 16-6），检查设定的相应参数和分子坐标，而后点击“run”按钮，选择存储“CH2O_SP. out”的文件夹，随后开始运行程序。如若程序顺利结束，可在程序窗口或“CH2O_SP. out”文件末尾处，看到名人名言和计算耗时等信息，以及“Normal termination of Gaussian 09”等字样，如图 16-7 所示。本步骤用于熟悉 G09W 输入文件的编辑，查看输出文件的内容等。

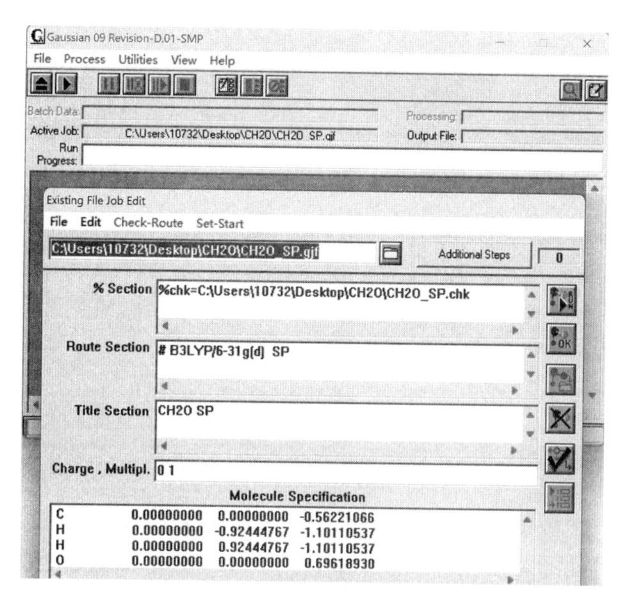

图 16-6 运行 G09W 读取输入文件

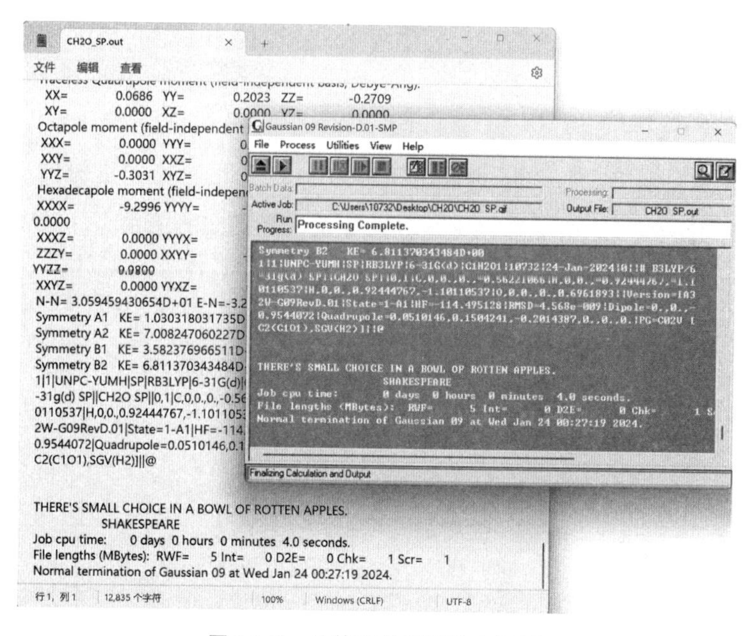

图 16-7 计算任务顺利结束示例

4. 几何构型优化和振动频率计算

对输入文件进行编辑，采用“B3LYP/6-31G(d) Opt Freq”命令，依次进行几何优化和振动频率计算，得到“CH2O_OPT.out”文件。若频率分析结果无虚频出现，则确认其为分子的稳定构型，如图 16-8 所示。

5. 计算电子结构和化学成键性质

使用上一步优化好的结构坐标，可采用两种方式：读取上一步“CH2O_OPT.chk”文件

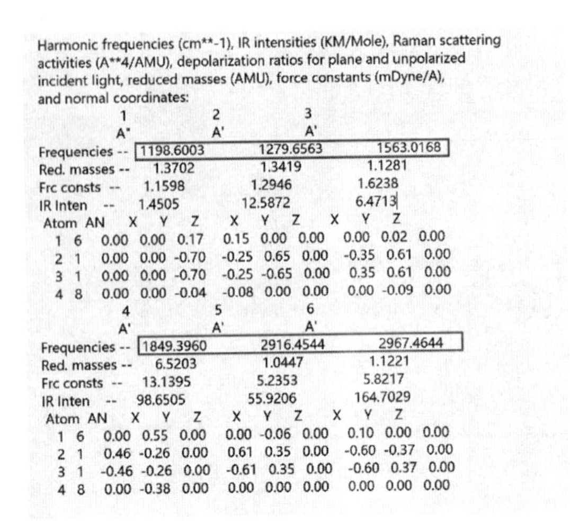

```
Harmonic frequencies (cm**-1), IR intensities (KM/Mole), Raman scattering
activities (A**4/AMU), depolarization ratios for plane and unpolarized
incident light, reduced masses (AMU), force constants (mDyne/A),
and normal coordinates:
                     1                      2                      3
                     A"                     A'                     A'
Frequencies --  1198.6003              1279.6563              1563.0168
Red. masses --     1.3702                 1.3419                 1.1281
Frc consts  --     1.1598                 1.2946                 1.6238
IR Inten    --     1.4505                12.5872                 6.4713
 Atom  AN      X      Y      Z        X      Y      Z        X      Y      Z
   1   6     0.00   0.00   0.17     0.15   0.00   0.00     0.00   0.02   0.00
   2   1     0.00   0.00  -0.70    -0.25   0.65   0.00    -0.35   0.61   0.00
   3   1     0.00   0.00  -0.70    -0.25  -0.65   0.00     0.35   0.61   0.00
   4   8     0.00   0.00  -0.04    -0.08   0.00   0.00     0.00  -0.09   0.00
                     4                      5                      6
                     A'                     A'                     A'
Frequencies --  1849.3960              2916.4544              2967.4644
Red. masses --     6.5203                 1.0447                 1.1221
Frc consts  --    13.1395                 5.2353                 5.8217
IR Inten    --    98.6505                55.9206               164.7029
 Atom  AN      X      Y      Z        X      Y      Z        X      Y      Z
   1   6     0.00   0.55   0.00     0.00  -0.06   0.00     0.10   0.00   0.00
   2   1     0.46  -0.26   0.00     0.61   0.35   0.00    -0.60  -0.37   0.00
   3   1    -0.46  -0.26   0.00    -0.61   0.35   0.00    -0.60   0.37   0.00
   4   8     0.00  -0.38   0.00     0.00   0.00   0.00     0.00   0.00   0.00
```

图 16-8 在 CH2O_OPT.out 文件中查看振动有无虚频

或将优化好的坐标编辑为新的输入文件“CH2O_NBO. gjf”。打印分子轨道的关键词为“Pop”。计算键级的关键词为“Pop=(NBORead, SaveNBOs)”。在分子坐标后空一行，输入自然键轨道（NBO）分析的关键词“ $NBO BNDIDX NLMO 3CBOND $END”，可进行 NBO 计算。

6. 查看输出信息

用文本编辑软件打开“. out”格式输出文件可查看全部计算结果数据。也可以使用 GaussView 对输出结果的基本信息、电荷布局、轨道图形等进行查看。下面简单介绍运用 GaussView 可以查看的一些输出结果。

（1）查看输出结果小结

在 GaussView 中打开“CH2O_OPT.out”文件，点击菜单栏“Results—Summary”（图 16-9）。

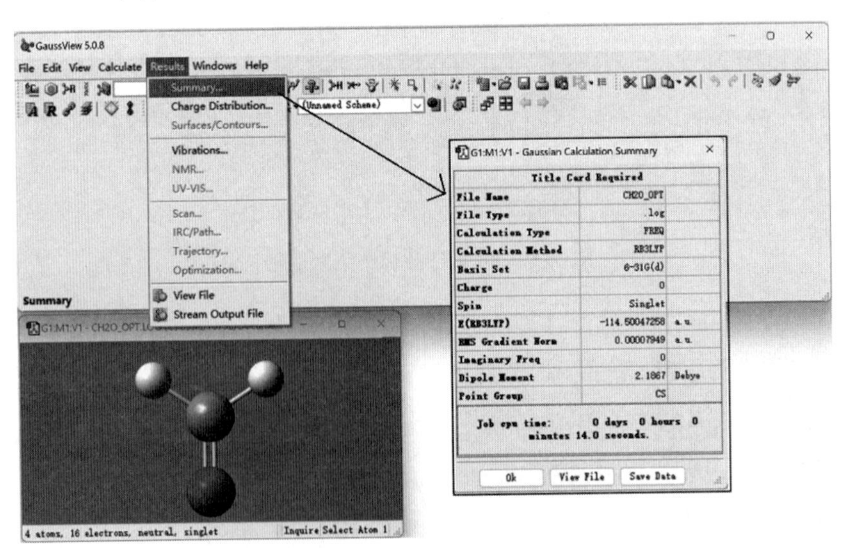

图 16-9 在 GaussView 中查看输出结果小结

（2）查看振动模式及红外光谱

在 GaussView 中打开“CH2O_OPT.out”文件，点击菜单栏“Results—Vibrations—Start Animation”，可观察不同振动频率的振动模式，如图 16-10(A)。点击该窗口中最下面一行的“Spectrum”，可以查看红外吸收光谱，如图 16-10(B)。

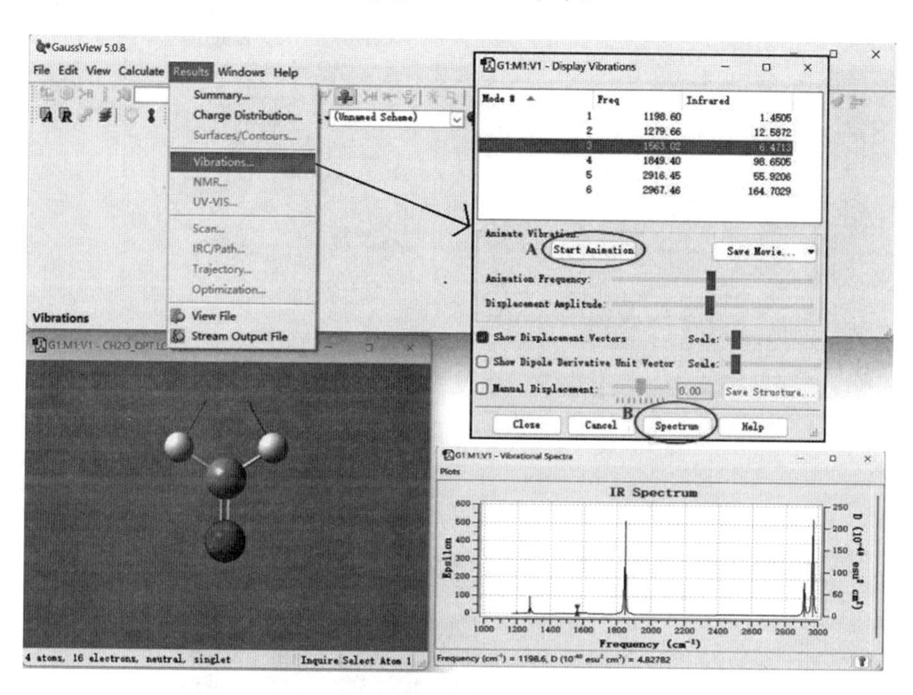

图 16-10 在 GaussView 中查看振动模式及红外光谱

（3）查看电荷和偶极矩

在 GaussView 中打开“CH2O_NBO. chk 文件”。而后单击“Results—Charge Distribution”可看到图 16-11 所示信息，勾选“Atomic Charges—Show Numbers”可查看原子净电荷，可以在 Type 菜单栏中选择 Mulliken 电荷或 NBO 电荷。勾选“Dipole Moment—Show Vector”可显示分子偶极矩大小及方向。

（4）查看分子前线轨道的三维图形

在 GaussView 中打开“CH2O_OPT.chk”文件，可浏览全部分子前线轨道的三维图形。点击菜单栏 Edit—MOs，左侧窗口为分子结构及轨道图形，右侧窗口为轨道序号、占据情况及能量，其中标黄的为选中轨道。点击 Visualize，选择 Update 可以生成轨道的三维图形，同时右侧窗口会出现红色和灰色方框，红框为目前显示的图形所属轨道，点击灰框可进行切换，如图 16-12 所示。自然键轨道三维图形的查看与上述方法相同。

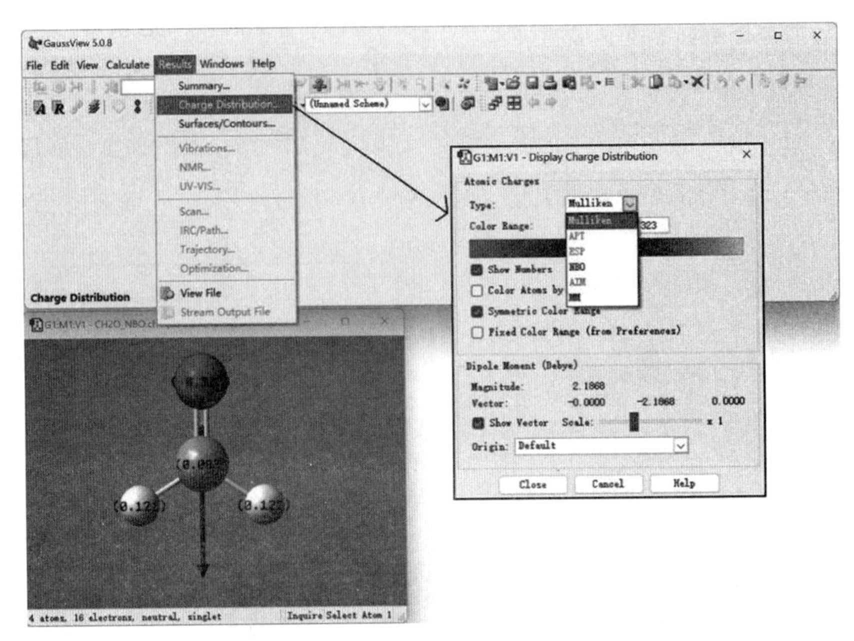

图 16-11　在 GaussView 中查看电荷和偶极矩

图 16-12　在 GaussView 中查看分子前线轨道的三维图形

注意事项

本实验的计算对象为真空中分子,故计算值与液态甲醛的实验测定值存在一定差异。量子化学计算的精度取决于所用的方法和基组。不同方法和基组得到的计算结果在某些情况下差异会很大,所以应查阅文献,并综合考虑实验条件、计算研究目标等因素,选择合适的方法和基组。

Gaussian 程序的功能十分强大。在实验过程中,应学会查阅在线帮助。量子力学计算是一个较为复杂的过程,初学者易出现各种问题,应学会查看输出文件给出的错误信息。Gaussian 程序的用户极多,所以初学者的常见问题,往往能在网络上找到答案或提示。

实验过程中安装软件、文件存储路径中必须使用全英文路径,不要出现中文字符和空格。

数据记录与处理

强制体系对称性为“C2v”,运行上述程序,并记录下列数据:

1. 记录计算所用的量子力学方法和基组。

2. 记录分子对称性、电子态和结构参数(键长、键角和二面角)。

3. 记录分子总能量及用零点能校正后的能量。

4. 记录每个原子上的净电荷,分子的偶极矩和化学键的键级。

5. 画出前线轨道图形,记录轨道能量和轨道占据情况。

6. 画出自然键轨道图形,记录轨道能量和轨道占据情况。

7. 记录振动的对称性、频率、红外吸收强度,描述振动模式特征。根据采用的理论计算方法,选用适当的校正因子(例如 B3LYP 泛函为 0.9613)进行理论振动频率校正。

8. 记录热力学函数(U、H、G、S)的理论预测值,验证热力学参数之间的关系。

思考与讨论

1. 查阅文献,将计算结果与实验值及其他理论计算值进行比较,讨论计算所用方法和基组的准确性。结合本实验计算结果和结构化学、物理化学、有机化学等专业化学的理论知识,对 CH_2O 分子的结构和性质进行分析和讨论。

2. 强制体系对称性为“Cs”,运行实验步骤中相关程序,并记录 Cs 对称性下,前述数据记录与处理(1)～(8)的实验数据。

3. 量子化学计算方法和基组种类繁多,且不同方法和基组的计算结果有时会有较大的差别。如何合理地选择计算方法和基组?

4. 试列举 Gaussian 软件能够计算的物质的性质参数。哪些参数能与实验值进行比对,从而互相验证?

参考资料

1. 王溢磊. 甲醛分子的结构和性质的计算化学研究——介绍一个计算化学实验. 大学化学, 2014, 29(5):66-70.

2. Frisch M J, Trucks G W, Schlegel H B, et al. Gaussian 09. Gaussian Inc., 2009.

Experiment 16

Computational chemistry study of the structure and properties of formaldehyde

Experimental purpose

1. Understand how to use the quantum chemistry software Gaussian and GaussView.

2. Through the relevant calculation of polar molecular formaldehyde, master the basic method of studying molecular structure and properties by quantum chemistry method.

Basical principles

With the improvement of research level in chemical science, people's understanding of the molecular level of the material world is also gradually deepening. Chemistry has traditionally relied mainly on experiments, while modern chemistry has a strong dependence on theory. Due to the rapid progress of computer science and technology and the rapid development of theoretical chemical calculation methods, computational chemistry with quantum chemistry as its mainstay, has become an indispensable and powerful tool for scientific research in the field of chemistry. Chemistry is no longer a purely experimental discipline in the traditional sense. The results of theoretical calculations can be compared and corroborated with the results of chemical experiments. The former can also provide reasonable explanations of chemical properties, reactions and laws at the atomic and molecular level, then provide predictions for experimental observations and rational design of materials and drugs. To date, experiments and theoretical

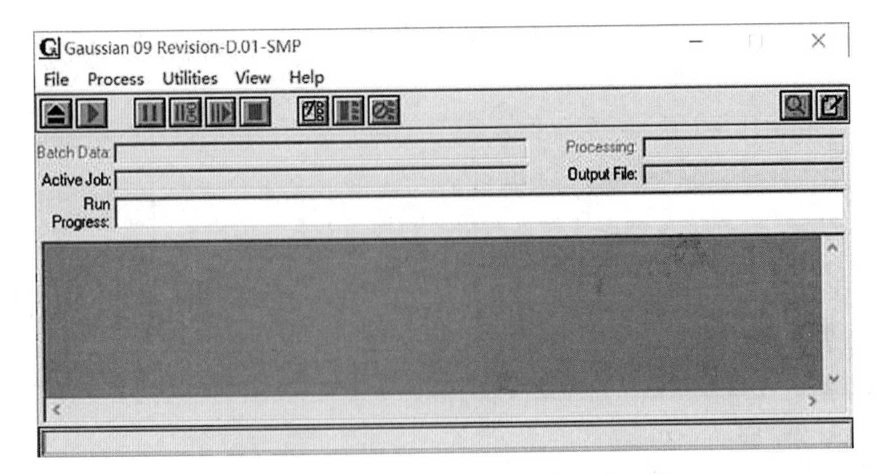

Fig. 16-1 Program interface of Gaussian 09

calculations have become two important pillars of chemistry. Therefore, learning and mastering the basic knowledge and methods of computational chemistry is essential for students of chemistry and related disciplines.

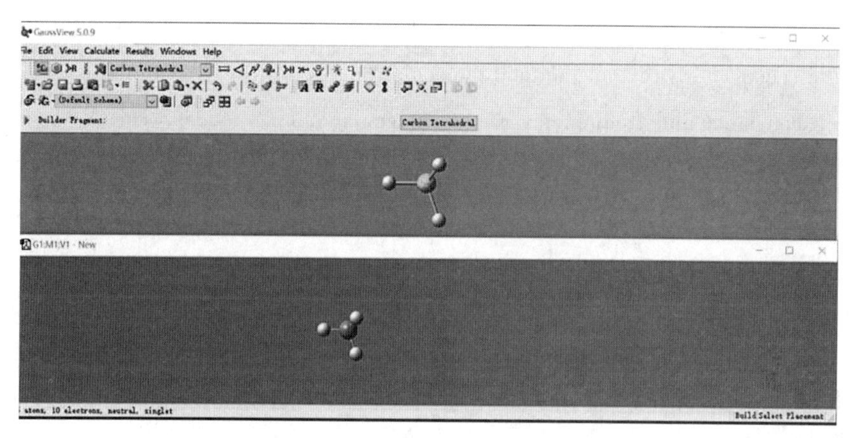

Fig. 16-2 GaussView interface

The Gaussian program is easy to learn, easy to operate and has powerful functions. It is the most popular and widely used comprehensive quantum chemistry calculation package in computational chemistry. It includes the current mainstream quantum mechanical calculations, such as *ab initio* calculations, semi-empirical methods, density functional theory, etc. It can calculate the energy, geometry, electronic structure, chemical thermodynamics parameters and kinetic parameters, spectral properties, excited state properties and crystal systems of molecules. It is suitable for chemists from the introduction to professional standards. In addition, GaussView is an intuitive and powerful graphical interface software that is compatible with Gaussian programs, used to observe, construct, modify molecular models, set up and submit computational tasks, and display computational results.

For a general system, the basic steps of calculating its molecular geometry, electronic structure, and properties are as follows:

(1) Build the initial geometric structure. Construct reasonable initial molecular model based on chemical theory or literature data.

(2) Choose the calculation method and basis. Refer to literature, select quantum mechanics calculation methods and basis sets from the aspects of research system properties, computer performance, etc., and attempt to do single point energy calculations. If accurate structural parameters are used for the molecular configuration, the results of the single point energy calculations can be used directly to study the electronic structure and properties of the molecule.

(3) Geometrical configuration optimization calculations. In each single-point calculation, the direction and magnitude of coordinate movement of each atom are determined according to the first derivative of the total energy of the molecule with respect to the coordinates.

(4) Vibrational frequency calculation. On the basis of the optimized configuration, calculate the second derivative of the total molecular energy with respect to the basis, and obtain the dynamic constants and frequencies of the molecular simple modes. The N-atom molecule should have $3N-5$ (linear molecules) or $3N-6$ (non-linear molecules) vibrational real frequencies (positive force constants) and no imaginary frequencies (negative force constants) in the stable energy state; the chemical transition state should have only one imaginary frequency.

(5) Calculation of chemical properties. Using the optimized configuration, calculate the electronic structure information of the molecule, such as charge distribution, orbital composition, bond level, partical physical and chemical properties of the molecule, etc.

The file types involved in the Gaussian programscommonly include: ① Input files with extensions of ".gjf", ".in", and ".com", which are used to pass various information to Gaussian programs. ②Output files with extensions of ".out" and ".log", containing calculated results, can be viewed using GaussView or file editor. ③ Checkpoint files with extensions of ".chk" and ".fchk", containing information related to the calculation process. ④Some temporary files.

The Gaussian program converts the input molecules into the Schrödinger equation and solves it to obtain necessary properties of molecules. Therefore, it is necessary to convert the studied system into an input file for the Gaussian program in order to call the operation of the Gaussian program. The input file content includes the initial structure of the molecule, the calculation method and basis set, job types, etc. The rationality of the input file content not only directly determines the rationality of the related Schrödinger equation, but also affects the accuracy of the results. Therefore, creating a reasonable input file is an important step in Gaussian program for computational simulation.

The basic structure of a Gaussian input file includes several different sections.

1. Link 0 commands

Locate and name scratch files (not blank line terminated). "%chk=" command is used to name the checkpoint file "*.chk" for process and result information, and provides a save path that must truly exist, otherwise an error will be reported in the subsequent calculation process.

2. Route section (# lines)

Specify desired calculation type, model chemistry and other options (blank line terminated).

Syntanx of route section

```
# Method/Basis Job Type/Keyword1=(option1,…) Keyword2=(option2,…)
```

"#" and the following letters are used for printing output control. Among them, "# N (default)" refers to printing output in default form; "# P" indicates outputting results in a more detailed form; "# T" represents outputting the basic information and results.

"Method/Bass" is the selected functional and basis set.

"Job Type/Keyword" refers to the task type/keyword used in the calculation. These task types are represented and executed using their respective keywords. The commonly used task types are: "SP" indicates that single point energy can be calculated by default task type, which can be left unspecified.; "Opt" represents geometric structure optimization; "Freq" represents frequency and thermochemical analysis, etc. Generally speaking, only one task type can be specified, with exceptions such as opt and freq being placed together.

3. Title section

Brief description of the calculation (blank line terminated). This section is required in the input, but is not interpreted in any way by Gaussian 09. It appears in the output for purposes of identification and description. Typically, this section might contain the compound name, symmetry, the electronic state, and any other relevant information. The title section cannot exceed five lines and must be followed by a terminating blank line. The following characters should be avoided in the title section: @ # ! - _ \ *control characters (especially* Ctrl-G)

4. Charge and spin multiplicity

Give the charge and spin state, define the net charge (positive or negative integer) and spin multiplicity (positive integer) carried by the molecular system. Two integers are separated by spaces or commas.

5. Molecular coordinates

Any calculation of molecular properties must input the molecular geometry configuration, which can be Cartesian coordinates or intra molecular coordinates. Given the coordinates of each atom in the molecule, the program can automatically recognize it. It can also be read from the checkpoint file ". chk" by adding the keyword "Geom=Check" to the Route section of the input file.

6. Optional additional sections

Additional input needed for specific job types (usually blank line terminated).

Ensure that there are empty lines at the end of the input file content, otherwise errors may occur during the calculation process.

Apparatus and software

Computer

Gaussian program

GaussView program

Experimental steps

1. Constructing the initial conformation of a molecule

(1) Draw a CH_2O molecule on GaussView 5.0. Tap the element fragment icon in the upper left corner, as shown in Fig. 16-3.

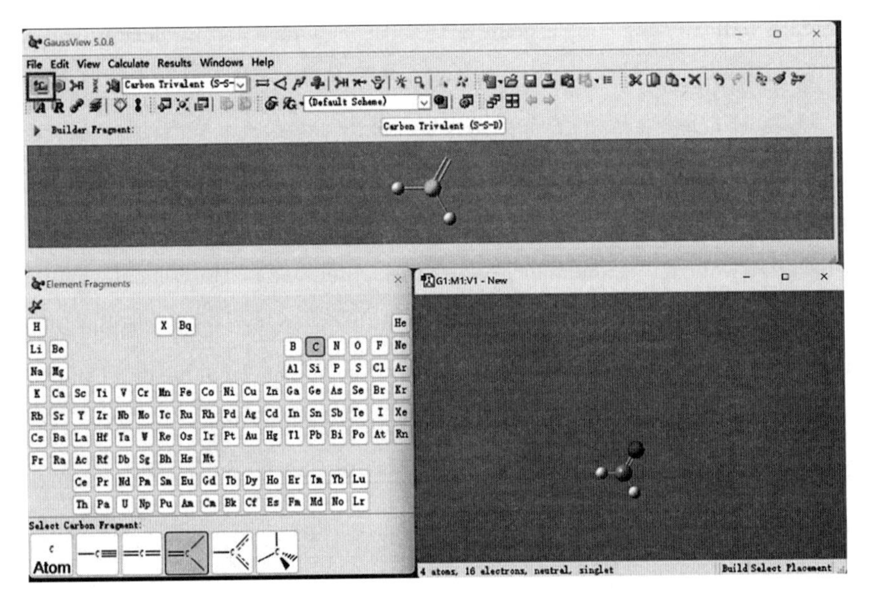

Fig. 16-3 Draw formaldehyde (CH_2O) molecule

(2) Tap the "Point Group" function in the Edit drop-down menu and select "Enable Point Group Symmetry" in the pop-up window, click the "Symmetrize" command, and force its symmetry to be "C2v", as shown in Fig. 16-4.

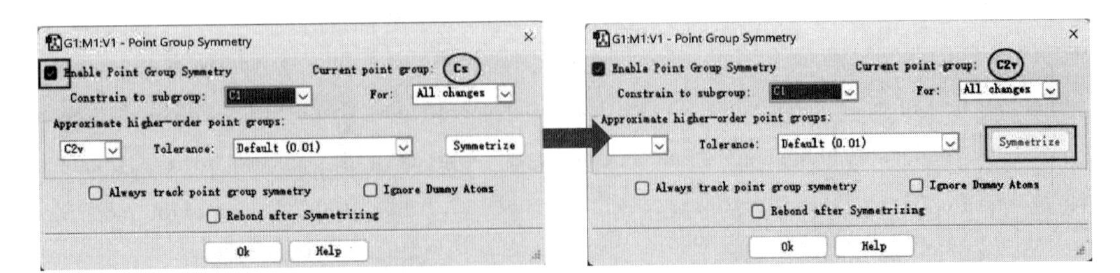

Fig. 16-4 Forced formaldehyde symmetry to "C2v"

2. Save input file

Save the file in ".gjf" format. Tap menu bar File-Save, name the file "CH2O.gjf", and save it to the specified folder. It should be noted that Chinese characters and spaces should not appear in the file name and path name (folder naming). The preliminary input file obtained is shown in Fig. 16-5 (A).

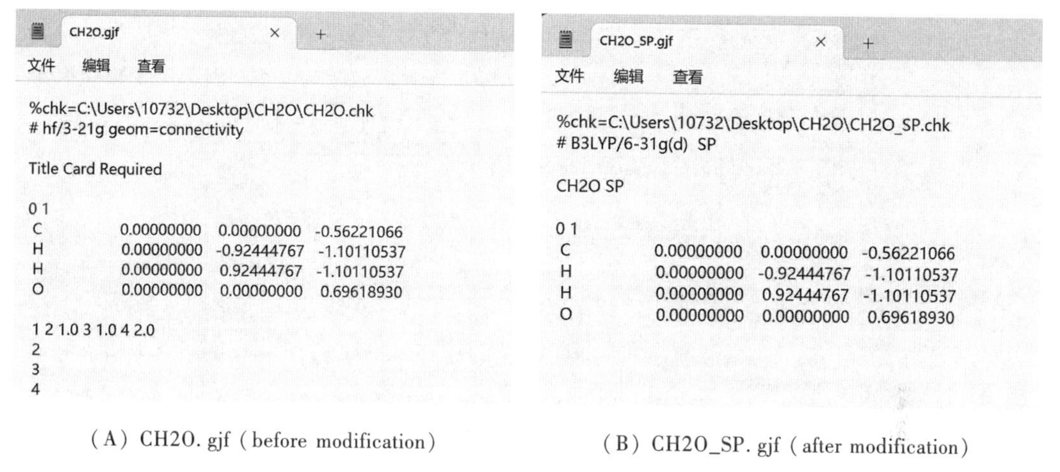

(A) CH2O. gjf (before modification)　　　(B) CH2O_SP. gjf (after modification)

Fig. 16-5 “. gjf” Input file format

3. Trial calculation of single point energy

(1) Edit the input file from the previous step, use the B3LYP/6-31G (d) command to calculate the single point energy (SP) of the molecule, as shown in Fig. 16-5 (B), and save it as a new input file “CH2O_SP. gjf”. Note that it is necessary to check and modify the path of “% chk=” and the file name of “. chk”.

(2) Start G09W. Tap menu bar File-open, select the “CH2O-SP. gjf” file in the target folder. In the pop-up window (Fig. 16-6), check the corresponding parameters and molecular coordinates, then click the “run” button, select the folder storing “CH2O-SP.out”, and start running the program. If the program ends smoothly, you can see celebrity quotes and calculation time information, as well as the words “Normal termination of Gaussian 09” at the end of the program window or “CH2O-SP.out” file, as shown in Fig. 16-7. This step is used to familiarize oneself with the editing of G09W input files, and to view the content of output files.

4. Geometry optimization and frequency analysis

Edit the input file using the “B3LYP/6-31G (d) Opt Freq” command, perform geometry optimization and vibration frequency calculation, and obtain the “CH2O_OPT.out” file. If there is no imaginary frequency in the frequency analysis result, the molecule is confirmed to be a stable configuration, as shown in Fig. 16-8.

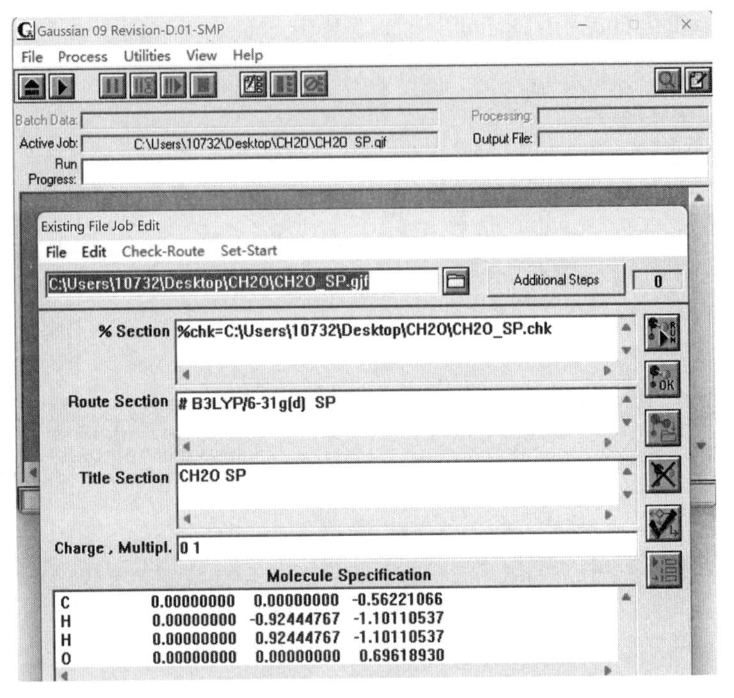

Fig. 16-6 Run G09W with input file

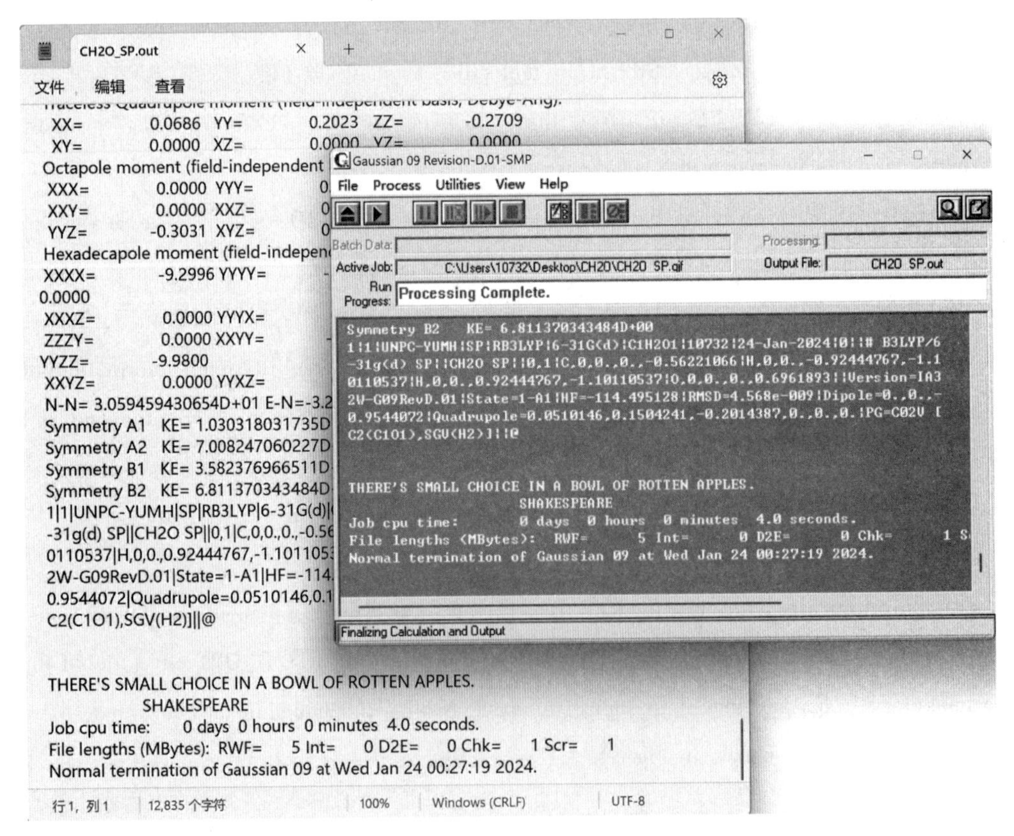

Fig. 16-7 Example of calculation job end

```
Harmonic frequencies (cm**-1), IR intensities (KM/Mole), Raman scattering
activities (A**4/AMU), depolarization ratios for plane and unpolarized
incident light, reduced masses (AMU), force constants (mDyne/A),
and normal coordinates:
                      1                      2                      3
                      A"                     A'                     A'
Frequencies --   1198.6003              1279.6563              1563.0168
Red. masses --      1.3702                 1.3419                 1.1281
Frc consts  --      1.1598                 1.2946                 1.6238
IR Inten    --      1.4505                12.5872                 6.4713
 Atom  AN      X      Y      Z        X      Y      Z        X      Y      Z
   1   6     0.00   0.00   0.17     0.15   0.00   0.00     0.00   0.02   0.00
   2   1     0.00   0.00  -0.70    -0.25   0.65   0.00    -0.35   0.61   0.00
   3   1     0.00   0.00  -0.70    -0.25  -0.65   0.00     0.35   0.61   0.00
   4   8     0.00   0.00  -0.04    -0.08   0.00   0.00     0.00  -0.09   0.00
                      4                      5                      6
                      A'                     A'                     A'
Frequencies --   1849.3960              2916.4544              2967.4644
Red. masses --      6.5203                 1.0447                 1.1221
Frc consts  --     13.1395                 5.2353                 5.8217
IR Inten    --     98.6505                55.9206               164.7029
 Atom  AN      X      Y      Z        X      Y      Z        X      Y      Z
   1   6     0.00   0.55   0.00     0.00  -0.06   0.00     0.10   0.00   0.00
   2   1     0.46  -0.26   0.00     0.61   0.35   0.00    -0.60  -0.37   0.00
   3   1    -0.46  -0.26   0.00    -0.61   0.35   0.00    -0.60   0.37   0.00
   4   8     0.00  -0.38   0.00     0.00   0.00   0.00     0.00   0.00   0.00
```

Fig. 16-8 Check the frequencies in CH2O_OPT.out

5. Calculation of electronic structure and chemical bonding properties

There are two ways to read the optimized structure coordinates from the previous step: read the "CH2O-OPT. chk" file or edit the optimized coordinates into a new input file "CH2O-NBO. gjf". The keyword for printing molecular orbitals is "Pop". The keyword for calculating the key level is "Pop = (NBORead, SaveNBOs)". After the molecular coordinates, leave a blank line and enter the keyword "$NBO BNDIDX NLMO 3CBOND $END" for natural bond orbital (NBO) analysis to perform NBO calculations.

6. View output information

Open the ". out" format output file with text editing software to view all calculation result data. GaussView can also be used to view the basic information of the output results, charge distribution, 3D graph of molecular orbitals, etc. A brief introduction to some output results that can be viewed using GaussView, as below.

(1) Summary

Open the "CH2O-OPT. out" file in GaussView and tap menu bar "Results—Summary" (Fig. 16-9).

(2) Vibration modes and infrared spectra

Open the "CH2O-OPT. out" file in GaussView, tap menu bar Results—Vibration—Start Animation to observe the vibration modes of different vibration frequencies, as shown in Fig. 16-10(A). Click on the "Spectrum" at the bottom row of the window to view the infrared spectra, as shown in Fig. 16-10(B).

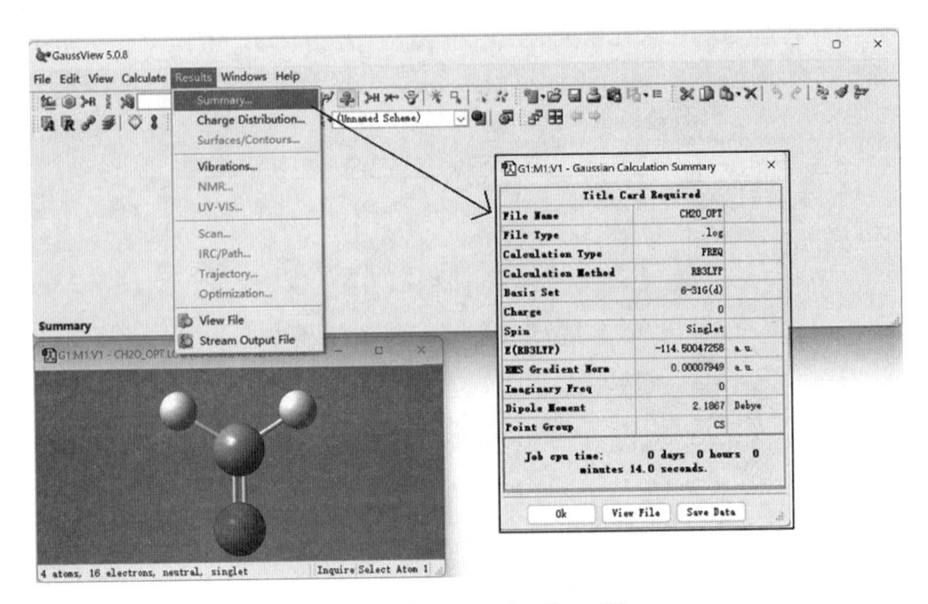

Fig. 16-9 Summary by GaussView

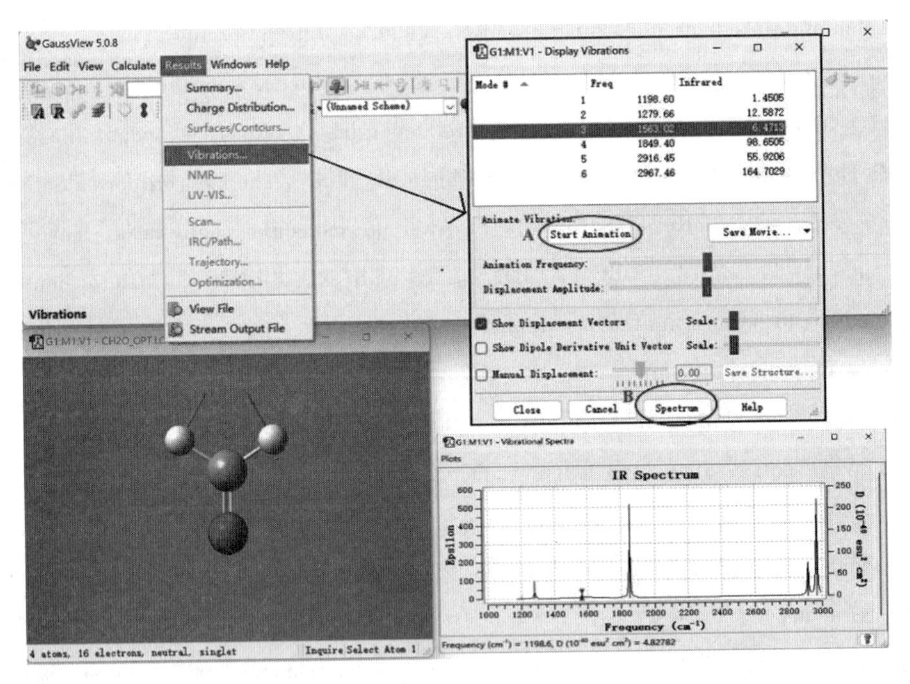

Fig. 16-10 Vibration modes and infrared spectra by GaussView

(3) Charge and dipole moment

Open the CH2O-NBO.chk file in GaussView. Then tap "Results—Charge Distribution" to see the information shown in Fig. 16-11. Check "Atomic Charges—Show Numbers" to view the net atomic charge. Use type menu bar to select Mulliken charge or NBO charge. Check "Dipole Moment—Show Vector" to display the dipole moment.

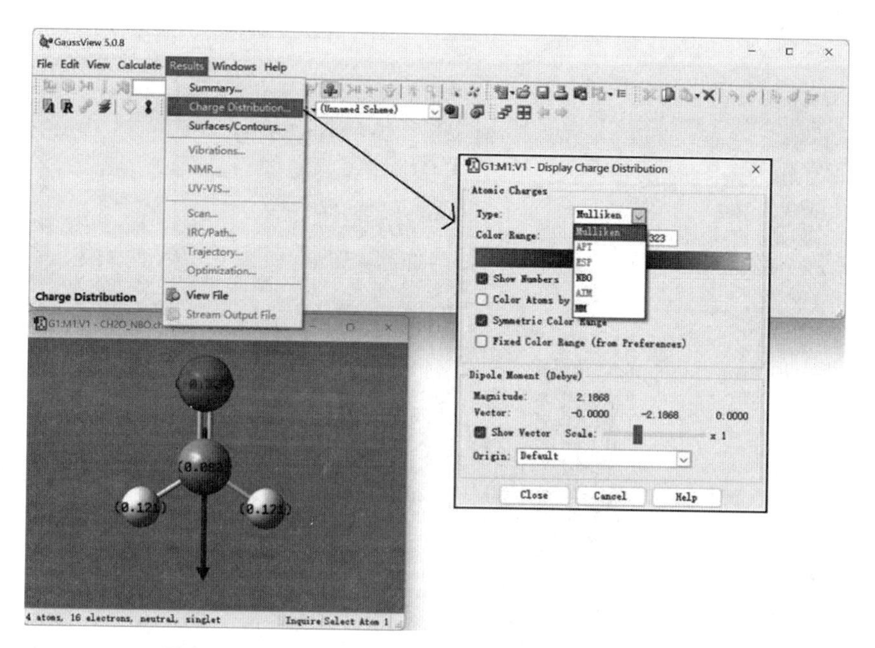

Fig. 16-11 Charge and dipole moment by GaussView

(4) 3D graph of molecular frontier orbitals

Open the "CH2O-OPT. chk" file in GaussView to browse the 3D graph of all molecular frontier orbitals. Tap the menu bar "Edit—MOs".The MOs dialog has its own embedded View window for displaying the molecule and surface corresponding to the currently selected MO. The left window displays the molecular structure and current surface, while the right of the current surface is the MO energy and occupancy diagram. MOs can be selected for visualization, inclusion in future calculation, and other operations by left-clicking on them. Selected MOs are highlighted in yellow. If a surface is available for an MO, a small square appears to the right of the MO. The square is red for the MO whose surface is currently being viewed. Clicking on a different MO's grey square will change the Current Surface View to show that MO's surface, as shown in Fig. 16-12. The same method 13 used to view the 3D graph of the natural bond orbitals.

Cautions

The calculations in this experiment are for molecules in vacuum, so the calculated values differ somewhat from the experimentally measured values for liquid formaldehyde. The accuracy of quantum chemical calculations depends on the method and the substrate used. The results of calculations obtained by different methods and substrates may vary considerably in some cases. Therefore, we should review literatures and choose a suitable method and base group, taking into account the experimental conditions, the computational study objectives and other factors.

Fig. 16-12 3D graph of molecular frontier orbitals by GaussView

Gaussian program is very powerful. During experiments, students should learn to consult the online help. Quantum-mechanical calculation is a complex process and beginners are prone to problems and should learn to check the output file for error messages. The Gaussian program has a large number of users, so answers to common problems for beginners can often be found on the web.

During the experiment, the software installation and file storage path must use the full English path, Chinese characters or spaces should be avoided.

Data recording and processing

Set the symmetry of the system to "C2v", run the program and record the following data.

1. Record the quantum mechanics method and base group used in the calculation.

2. Record molecular symmetry, electronic states and structural parameters (recording bond lengths, bond angles and dihedral angles).

3. Record the total energy of the molecule and the energy corrected with zero point energy.

4. Record the net charge on each atom, the dipole moment of the molecule and the bond level of the chemical bond.

5. Draw molecular frontier orbitals, record the corresponding MOs occupancies and energies.

6. Draw natural bond orbitals, record the corresponding MOs occupancies and energies.

7. Record the symmetry, frequency and infrared absorption intensity of vibrations, describe the vibration pattern characteristics. According to the theoretical calculation method, appropriate correction factors (such as B3LYP functional 0.9613) are selected for theoretical vibration frequency correction.

8. Record the theoretical predicted values of thermodynamic functions (U, H, G, S) to verify the relationship between thermodynamic parameters.

Thinking and discussing

1. Reviewing the literature, compare the results of the calculations with experimental values and other theoretical calculations, discuss the accuracy of the methods and basis sets used in the calculations. Analyze and discuss the structure and properties of the CH_2O molecule in combination with the results of this experimental calculation and theoretical knowledge of specialised chemistry such as structural chemistry, physical chemistry and organic chemistry.

2. Force the system symmetry to "Cs". Run the relevant procedures in the experimental steps, and record the experimental data from the previous data recording and processing (1) ~ (6) under Cs symmetry.

3. There is a wide variety of quantum chemistry calculation methods and basis sets, and the results of different methods and basis sets sometimes vary considerably. How to choose the right method and basis set?

4. Try to list the parameters of the properties of substances that the Gaussian software can calculate. Which parameters can be compared with experimental values thus verified against each other?

Reference materials

1. Wang Y. A computational chemistry study on the structure and properties of formaldehyde molecule—an experiment of computational chemistry is introduced. University Chemistry, 2014, 29(5): 66-70.

2. Frisch M J, Trucks G W, Schlegel H B, et al. Gaussian 09. Gaussian Inc., 2009.

实验 17　甲烷分子的结构和性质的计算化学研究

实验目的

1. 进一步了解、掌握量子化学软件 Gaussian 和 GaussView 的使用方法。

2. 通过非极性分子甲烷的相关计算，进一步掌握用 Gaussian 程序研究分子结构和性质的基本方法。

基本原理

Gaussian 程序是一款功能非常强大的量子化学计算软件包。它包含了目前主流的计算方法，如从头算、密度泛函理论、半经验方法等。通过 Gaussian 可以计算获得研究对象的能量、几何结构、振动模式、热力学量、静电势、电子结构、光谱性质和激发态性质等信息，也可用于寻找反应的过渡态和反应路径。而 GaussView 程序是辅助 Gaussian 程序强有力的可视化工具软件。两者结合使用，可满足从入门到专业级化学工作者的研究需求。

本实验主要内容包括：

单点能计算：用来预测给定结构的分子能量等信息。若分子构型采用了准确的结构参数，单点能的计算结果可提供分子非常精确的能量和相关性质。

几何构型优化计算：通过搜寻分子势能面上的极小点，预测分子的平衡几何构型，也称为优化构型。

振动频率分析：在优化构型的基础上，使用与结构优化相同的计算方法和基组可得到分子各简振模式的振动力常数和频率。对于稳定结构，N 个原子的线性分子和非线性分子分别有 $3N-5$ 和 $3N-6$ 个振动实频（力常数为正），且无虚频；化学反应过渡态（鞍点）有且仅有一个虚频，即有且只有一个负的力常数。同时该计算可获得分子的 IR 和 Raman 光谱，零点振动能和相关热力学数值。

在获取优化构型后，可进一步计算分子的电荷分布、轨道成分、键级等信息。

仪器和软件

计算机　　　　Gaussian 程序

Gaussview 程序

实验步骤

1. 构建分子初始构型

使用 GaussView5.0 软件画出一个甲烷分子（CH_4）的结构模型，如图 17-1 所示。而后点击“Edit”下拉菜单，选择“Point Group”，在弹出的窗口上用“Enable Point Group Symmetry”强制其对称性为“Td”。

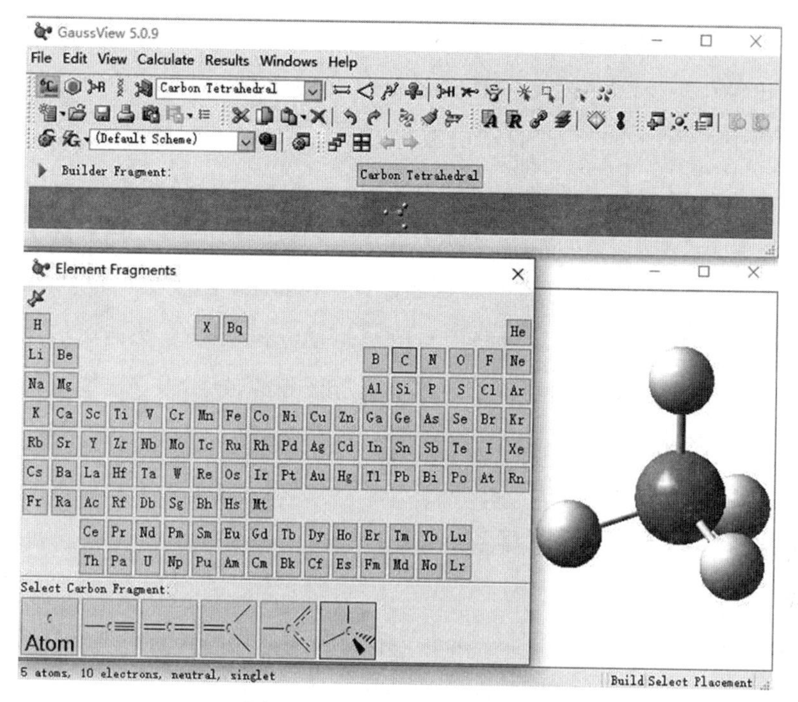

图 17-1 甲烷分子的 3D 结构

2. 保存输入文件

鼠标左击“File—Save”，将文件命名为 CH4. gjf，并保存至指定的文件夹中。需要注意的是文件名和路径名（文件夹命名）中不要出现中文字符和空格。这样我们就得到了一个初步的输入文件，如图 17-2(a)所示。而后用记事本或其他文本编辑软件打开 CH4. gjf，对输入文件进行编辑、修改后得到单点能计算的输入文件，如图 17-2(b)所示。其中，数字(1)～(4)标注出输入文件主要的四部分结构，将在下文做详细说明。

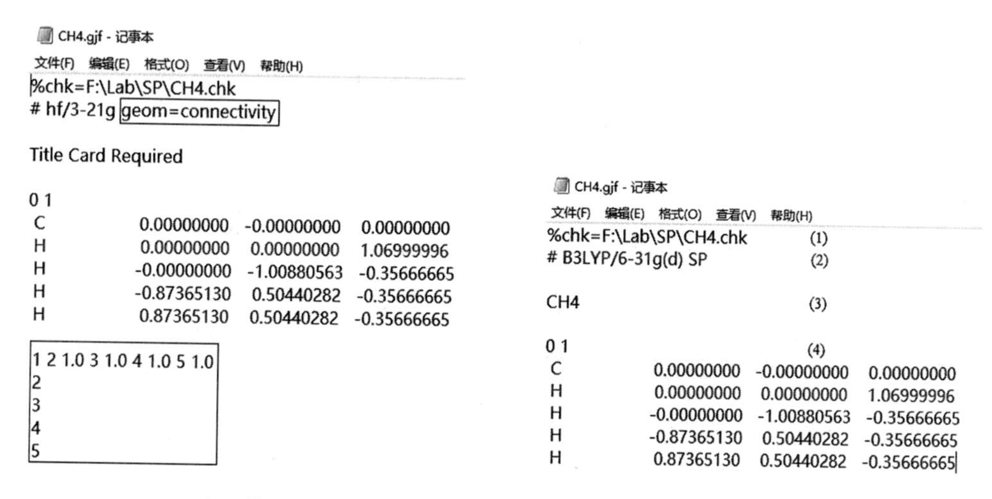

(a) 修改前

(b) 修改后

图 17-2 手动生成 CH4. gjf 文件

(1)“%chk=”命令用于对过程和结果信息的检查点文件“*.chk”进行命名(CH4.chk),同时给出保存路径,该路径必须真实存在,否则在后续计算过程中会报错。

(2)“#”及其后面的字母用于打印输出的控制。其中,“#N(缺省)”指以默认形式打印输出;“#P”表示以更详细的形式输出结果;“#T”表示输出最基本的信息和结果。“B3LYP”指本次计算采用DFT方法并选用B3LYP泛函进行计算。“6-31g(d)”是本次计算所用的基组。“SP”表示具体任务为单点能计算。Gaussian的关键词较多,除SP外使用较多的包括opt(结构优化)、irc(反应路径)、freq(振动分析)、scan(势能面扫描)等。

(3)对运行任务的命名行。可以修改本次计算任务的名字,根据需要进行修改即可。

(4)体系的总电荷和自旋多重度。“0”指该分子的总电荷为0。“1”指该分子的自旋多重度为0。该体系中包含1个C原子和4个H原子,这5个原子各自的笛卡儿坐标随之给出。输入文件*.gjf内容的最末尾务必保证存在空行,否则在计算过程中会报错。

3. 单点能计算

启动G09W程序,鼠标左击“File—open”,在目标文件夹内找到刚刚生成的CH4.gjf文件,可看到已经设定的相应参数和坐标,如图17-3,而后点击“run”按钮,选择存储CH4.out的文件夹后程序开始运行。如若程序顺利结束,打开CH4.out文件移动到末尾处可看到一句名人名言以及计算耗时等信息,而后出现“Normal termination of Gaussian 09”等字样(图17-4)。

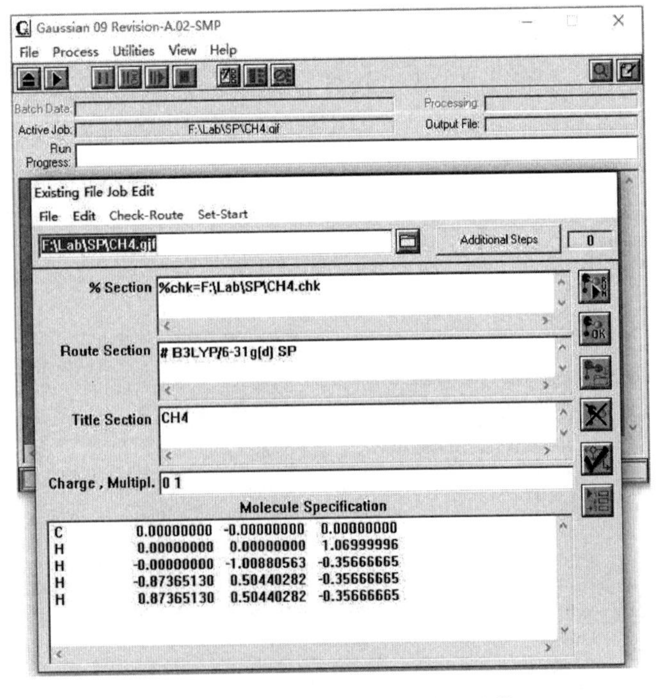

图17-3 运行G09W读取输入文件

```
1|1|UNPC-DESKTOP-2PQPC9R|SP|RB3LYP|6-31G(d)|C1H4|PCLWANG|23-Mar-2023|0
||# B3LYP/6-31g(d) SP||CH4||0,1|C,0,0.,0.,0.|H,0,0.,0.,1.06999996|H,0,
0.,-1.00880563,-0.35666665|H,0,-0.8736513,0.50440282,-0.35666665|H,0,0
.8736513,0.50440282,-0.35666665||Version=IA32W-G09RevA.02|State=1-A1|H
F=-40.5169541|RMSD=8.963e-009|Dipole=0.,0.,0.|Quadrupole=0.,0.,0.,0.,0
.,0.|PG=TD [O(C1),4C3(H1)]||@


ANYONE WHO IS NOT SHOCKED BY QUANTUM THEORY HAS
NOT UNDERSTOOD IT. -- NIELS BOHR(1885-1962)
Job cpu time:  0 days  0 hours  0 minutes  2.0 seconds.
File lengths (MBytes):  RWF=      5 Int=      0 D2E=      0 Chk=      1 Scr=      1
Normal termination of Gaussian 09 at Thu Mar 23 10:47:44 2023.
```

图 17-4　计算任务顺利结束示例

用 GaussView 打开“＊.out”格式输出文件，单击“Results—Summary”可查看本次计算得到的能量和偶极矩等基本信息。而后单击“Results—Charge Distribution”可看到图 17-5 所示信息，勾选“Atomic Charges—Show Numbers”可查看原子净电荷；勾选“Dipole Moment—Show Vector”可显示分子偶极矩。由于甲烷分子偶极矩为零，所以图例中并未看到偶极矩的示例。

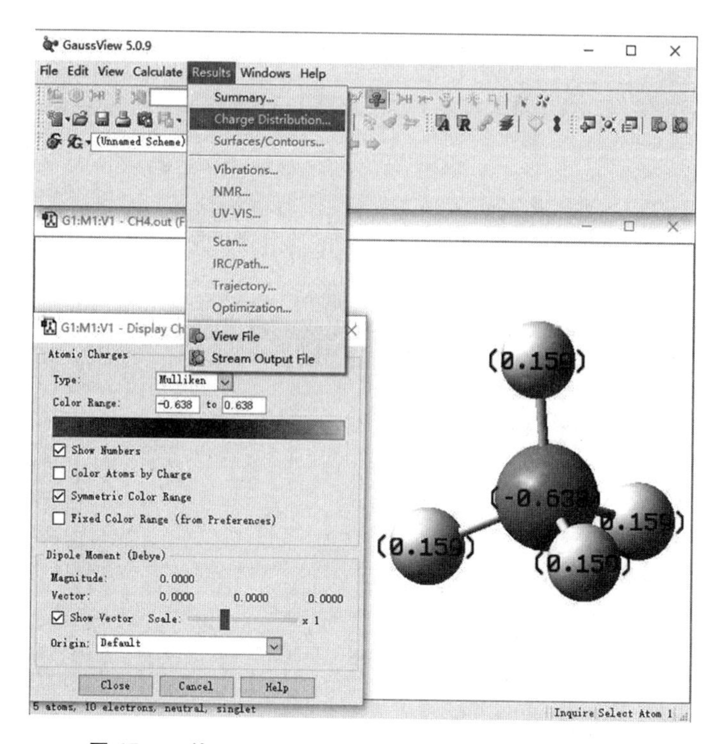

图 17-5　使用 GaussView 查看原子净电荷和分子偶极

4. 几何构型优化和振动频率计算

计算方法和基组继续采用 B3LYP/6-31G(d)，关键词使用 Opt 结合 Freq，依次进行几何优化和振动频率计算。若输出文件中出现四个 YES(图 17-6)，同时频率分析结果无虚频(图 17-7)，可确认分子为稳定构型。

```
         Item               Value     Threshold  Converged?
Maximum Force            0.000165     0.000450     YES
RMS     Force            0.000088     0.000300     YES
Maximum Displacement     0.000465     0.001800     YES
RMS     Displacement     0.000248     0.001200     YES
Predicted change in Energy=-1.430245D-07
Optimization completed.
   -- Stationary point found.
                           ----------------------------
                           !   Optimized Parameters   !
                           ! (Angstroms and Degrees)  !
 --------------------------                            --------------------------
 ! Name  Definition              Value          Derivative Info.                !
 --------------------------------------------------------------------------------
 ! R1    R(1,2)                  1.0935         -DE/DX =   -0.0002              !
 ! R2    R(1,3)                  1.0935         -DE/DX =   -0.0002              !
 ! R3    R(1,4)                  1.0935         -DE/DX =   -0.0002              !
 ! R4    R(1,5)                  1.0935         -DE/DX =   -0.0002              !
 ! A1    A(2,1,3)              109.4712         -DE/DX =    0.0                 !
 ! A2    A(2,1,4)              109.4712         -DE/DX =    0.0                 !
 ! A3    A(2,1,5)              109.4712         -DE/DX =    0.0                 !
 ! A4    A(3,1,4)              109.4712         -DE/DX =    0.0                 !
 ! A5    A(3,1,5)              109.4712         -DE/DX =    0.0                 !
 ! A6    A(4,1,5)              109.4712         -DE/DX =    0.0                 !
 ! D1    D(2,1,4,3)           -120.0            -DE/DX =    0.0                 !
 ! D2    D(2,1,5,3)            120.0            -DE/DX =    0.0                 !
 ! D3    D(2,1,5,4)           -120.0            -DE/DX =    0.0                 !
 ! D4    D(3,1,5,4)            120.0            -DE/DX =    0.0                 !
 --------------------------------------------------------------------------------
```

图 17-6 结构优化对稳定构型的确认

用 GaussView 打开“*.out”格式输出文件,点击“Results—Vibrations”,可查看红外吸收光谱以及分子的简振模式。同时可通过“Start Animation”按钮查看各振动模式的动画演示(图 17-8)。

```
Harmonic frequencies (cm**-1), IR intensities (KM/Mole), Raman scattering
activities (A**4/AMU), depolarization ratios for plane and unpolarized
incident light, reduced masses (AMU), force constants (mDyne/A),
and normal coordinates:
                      1                      2                      3
                     T2                     T2                     T2
Frequencies --  1373.9889              1373.9889              1373.9889
Red. masses --     1.1787                 1.1787                 1.1787
Frc consts  --     1.3110                 1.3110                 1.3110
IR Inten    --    15.3929                15.3929                15.3929
 Atom  AN      X      Y      Z        X      Y      Z        X      Y      Z
    1   6     0.00   0.00   0.12     0.03   0.12   0.00     0.12  -0.03   0.00
    2   1     0.22   0.23  -0.36     0.13  -0.31   0.29    -0.42   0.31   0.18
    3   1    -0.24  -0.23  -0.38     0.14  -0.30  -0.27    -0.41   0.32  -0.16
    4   1     0.22  -0.25  -0.37    -0.32  -0.41  -0.16    -0.31  -0.13   0.29
    5   1    -0.25   0.22  -0.37    -0.31  -0.42   0.18    -0.30  -0.14  -0.27
                      4                      5                      6
                      E                      E                     A1
Frequencies --  1594.1193              1594.1193              3051.5424
Red. masses --     1.0078                 1.0078                 1.0078
Frc consts  --     1.5090                 1.5090                 5.5294
IR Inten    --     0.0000                 0.0000                 0.0000
 Atom  AN      X      Y      Z        X      Y      Z        X      Y      Z
    1   6     0.00   0.00   0.00     0.00   0.00   0.00     0.00   0.00   0.00
    2   1     0.40  -0.28  -0.12    -0.09  -0.30   0.39    -0.29  -0.29  -0.29
    3   1    -0.40   0.28  -0.12     0.09   0.30   0.39     0.29   0.29  -0.29
    4   1    -0.40  -0.28   0.12     0.09  -0.30  -0.39     0.29  -0.29   0.29
    5   1     0.40   0.28   0.12    -0.09   0.30  -0.39    -0.29   0.29   0.29
                      7                      8                      9
                     T2                     T2                     T2
Frequencies --  3161.6635              3161.6635              3161.6635
Red. masses --     1.1019                 1.1019                 1.1019
Frc consts  --     6.4898                 6.4898                 6.4898
IR Inten    --    26.4653                26.4653                26.4653
 Atom  AN      X      Y      Z        X      Y      Z        X      Y      Z
    1   6     0.09   0.00   0.00     0.00   0.09   0.00     0.00   0.00   0.09
    2   1    -0.28  -0.30  -0.30    -0.28  -0.27  -0.28    -0.30  -0.30  -0.28
    3   1    -0.28  -0.30   0.30    -0.29  -0.27   0.29     0.29   0.29  -0.27
    4   1    -0.27   0.29  -0.29     0.30  -0.29   0.30    -0.29   0.29  -0.27
    5   1    -0.27   0.29   0.29     0.30  -0.28  -0.30     0.30  -0.29  -0.28
```

图 17-7 *.out 文件中查看振动有无虚频

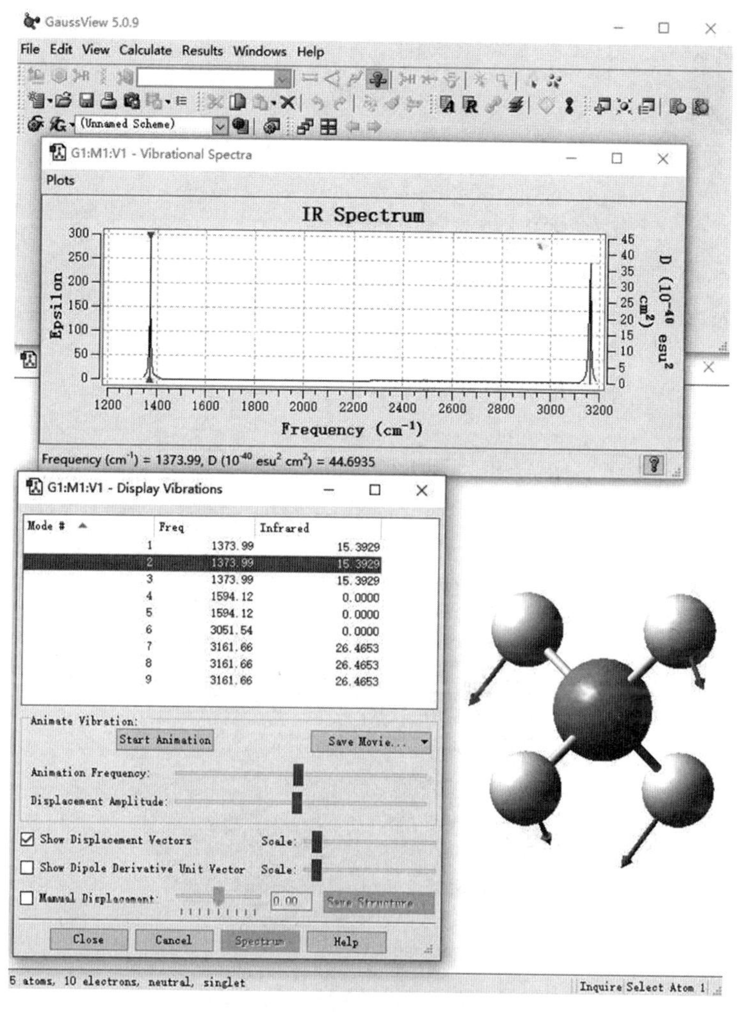

图 17-8　振动模式和 IR 光谱

5. 计算电子结构和化学键性质

使用优化好的结构坐标和步骤 4 的方法加基组，计算键级的关键词为“Pop=(NBORead,SaveNBOs)”。在分子坐标后空一行，输入 NBO 计算的关键词“$NBO BNDIDX NLMO 3CBOND $END”，可进行 NBO 分析。

计算任务结束后，用 GaussView 打开 *.chk 文件，点击“Results—Surfaces/Contours”，可浏览全部分子轨道的三维图形(图 17-9)。

数据记录与处理

1. 记录计算使用的方法和基组。
2. 记录分子对称性和结构参数(键长：保留 3 位小数；键角和二面角：保留 1 位小数)。
3. 记录分子总能量、零点能校正后的能量、原子净电荷、分子偶极矩和化学键的键级。
4. 选用适当的校正因子[例如 B3LYP/6-31G(d)为 0.9613]进行振动频率的理论值校

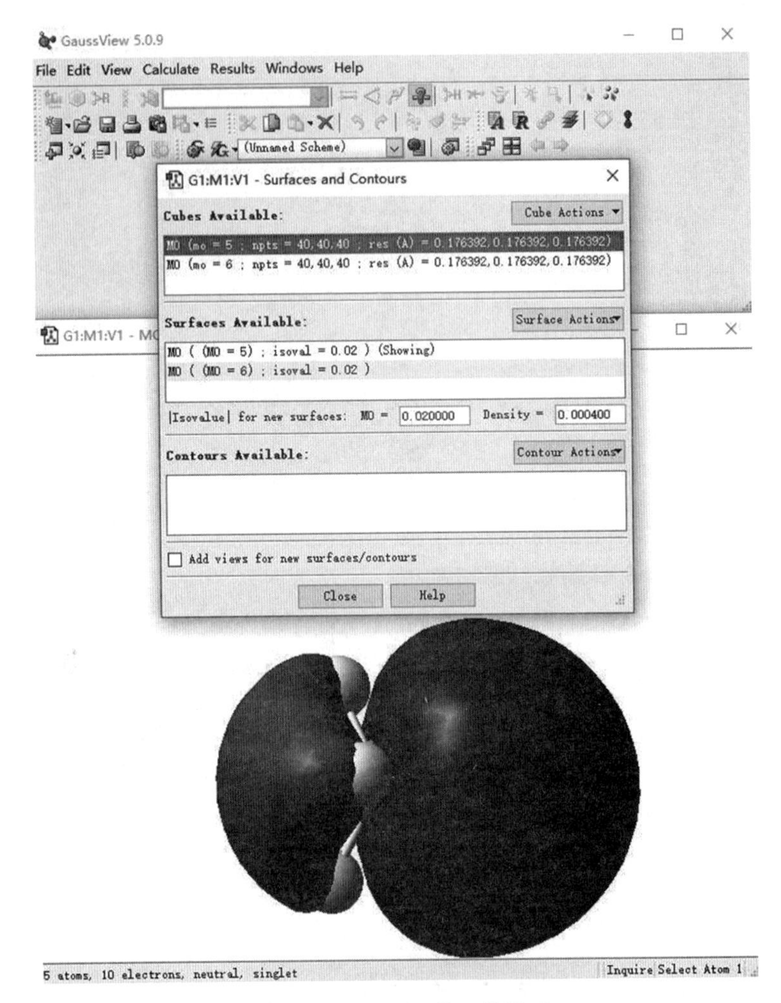

图 17-9　HOMO 的三维图形

正，而后记录振动的频率、红外吸收强度，并尝试描述振动模式的特征。

思考与讨论

1. 查阅文献，列举 Gaussian 软件能够计算的物质的性质参数，将计算结果与实验值及其他理论值进行比较，是否相同？同时讨论计算所用方法和基组的准确性。

2. 为什么所有振动频率为正值的结构是稳定结构？如果有一个振动频率为负值，这样的结构通常对应于什么结构？

参考资料

1. 华彤文，王颖霞. 普通化学原理（第四版）. 北京：北京大学出版社，2013.

2. 周公度，段连运. 结构化学基础（第五版）. 北京：北京大学出版社，2017.

3. 胡红智，马思渝. 计算化学实验. 北京：北京师范大学出版社，2008.

4. https://www.gaussian.com/

Experiment 17

Computational chemistry study of the structure and properties of methane

Experimental purpose

1. Understand and master the use of quantum chemistry software Gaussian and GaussView.

2. Master the basic methods of studying molecular structure and properties with Gaussian program via the calculation of non-polar methane.

Basical principles

Gaussian program is a very powerful quantum chemistry software package, including the current mainstream methods in computational chemistry, such as ab-initio method, density functional theory, semi-empirical method, etc. Through Gaussian calculation, we can obtain information of molecular, such as the energy, geometric structure, vibrational frequency, thermodynamic property, electrostatic potential, electronic structure, spectral properties, excited state properties and other information, meanwhile it could be used to find the transition state and the reaction path. GaussView program is a powerful visualization software that assists the use of Gaussian program. The combination of these two programs can meet the requirements of researchers from entry-level to professional level.

The content of this experiment includes:

Single-point energy calculation: In general, it is used to predict the electronic energy and other related information of a molecule at a given structure. If the molecule structure is from accurate experimental parameters, the results of single-point energy calculation can provide very accurate values of energy and properties of the molecule.

Geometric configuration optimization: Search the minimum on the potential energy surface, so as to predict the geometric configuration of molecule at equilibrium (optimized configuration).

Calculation of vibrational frequency: Beyond the optimized configuration, the vibrational force constants and frequencies of each vibrational mode of the molecule can be calculated using the same method and basis to geometric configuration optimization. N atomic molecules should have $3N-5$ (linear molecules) or $3N-6$ (non-linear molecules) real vibrational frequencies (with a positive force constant) and no imaginary frequencies at the stable states; the transition state (saddle point) of chemical reaction should have only one imaginary frequency, which means the force constant is negative. Simultaneously, the IR and Raman spectra, zero vibration energy, and related thermodynamic values of the molecule can be obtained.

After the optimized configuration, the information, such as charge distribution, orbital composition and bond order of the molecule, can be further calculated.

Apparatus and software

Computer　　　　Gaussian program

GaussView program

Experimental procedures

1. Construction of the initial configuration of the molecule

As shown in Fig. 17-1, draw the structure of methane molecule (CH_4) using GaussView 5.0. Then click on the "Edit" menu, select "Point Group", and use the "Enable Point Group Symmetry" command on the pop-up window to force its symmetry to "Td".

2. Saving the input files

Left click "File—Save", name the file CH4. gjf and save it to the selected folder. It is important to note that Chinese characters and spaces should not appear in all names (including folder name). In this way, a preliminary input file is obtained, which is shown in Fig. 17-2 (a). Then, Notepad is used to open CH4. gjf and modify the input file for single-point energy calculation, which is shown in Fig. 17-2 (b). The labels (1) ~(4) in Fig. 17-2 (b) indicate the main four part of the input file, and it will be explained in detail later.

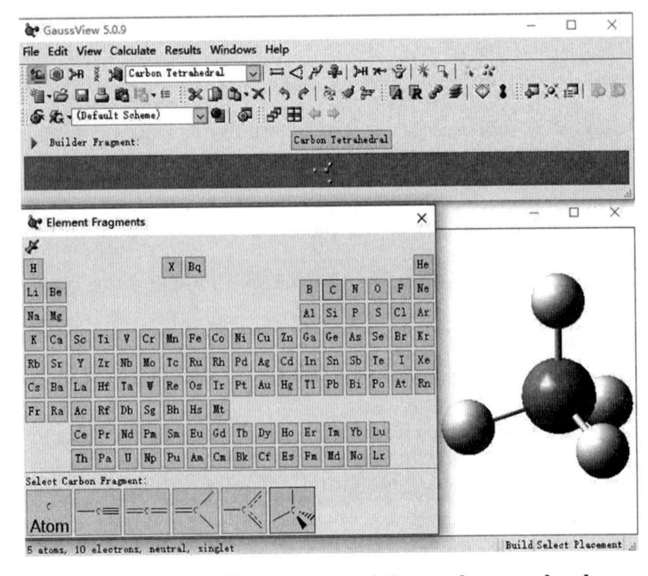

Fig. 17-1　The 3D structure of the methane molecule

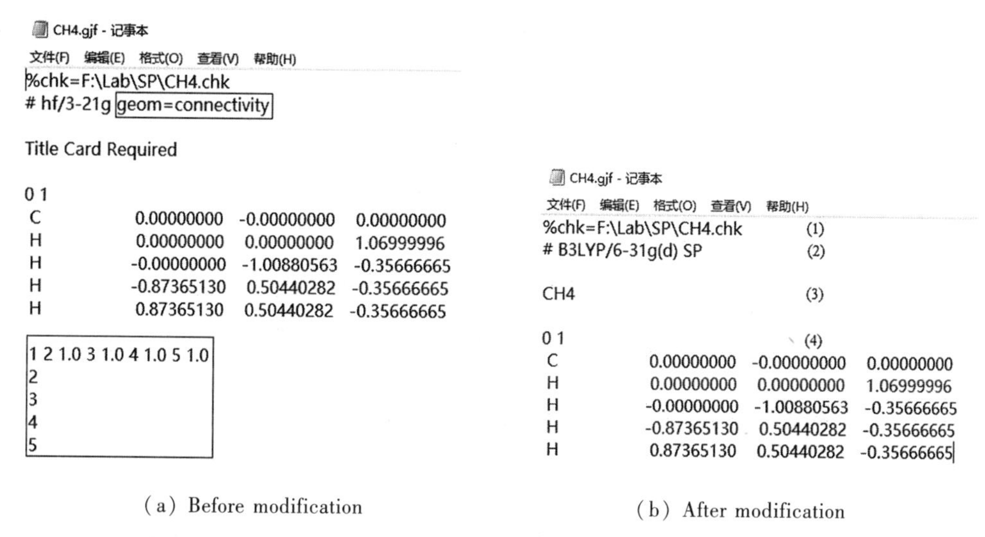

(a) Before modification (b) After modification

Fig. 17-2 Generating CH4. gjf file manually

(1) The "% chk =" command is used to specify the " * . chk" file name (CH4. chk) which contains the information of calculation process and result, meanwhile it specifies the file path. This path must be real, otherwise an error will be reported in the subsequent calculation.

(2) "#" and the letter after is used to control the output, where "#N(omitting)" refers to outputting results in normal form. "#P" will output the results in detail, and "#T" will output in simplified form. "B3LYP" indicates that the DFT method is used for this calculation and the B3LYP functional is chosen. "6-31g (d)" refers to the basis set used in this calculation. "SP" refers to single-point energy calculation. There are many keywords in Gaussian, such as opt (structure optimization), irc (reaction path), freq (frequency analysis), scan (potential energy surface scanning), etc.

(3) The name of the task for running. You can modify this name as needed.

(4) The total charge and spin multiplicity of the system. "0" refers to the total charge of the system, which is 0. "1" refers to the spin multiplicity of the system, which is 0. The file indicates 1 C atom and 4 H atoms are included, as well as their Cartesian coordinates. At the same time, there should be blank lines at the end of the input file * . gjf, otherwise an error will be reported during the calculation.

3. Single-point energy calculation

Run the G09W program, left click on "File—open", and find the newly generated CH4. gjf file in the target folder. Then, the corresponding parameters and coordinates that have been set in Fig. 17-3 can be seen. Click on the "run" button, select the folder where CH4.out is stored, and the program will start. If the job finishes normally, a famous quote and calculation time information will be shown at the end of CH4.out, which is followed by the words "Normal termination of Gaussian 09" (Fig. 17-4).

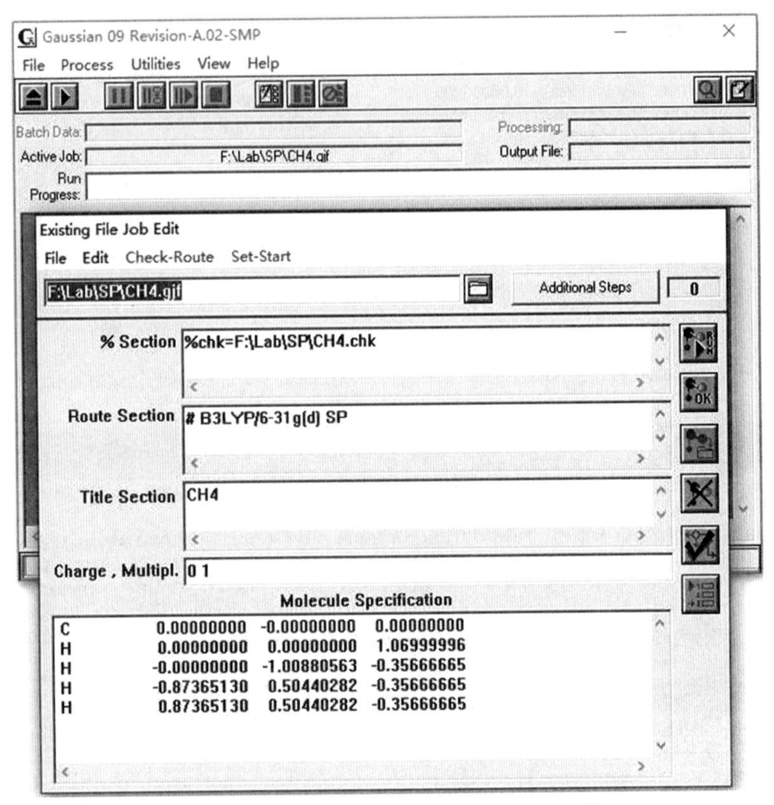

Fig. 17-3 Running G09W to read input files

```
1|1|UNPC-DESKTOP-2PQPC9R|SP|RB3LYP|6-31G(d)|C1H4|PCLWANG|23-Mar-2023|0
||# B3LYP/6-31g(d) SP||CH4||0,1|C,0,0.,0.,0.|H,0,0.,0.,1.06999996|H,0,
0.,-1.00880563,-0.35666665|H,0,-0.8736513,0.50440282,-0.35666665|H,0,0
.8736513,0.50440282,-0.35666665||Version=IA32W-G09RevA.02|State=1-A1|H
F=-40.5169541|RMSD=8.963e-009|Dipole=0.,0.,0.|Quadrupole=0.,0.,0.,0.,0
.,0.|PG=TD [O(C1),4C3(H1)]||@


ANYONE WHO IS NOT SHOCKED BY QUANTUM THEORY HAS
NOT UNDERSTOOD IT. -- NIELS BOHR(1885-1962)
Job cpu time:  0 days  0 hours  0 minutes  2.0 seconds.
File lengths (MBytes):  RWF=      5 Int=      0 D2E=      0 Chk=      1 Scr=      1
Normal termination of Gaussian 09 at Thu Mar 23 10:47:44 2023.
```

Fig. 17-4 Demo of successful termination

Open the " * . out" output file with GaussView and left click on "Results—Summary", the results such as energy and dipole moment will be shown. The content of Fig. 17-5 can be displayed under the command "Results—Charge Distribution". Then checking "Atomic Charges—Show Numbers" and "Dipole Moment—Show Vector", the numbers of atomic net charge and the vector of dipole moment can be seen. Since the dipole moment of methane molecule is zero, no arrow of dipole moment is shown in the figure.

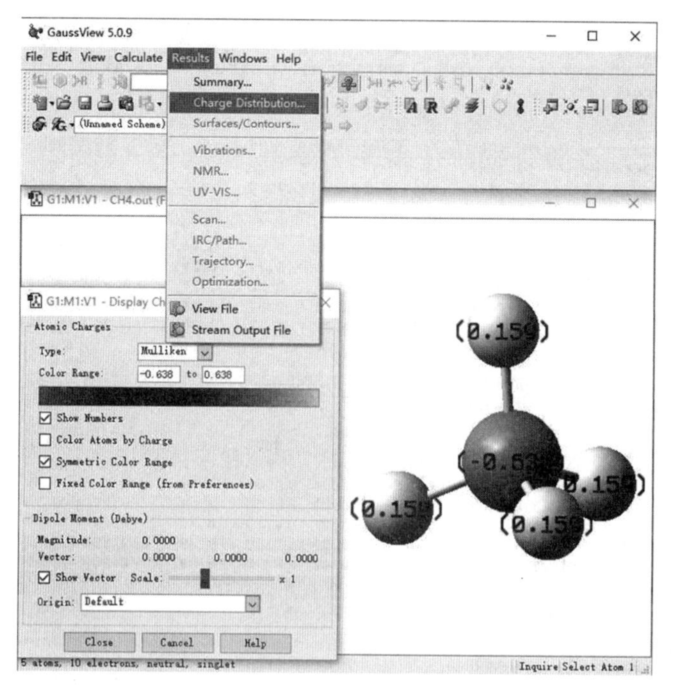

Fig. 17-5 Atomic net charge and molecular dipole under GaussView

4. Geometry optimization and vibrational frequency analysis

The method and the basis set are chosen as B3LYP/6-31G(d), "Opt" with "Freq" keywords are used. Then geometry optimization and vibrational frequency analysis is performed. If there are four "YES" (Fig. 17-6) with no imaginary frequency in the results of frequency analysis (Fig. 17-7), the stable configuration of the molecule is confirmed.

```
         Item               Value     Threshold  Converged?
 Maximum Force            0.000165     0.000450     YES
 RMS     Force            0.000088     0.000300     YES
 Maximum Displacement     0.000465     0.001800     YES
 RMS     Displacement     0.000248     0.001200     YES
 Predicted change in Energy=-1.430245D-07
 Optimization completed.
    -- Stationary point found.
                           ----------------------------
                           !   Optimized Parameters   !
                           ! (Angstroms and Degrees)  !
 --------------------------                            --------------------------
 ! Name  Definition              Value          Derivative Info.                !
 --------------------------------------------------------------------------------
 ! R1    R(1,2)                  1.0935         -DE/DX =   -0.0002              !
 ! R2    R(1,3)                  1.0935         -DE/DX =   -0.0002              !
 ! R3    R(1,4)                  1.0935         -DE/DX =   -0.0002              !
 ! R4    R(1,5)                  1.0935         -DE/DX =   -0.0002              !
 ! A1    A(2,1,3)              109.4712         -DE/DX =    0.0                 !
 ! A2    A(2,1,4)              109.4712         -DE/DX =    0.0                 !
 ! A3    A(2,1,5)              109.4712         -DE/DX =    0.0                 !
 ! A4    A(3,1,4)              109.4712         -DE/DX =    0.0                 !
 ! A5    A(3,1,5)              109.4712         -DE/DX =    0.0                 !
 ! A6    A(4,1,5)              109.4712         -DE/DX =    0.0                 !
 ! D1    D(2,1,4,3)           -120.0            -DE/DX =    0.0                 !
 ! D2    D(2,1,5,3)            120.0            -DE/DX =    0.0                 !
 ! D3    D(2,1,5,4)           -120.0            -DE/DX =    0.0                 !
 ! D4    D(3,1,5,4)            120.0            -DE/DX =    0.0                 !
 --------------------------------------------------------------------------------
```

Fig. 17-6 Confirmation of the stable configuration after geometry optimization

```
Harmonic frequencies (cm**-1), IR intensities (KM/Mole), Raman scattering
activities (A**4/AMU), depolarization ratios for plane and unpolarized
incident light, reduced masses (AMU), force constants (mDyne/A),
and normal coordinates:
                      1                      2                      3
                     T2                     T2                     T2
Frequencies --   1373.9889              1373.9889              1373.9889
Red. masses --      1.1787                 1.1787                 1.1787
Frc consts  --      1.3110                 1.3110                 1.3110
IR Inten    --     15.3929                15.3929                15.3929
 Atom  AN      X      Y      Z        X      Y      Z        X      Y      Z
   1   6     0.00   0.00   0.12     0.03   0.12   0.00     0.12  -0.03   0.00
   2   1     0.22   0.23  -0.36     0.13  -0.31   0.29    -0.42   0.31   0.18
   3   1    -0.24  -0.23  -0.38     0.14  -0.30  -0.27    -0.41   0.32  -0.16
   4   1     0.22  -0.25  -0.37    -0.32  -0.41  -0.16    -0.31  -0.13   0.29
   5   1    -0.25   0.22  -0.37    -0.31  -0.42   0.18    -0.30  -0.14  -0.27
                      4                      5                      6
                      E                      E                     A1
Frequencies --   1594.1193              1594.1193              3051.5424
Red. masses --      1.0078                 1.0078                 1.0078
Frc consts  --      1.5090                 1.5090                 5.5294
IR Inten    --      0.0000                 0.0000                 0.0000
 Atom  AN      X      Y      Z        X      Y      Z        X      Y      Z
   1   6     0.00   0.00   0.00     0.00   0.00   0.00     0.00   0.00   0.00
   2   1     0.40  -0.28  -0.12    -0.09  -0.30   0.39    -0.29  -0.29  -0.29
   3   1    -0.40   0.28  -0.12     0.09   0.30   0.39     0.29   0.29  -0.29
   4   1    -0.40  -0.28   0.12     0.09  -0.30  -0.39     0.29  -0.29   0.29
   5   1     0.40   0.28   0.12    -0.09   0.30  -0.39    -0.29   0.29   0.29
                      7                      8                      9
                     T2                     T2                     T2
Frequencies --   3161.6635              3161.6635              3161.6635
Red. masses --      1.1019                 1.1019                 1.1019
Frc consts  --      6.4898                 6.4898                 6.4898
IR Inten    --     26.4653                26.4653                26.4653
 Atom  AN      X      Y      Z        X      Y      Z        X      Y      Z
   1   6     0.09   0.00   0.00     0.00   0.09   0.00     0.00   0.00   0.09
   2   1    -0.28  -0.30  -0.30    -0.28  -0.27  -0.28    -0.30  -0.30  -0.28
   3   1    -0.28  -0.30   0.30    -0.29  -0.27   0.29     0.29   0.29  -0.27
   4   1    -0.27   0.29  -0.29     0.30  -0.29   0.30    -0.29   0.29  -0.27
   5   1    -0.27   0.29   0.29     0.30  -0.28  -0.30     0.30  -0.29  -0.28
```

Fig. 17-7　Frequency check in the * . out file

Open the " * . out" file with GaussView, click "Results—Vibrations" to view the infrared absorption spectrum and various vibrational modes of the molecule. Moreover, the "Start Animation" button can be used to view demonstrations of various vibrational modes (Fig. 17-8).

5. Electronic structure and chemical bonding properties

The coordinates of optimized structure, the method and basis set in step 4 are used here again. The keyword for calculating the bond order is "Pop=(NBORead, SaveNBOs)", leave a blank line after the molecule coordinates, and enter the keyword " $NBO BNDIDX NLMO 3CBOND $END" for NBO calculation.

After the job is done, open the * . chk file with GaussView and click "Results—Surfaces/Contours" to browse the 3D graphics of all molecule orbital (Fig. 17-9).

Data recording and processing

1. Record the methods and basis sets used in the calculation.

2. Record molecule symmetry and structure parameters (bond length: keep 3 decimal places; bond angle and dihedral angle: keep 1 decimal place).

3. Record the total energy and the energy corrected with zero point energy, the net charge

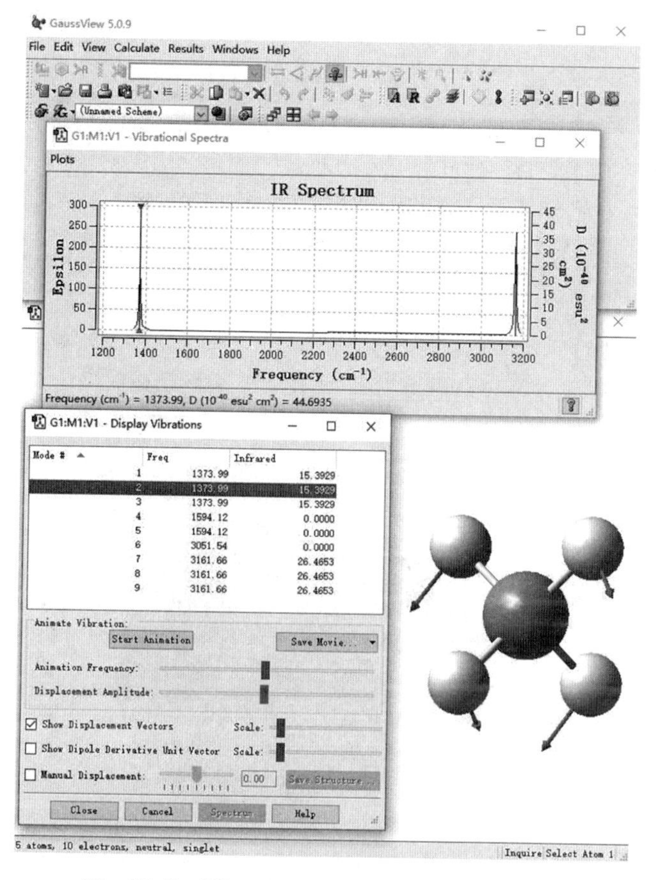

Fig. 17-8 Vibrational modes and IR spectrum

on each atom, the dipole moment and the bond order of chemical bond.

4. Select appropriate correction factors [0.9613 for B3LYP/6-31G(d)] to calculate the true vibrational frequency. Later, record the vibrational frequency and infrared absorption intensity, then try to describe the vibration modes.

Thinking and discussing

1. Referring to the literature, list all the properties that can be predicted by Gaussian. Compare the results with experimental and other theoretical values. Are they the same? Discuss the accuracy of the methods and basis sets which are used for calculation.

2. Why is the structure with all positive vibrational frequencies stable? If there is a negative vibrational frequency, what structure does it correspond to?

Reference materials

1. Hua T, Wang Y, et al. Principles of General Chemistry (4th ed). Beijing: Peking University Press, 2013.

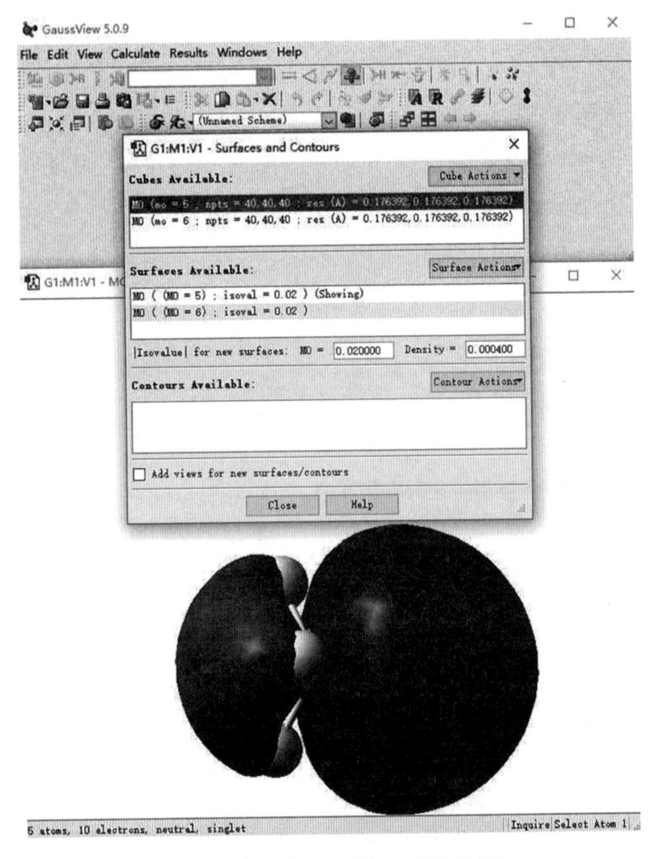

Fig. 17-9　3D graphics of HOMO

2. Zhou G, Duan L. Fundamentals of Structural Chemistry (5th ed). Beijing: Peking University Press, 2017.

3. Hu H, Ma S. Experiment of Computational Chemistry. Beijing: Beijing Normal University Press, 2008.

4. https://www.gaussian.com/

实验 18　富勒烯在不同工作介质中的电化学行为研究(虚拟仿真实验)

实验目的

1. 进一步熟悉电化学分析仪的使用方法。

2. 通过 C_{60}、C_{70}在不同条件下的循环伏安及示差脉冲伏安测定,了解多电子转移体系的电化学行为。

3. 通过电器着火的安全演练,提升学生安全意识,提高学生自我防护能力。

基本原理

1. 循环伏安法

循环伏安法的基本原理及应用见本书实验10。

2. 示差脉冲伏安法

示差脉冲伏安法(Differential pulse voltammetry, DPV)也是一种常用的伏安分析方法,其原理是以线性电压或阶梯电压和幅值固定的脉冲加和为激励信号,如图18-1所示,在即将应用脉冲之前和脉冲末期,对电流两次取样,以电流差对电压作图,可得图18-2的示差脉冲实验结果。加入脉冲电压后,一方面对电极充电,可产生相应的充电电流,充电电流会很快衰减至零;另一方面,加入的脉冲电压可引起被测物质发生电极反应,产生电解电流(即法拉第电流),受电极反应物质的扩散所控制,电解电流将随着被测物质在电极上的反应而慢慢衰减,但速度比充电电流的衰减慢得多,因此在施加脉冲电压的后期进行电流取样,测得的几乎是电解电流。由于电流差减的缘故,DPV杂质的氧化还原电流导致的背景电流被大大地消除,因此,DPV具有更高的检测灵敏度及更低的检测限,在实验条件控制良好的条件下,其检测限可低至10^{-8} mol · L^{-1}。在定量测试方面,甚至比大部分分子或者原子吸收光谱更加灵敏。

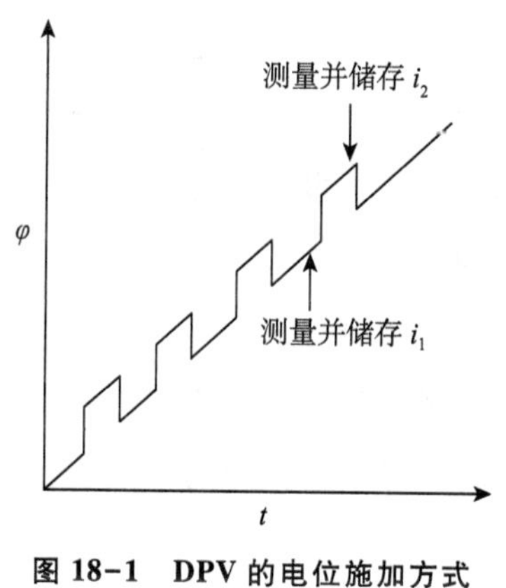

图18-1 DPV的电位施加方式

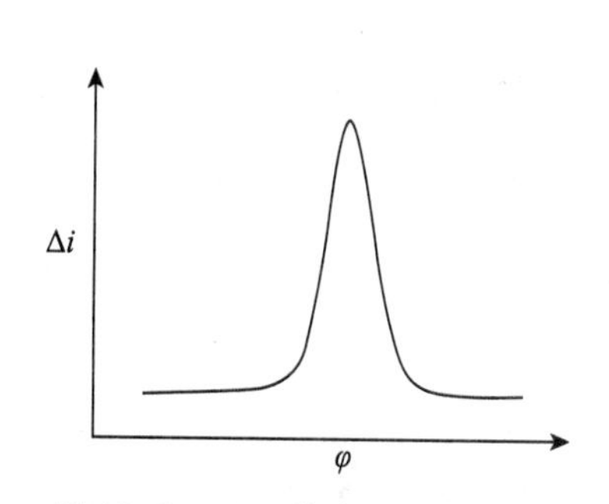

图18-2 DPV的电流-电位关系

3. 富勒烯C_{60}、C_{70}概述

富勒烯是由碳原子组成的一系列笼形单质分子的总称,是Smalley、Kroto和Curl三位化学家于1985年发现的继金刚石、石墨之后碳单质的第三种稳定存在形式,发现者也因此被授予1996年诺贝尔化学奖。由于其特殊的结构和性能,富勒烯在材料、化学、超导与半导体物理、生物等学科和激光防护、催化剂、燃料、润滑剂、合成、化妆品、量子计算机等工程领域具有重要的研究价值和应用前景。

C_{60}分子具有圆球形笼状结构以及缺电子烯烃的化学特性,六元环间的6∶6双键为反应的活性部位，可发生诸如氢化、卤化、氧化还原、环加成、光化与催化、配合和自由基加成等多种化学反应。共价富勒烯化学的系统发展使得 C_{60} 这一三维建筑块具有前所未有的多变性，为合成新材料开辟了广阔的天地。与 C_{60} 分子不同，具有 D_5h 点群对称性的 C_{70} 分子形状类似于橄榄球，具有比 C_{60} 更强的电子亲和力，更易于通过电荷转移形成阴离子或自由基。它不但具有亲核加成和氧化加成、还原低价过渡金属的能力，而且能进行亲电加成、卤化和自由基反应，以及以阴离子为中间体的烃基化、烷氧基化和以 Lewis 酸为中间体的芳基化反应。

C_{60}、C_{70} 的电化学行为研究具有重要意义，它能反映加成或者化学反应等过程对富勒烯电子结构的影响，但 C_{60}、C_{70} 的电化学性质比较稳定，常温常压下反应很难进行，观察到相关实验现象就更难了，因此，本实验通过虚拟仿真软件进行。

仪器和软件

电脑　　　　　　　　　　　　虚拟仿真实验软件

实验步骤

1. C_{60} 在 KI/液氨混合溶液中的电化学行为

（1）利用循环伏安法研究 C_{60} 在 KI/液氨混合溶液中的电化学行为。

（2）假设实验中电器设备着火，进行灭火演练。

2. C_{60}、C_{70} 在乙腈/甲苯混合溶液中的电化学行为

利用循环伏安法研究 C_{60}、C_{70} 在乙腈/甲苯混合溶液中的电化学行为。

3. C_{60}、C_{70} 在离子液体/甲苯溶液中的电化学行为

（1）利用四己基氯化铵和三氟甲基磺酰亚胺锂制备离子液体。

（2）利用循环伏安法研究 C_{60}、C_{70} 在离子液体/甲苯溶液中的电化学行为。

（3）利用示差脉冲伏安法研究 C_{60}、C_{70} 在离子液体/甲苯溶液中的电化学行为。

具体虚拟实验操作请登录下面链接进行。

http://lnnu-flx.dlvrtec.com

数据记录与处理

1. 记录上述实验过程的循环伏安图或示差脉冲伏安图。

2. 比较 C_{60}、C_{70} 的还原性。

思考与讨论

1. 比较示差脉冲伏安法和循环伏安法的优点和特点。

2. 通过查阅文献，结合本实验结果，分析总结工作介质对富勒烯电化学行为的影响。

参考资料

1. Oswald J, Beretta D, Stiefel M, et al. Field and thermal emission limited charge injection in Au-C60-graphene van der waals vertical heterostructures for organic electronics. ACS Applied Nano Materials, 2023, 6(11): 9444-9452.

2. Shukla M K, Dubey M, Leszczynski J. Theoretical investigation of electronic structures and properties of C60-bgold nanocontacts. ACS Nano, 2008, 2(2): 227-234.

3. Zhou F, Jehoulet C, Bard A J. Reduction and electrochemistry of fullerene C60 in liquid ammonia. Journal of the American Chemical Society, 1992, 114(27): 11004-11006.

4. Xie Q, Perez-Cordero E, Echegoyen L. Electrochemical detection of C_{60}^{6-} and C_{70}^{6-}: enhanced stability of fullerides in solution. Journal of the American Chemical Society, 1992, 114(10): 3978-3980.

5. 张美芹，刘卉，胡虎，等. 室温下甲苯中 C_{60}^{6-} 和 C_{70}^{6-} 的电化学检测. 高等学校化学学报，2007，28(4)：727-730.

Experiment 18

Study on the electrochemical behavior of fullerene in different working media (virtual simulation experiment)

Experimental purpose

1. Learn more about electrochemical analysis.

2. Understand the electrochemical behavior of multi-electron transfer systems by cyclic voltammetry and differential pulse voltammetry measurements of C_{60} and C_{70} under different conditions.

3. Raise students' experiment safety consciousness and improve students' self-protection ability through the safety drill of electric appliance fire.

Basical principles

1. Cyclic voltammetry (CV)

The basic principles and applications of CV can be found in Experiment 10 of this book.

2. Differential pulse voltammetry (DPV)

DPV is also a commonly used voltammetric analysis method, which is based on a linear or step voltage and a fixed amplitude pulse addition as the excitation signal, as shown in Fig. 18-1. Before and at the end of the pulse, the current is sampled twice and the voltage is plotted by the current difference. The experimental results of the DPV can be shown in Fig. 18-2. After adding the pulse voltage, on one hand, charging the electrode can produce the corresponding charging current, and the charging current will quickly decay to zero. On the other hand, the added pulse voltage can cause the electrode reaction of the tested substance, to produce an electrolytic current (the Faraday current). Controlled by the diffusion of the reactant, the electrolytic current will decay slowly with the reaction of the measured substance on the electrode, but at a much slower rate than the decay of the charging current. Therefore, the current sampling at the later stage of the pulse voltage is almost the electrolytic current. Because of the current difference, the background current is largely eliminated which is caused by the redox current of the DPV impurity. As a result, the DPV has higher sensitivity and lower detection limit, the detection limit can be as low as 10^{-8} mol · L^{-1}. In quantitative tests, it is even more sensitive than most molecules or atomic absorption spectroscopy.

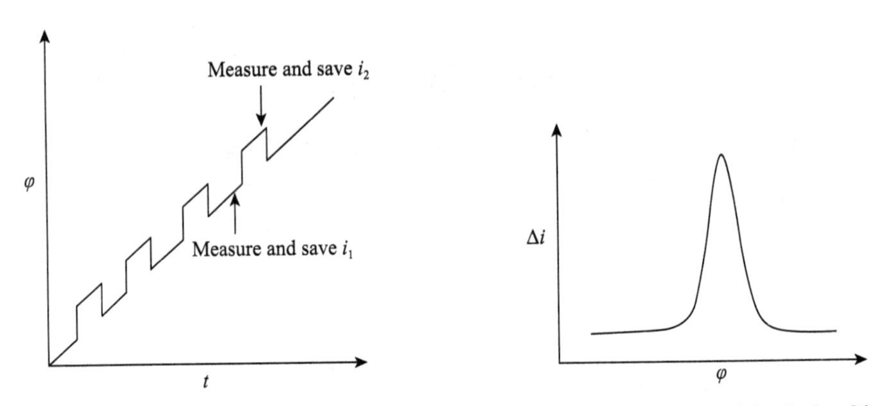

Fig. 18-1 Potential application of DPV **Fig. 18-2 The current-potential relationship of DPV**

3. Overview of fullerene C_{60}、C_{70}

Fullerenes are a group of cage-like molecules made up of carbon atoms. They are the third stable form of carbon found by chemists Smalley, Kroto and Curl in 1985, after diamond and graphite, and the discoverer was awarded the 1996 Nobel Prize for Chemistry. Because of its special structure and performance, fullerene has important research value and application prospect in materials, chemistry, superconducting and semiconductor physics, biology and engineering fields such as laser protection, catalyst, fuel, lubricant, synthesis, cosmetics, quantum computer and so on.

The C_{60} molecule has spherical cage structure and the chemical characteristics of electron-deficient olefins. The 6 : 6 double bond between the six-membered ring is the active part of the reaction, which can occur many chemical reactions such as hydrogenation, halogenation, redox, cycloaddition, photocatalysis, complexation and free-radical addition. The systematic development of covalent fullerene chemistry has made C_{60}, the three-dimensional building block, more versatile than ever before, opening up a wide field for the synthesis of new materials.

Unlike C_{60}, the D_5h point group symmetry of C_{70} is similar in shape to that of a rugby, with a stronger electron affinity than C_{60} and easier to form anions or free radicals through charge transfer. It is capable of nucleophilic addition, oxidative addition, reduction of low-valent transition metals, electrophilic addition, halogenation, radical reaction, alkylation, alkoxylation with anions as intermediates and arylation with Lewis acid as intermediates.

The study of the electrochemical behavior of C_{60} and C_{70} is of great significance. It can reflect the effect of addition or chemical reaction on the electronic structure of fullerene. However, the electrochemical properties of C_{60} and C_{70} are relatively stable, it is difficult to carry out the reaction under normal temperature and pressure, and it is even more difficult to observe the related experimental phenomena. Therefore, this experiment is carried out by virtual simulation software.

Apparatus and software

Computer　　　　Virtual simulation experiment software

Experimental procedures

1. Electrochemical behavior of C_{60} in KI/liquid ammonia mixed solution

(1) Investigate the electrochemical behavior of C_{60} in KI/liquid ammonia mixed solution by CV.

(2) In the experiment, suppose that the electrical equipment is on fire, then carry out fire-fighting drill.

2. Electrochemical behavior of C_{60} and C_{70} in acetonitrile/toluene mixed solution

Investigate the electrochemical behavior of C_{60} and C_{70} in acetonitrile/toluene mixed solution by CV.

3. Electrochemical behavior of C_{60} and C_{70} in ionic liquids/toluene solutions

(1) Prepare ionic liquids using tetrahexyl ammonium chloride and lithium trifluoromethyl sulfonimide.

(2) Investigate the electrochemical behavior of C_{60} and C_{70} in ionic liquid/toluene by CV.

(3) Investigate the electrochemical behavior of C_{60} and C_{70} in ionic liquid/toluene by DPV.

Please click on the link below for the specific virtual experiment.

http://lnnu-flx.dlvrtec.com

Data recording and processing

1. Record and analyze the cyclic voltammogram and differential pulse voltammogram obtained.

2. Compare the reducibility of C_{60} and C_{70}.

Thinking and discussing

1. Compare the advantages and characteristics of DPV and CV.

2. By consulting literatures and combining with the experimental results, analyze and summarize the influence of working medium on the electrochemical behavior of fullerene.

Reference materials

1. Oswald J, Beretta D, Stiefel M, et al. Field and thermal emission limited charge injection in Au-C60-graphene van der Waals vertical heterostructures for organic electronics. ACS Applied Nano Materials 2023, 6(11), 9444-9452.

2. Shukla M K, Dubey M, Leszczynski J. Theoretical investigation of electronic structures and properties of C60-bgold nanocontacts. ACS Nano, 2008, 2(2): 227-234.

3. Zhou F, Jehoulet C, Bard A J. Reduction and electrochemistry of fullerene C60 in liquid ammonia. Journal of the American Chemical Society, 1992, 114(27): 11004-11006.

4. Xie Q, Perez-Cordero E, Echegoyen L. Electrochemical detection of C_{60}^{6-} and C_{70}^{6-}: enhanced stability of fullerides in solution. Journal of the American Chemical Society, 1992, 114(10): 3978-3980.

5. Zhang M., Liu H., Hu H., et al. Electrochemical determination of C_{60}^{6-} and C_{70}^{6-} in toluene at room temperature. Journal of Chemistry in Colleges and Universities, 2007, 28(4): 727-730.

第三章

综合实验

实验 19 气相色谱法测定非电解质溶液的热力学函数

实验目的

1. 了解气相色谱法在化学热力学方面的一些应用。

2. 用气相色谱法测定环己烷在邻苯二甲酸二壬酯溶液中的无限稀释活度系数,并求出偏摩尔溶解焓、偏摩尔超额溶解焓和摩尔气化焓。

基本原理

气相色谱法是用气体作流动相的色谱方法,是分离分析应用最广泛的技术之一。气相色谱有两个相,流动相和固定相。固定相可以是固体吸附剂,也可以是涂敷在惰性多孔担体上的液体。流动相(又称为载气)是一些不会与固定相和待测样品发生化学反应的气体。显然,试样(或称第三组分)在两相间的分配情况与试样和固定相之间相互作用的热力学和动力学性质密切相关。试样由流动相带动,通过大比表面积的固定相的空隙,并在气相和固定相之间进行反复多次连续的热力学分配,由于试样中各组分之间性质的微小差异,可达到分离的目的,并可测定无限稀释的非电解质溶液热力学函数。

利用脉冲进样方法可以测定某些溶液体系的热力学函数。作为溶质的色谱试样通过进样口进入色谱仪后,部分留于气相;部分溶解在色谱柱固定液中与固定液组成溶液。随着载气的流动,经过一段时间后溶质样品将被带出色谱柱。图 19-1 就是一个较为理想的色谱图。图中在 t_d 处的峰意味着色谱仪的气路上有“死空间”存在。真正的样品峰则出现在 t_r 处。本实验选用氮气作为流动相(载气)。

从溶质进样到检测器出现浓度极大值所需的时间 t_r 称为保留时间。以皂膜流量计测得的载气流量 F 乘以 t_r 即为保留体积 V_r。F 与 t_d 的乘积 V_d 则称为死体积,它与溶解过程无关,只与色谱仪的进样器、色谱柱和检测器这三部分的空间大小有关,所以 t_r 与 t_d 之差

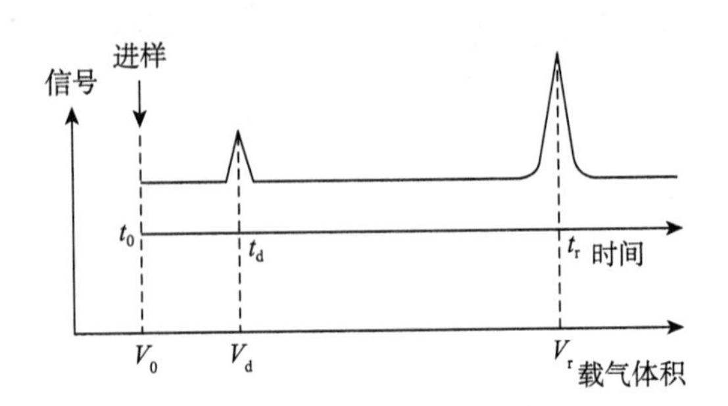

图 19-1　脉冲进样色谱示意图

表征了溶质的溶解或溶液的性质。另外，应以色谱柱内载气的流速 F_c 来讨论保留体积才较合理。然而，从色谱柱的性质可知，柱内压力是柱长的函数，因而柱内各部位的实际流量也不是常量，对此可用压力校正因子 j 加以校正。如再考虑其他因素的影响，则应以单位质量固定液上样品比保留体积 V_g^0 来表示，才能真正反映溶质与作为溶剂的固定液之间相互作用的特性

$$V_g^0/\mathrm{dm}^3\cdot\mathrm{kg}^{-1}=(t_r-t_d)\cdot j\cdot\frac{(p_0-p_w)}{p_0}\cdot\frac{273}{T_r}\cdot\frac{F}{m_1}\tag{1}$$

式中

$$j=\frac{3}{2}\cdot\frac{[(p_i/p_0)^2-1]}{[(p_i/p_0)^3-1]}\tag{2}$$

T_r 为皂膜流量计所处的温度（K，T_r = 室温+273.15）；p_i、p_0、p_W 分别为色谱柱前压力、出口压力（大气压力）以及 T_r 时水的饱和蒸气压；m_1 为固定液质量。

若溶质在气、液两相的浓度分别用不同的概念来定义，则 273 K 时溶质在两相间的分配系数可表示如下

$$K_D=\frac{\text{固定液上溶质质量 / 固定液质量}}{\text{流动相中溶质质量 / 流动相体积}}=\frac{m_2^s/m_1}{m_2^g/V_d}$$

式中下标 1、2 分别表示固定液和溶质；上标 s、g 分别表示固定液相和气相。

在理想条件下，色谱峰峰形应该是对称的。那么，在 t_r 时，恰好有一半溶质被载气带离检测器，另一半还留在色谱柱内。两部分质量相等，色谱柱内的溶质又分别处于气相和液相中。因此

$$V_r^c\cdot\frac{m_2^g}{V_d}=V_d^c\cdot\frac{m_2^g}{V_d}+V_s\cdot\frac{m_2^s}{m_1}\cdot\rho_1\tag{3}$$

这里 ρ_1 是固定液的密度，V_r^c 和 V_d^c 分别表示柱温柱压条件下的保留体积和死体积。移项并作压力和温度校正，得

$$(t_r-t_d)\cdot j\cdot\frac{(p_0-p_w)}{p_0}\cdot\frac{273}{T_r}\cdot F\cdot\frac{m_2^g}{V_d}=V_s\cdot\rho_1\cdot\frac{m_2^s}{m_1}\tag{4}$$

因 $V_s \cdot \rho_1 = m_1$，再分别与式(1)和式(2)比较，即得

$$V_g^0 = \frac{m_2^s/m_1}{m_2^g/V_d} = K_D \tag{5}$$

由于脉冲进样量非常小，样品在气液两相的行为可分别用理想气体方程和拉乌尔定律作近似处理

$$p_2 V_d = nRT_c$$

$$p_2 = \frac{m_2^g RT}{V_d M_2} \tag{6}$$

$$p_2^* = \frac{p_2}{x_2} = p_2\left(\frac{n_1 + n_2}{n_2}\right) \approx p_2 \cdot \frac{n_1}{n_2} = p_2 \cdot \frac{M_2}{M_1} \cdot \frac{m_1}{m_2^s} \tag{7}$$

式中 p_2^* 和 p_2 分别为纯溶质和溶液中溶质的蒸气压；x_2 为溶质在溶液中的摩尔分数；M_1 和 M_2 分别为固定液和溶质的摩尔质量；n_1 和 n_2 分别为固定液和溶质在溶液中的物质的量。将蒸气压由柱压 T_c 校正至 273 K，并以式(5)和式(6)代入式(7)，得

$$p_2^* = p_2 \cdot \frac{273}{T_c} \cdot \frac{M_2}{M_1} \cdot \frac{V_d}{K_D \cdot m_2^g} = \frac{273R}{K_D \cdot M_1} \tag{8}$$

结合式(5)得

$$V_g^0 = \frac{273R}{p_2^* \cdot M_1} \tag{9}$$

实际上，色谱柱固定液的沸点都较高，蒸气压很低，且摩尔质量和摩尔体积都较大；然而，适用于作溶质的样品，其物理性质则与之相去甚远。所以溶液性质往往会偏离拉乌尔定律。不过，在此稀溶液中，溶质分子的实际蒸气压主要取决于溶质与溶剂分子之间的相互作用力，故可用亨利定律来处理。所以式(9)可表示为

$$V_g^0/\mathrm{m^3 \cdot kg^{-1}} = \frac{273R}{\gamma_2^\infty \cdot p_2^* \cdot M_1} \tag{10}$$

$$\gamma_2^\infty = \frac{273R}{V_2^\infty \cdot p_2^* \cdot M_1} \tag{11}$$

上式将色谱的特有概念——比保留体积 V_g^0 与溶液热力学的重要参数——无限稀释的活度系数 γ_2^∞ 相关联。

根据克劳修斯-克拉佩龙(Clausius-Clapeyron)方程并结合亨利定律，可得

$$\mathrm{d}(\ln p_2^*/\mathrm{kPa}) = \frac{\Delta_{\mathrm{vap}} H_{\mathrm{m}}}{RT^2}\mathrm{d}T \tag{12}$$

$$\mathrm{d}[\ln(p_2^* x_2 \gamma_2^\infty/\mathrm{kPa})] = \frac{\Delta_{\mathrm{vap}} H_{2,\mathrm{m}}}{RT^2}\mathrm{d}T \tag{13}$$

$\Delta_{vap}H_{2,m}$ 表示溶质从一溶液中气化的偏摩尔气化焓。对于理想溶液，$\gamma_2^{\infty}=1$，溶质的分压可用 $p_2^*\cdot x_2$ 表示，而其偏摩尔气化焓与纯溶质的偏摩尔气化焓相等，偏摩尔溶解焓等于气化焓，即 $\Delta_{vap}H_{2,m}=\Delta_{vap}H_m=-\Delta_{sol}H_{2,m}=-\Delta_{sol}H_m$。非理想溶液的偏摩尔溶解焓 $\Delta_{sol}H_{2,m}$ 虽然也等于 $-\Delta_{vap}H_{2,m}$，但它们与活度系数有关。

将式(10)取对数并对 T 微分，再以式(13)代入可得

$$\frac{d(\ln V_g^0/m^3\cdot kg^{-1})}{dT}=-\frac{d\ln(p_2^*\cdot\gamma_2^{\infty}/kPa)}{dT}=-\frac{\Delta_{vap}H_{2,m}}{RT^2} \tag{14}$$

设在一定温度范围内，$\Delta_{vap}H_{2,m}$ 可视为常数，积分可得

$$d(\ln V_g^0/m^3\cdot kg^{-1})=\frac{\Delta_{vap}H_{2,m}}{RT}+C \tag{15}$$

将式(13)与式(12)两式相减(无限稀释溶液 $x_2\rightarrow 0$)，并代之以溶解焓，则得

$$d(\ln\gamma_2^{\infty})=\frac{(\Delta_{vap}H_{2,m}-\Delta_{vap}H_m)}{RT^2}dT=-\frac{(\Delta_{sol}H_{2,m}-\Delta_{sol}H_m)}{RT^2}dT \tag{16}$$

与式(15)一样，积分可得

$$\ln\gamma_2^{\infty}=-\frac{(\Delta_{vap}H_{2,m}-\Delta_{vap}H_m)}{RT}+D=\frac{(\Delta_{sol}H_{2,m}-\Delta_{sol}H_m)}{RT}+D=\frac{\Delta_{sol}H^E}{RT}+D \tag{17}$$

式(15)和式(17)中的 C、D 均为积分常数。$\Delta_{sol}H^E$ 为非理想溶液与理想溶液中溶质的溶解焓之差，称偏摩尔超额溶解焓

$$\Delta_{sol}H^E=\Delta_{sol}H_{2,m}-\Delta_{sol}H_m=-(\Delta_{vap}H_{2,m}-\Delta_{vap}H_m) \tag{18}$$

$\gamma_2^{\infty}>1$ 时，溶液对拉乌尔定律产生正偏差，溶质与溶剂分子之间的作用力小于溶质之间的作用力，$\Delta_{sol}H^E>0$；反之则相反。

仪器和试剂

气相色谱仪
色谱工作站(包括计算机)
红外加热灯
邻苯二甲酸二壬酯
丙酮(分析纯)
氮气钢瓶
微量进样器
秒表
环己烷(分析纯)

实验步骤

1. 实验前准备

准确称取一定量的邻苯二甲酸二壬酯固定液于蒸发皿中，加适量丙酮稀释固定液，按固定液与载体比为 1∶5 来称取红色载体，倒入蒸发皿浸泡，用吹风机吹热风使丙酮蒸发干。

将涂好固定液的载体小心装入已洗净干燥的色谱柱中，柱的一端塞以少量玻璃棉，接

上真空泵,用小漏斗由柱的另一端加入载体,同时不断振动柱管,填满后同样塞以少量玻璃棉,准确计算装入色谱柱内固定液的质量。

按操作规程检查色谱仪,确保色谱仪气路及电路正常。打开氮气钢瓶阀门,利用减压阀和色谱仪的针形阀调节气流流量至 50 $mL \cdot min^{-1}$左右。将载气出口处堵死,柱前转子流量计的标示应下降至零,这表示气密性良好。如流量计显示有气流,则表示系统漏气。通常可用肥皂水顺次检查各接头处,必要时应再旋紧接头,直至整个气路不漏气。

保持氮气流量,将柱温(或称层析室温度)调到 130 ℃,恒温约 4 h,使固定相老化。注意,切勿超过 150 ℃。

2. 测试

(1) 测定保留时间

将柱温调到大约 70 ℃,柱前压力约为 2×10^5 Pa(表压为 1.1 $kgf \cdot cm^{-2}$)。打开热导电源并调节桥路电流至 120 mA,气化室温度约为 130 ℃。待色谱工作站记录的基线稳定后便可进样。

用 10 μL 微量注射器吸取 1 μL 环己烷进样,进样前按下色谱工作站的启动键,进样的同时立即按下秒表,计算机将自动记录曲线。注意观察,样品峰达到最高点停表,此时间即为保留时间。

每一柱温下进样 2 次,取保留时间的平均值。记下柱前压力、大气压力、层析室温度(柱温)、皂膜流量计温度,并用皂膜流量计测定载气流量。

(2) 保留时间与柱温的关系

升高层析室温度,重复上段(1)的操作,测定不同柱温下的保留时间及其他数据。每次升温幅度可控制在 2 ℃,从 70 ℃测到 78 ℃共测定 5 组数据。每个温度测定 2 次,每次的时间误差不超过 0.5 s。

3. 实验完毕后,逐一关闭各个部分开关,再关闭电源,待层析室温度接近室温后再关闭气源。

数据记录与处理

1. 设计合理的表格并将原始数据和计算结果列入表中。

(1) 环己烷的饱和蒸气压的计算

$$p_2^*/\text{mmHg} = \exp[15.957 - 2879.9/(228.20 + t/℃)]$$

$$p_2^*/\text{Pa} = 133.3 \times \exp[15.957 - 2879.9/(228.20 + t/℃)]$$

(2) 水的饱和蒸气压的计算

$$p_w/\text{mmHg} = 4.5829 + 0.33173t/℃ + 1.1113 \times 10^{-2}(t/℃)^2 + 1.6196 \times 10^{-4}(t/℃)^3 + 3.5957 \times 10^{-6}(t/℃)^4$$

$$p_w/\text{Pa} = 6.1100 \times 10^2 + 4.4227 \times 10\ t/℃ + 1.4816(t/℃)^2 + 2.1593 \times 10^{-2}(t/℃)^3 + 4.7939 \times 10^{-4}(t/℃)^4$$

适用温度范围:0～40 ℃。

(3) 比保留体积和活度系数的计算

根据式(1)可计算出不同柱温时的 V_g^0,由式(11)计算出 γ_2^∞。

2. 求偏摩尔溶解焓和偏摩尔超额溶解焓

以环己烷在不同柱温下测得的 $\ln[V_g^0/(\text{m}^3 \cdot \text{kg}^{-1})]$ 和 $\ln\gamma_2^\infty$ 分别对 $1/T$ 作图,得两条直线。由其斜率可按式(15)和式(17)分别求出环己烷的 $\Delta_{vap}H_{2,m}$ 和 $\Delta_{sol}H^E$。

3. 求摩尔气化焓

由式(18)计算纯态环己烷的摩尔气化焓 $\Delta_{vap}H_m$。

思考与讨论

1. 实验中若采用氢气作为载气,应注意哪些问题?

2. 什么样的溶液体系才适于用气相色谱法测定其热力学函数?

3. 试从热力学函数对温度的依赖关系与实验测量误差两个角度讨论测定温度范围的合理选择。

参考资料

1. 李民,刘衍光,傅伟康,等.气相色谱法测定非电解质溶液热力学函数值的实验条件选择.化学通报,1988,(4):54.

2. 曹才放,李小文,聂华平,等.离子色谱法研究钨冶金体系阴离子动态交换的热力学.有色金属科学与工程,2019,10(04):1-6.

3. 邓丽霜,王强,张正方,等.反气相色谱法测定离子液体 1-己基-3-甲基咪唑三氟甲磺酸盐的热力学参数.色谱,2014,32(02):169-173.

Experiment 19

Determination of thermodynamic function of non-electrolyte solution by gas chromatography

Experimental purpose

1. Understand some applications of gas chromatography in chemical thermodynamics.

2. Determinate of the infinite dilution activity coefficient of cyclohexane in dinonyl phthalate

solution by gas chromatography, and obtain partial molar solution enthalpy, excess partial molar enthalpy, and molar enthalpy of vaporization.

Basical principles

Gas chromatography is a chromatographic method using gas as mobile phase, which is the most widely used technology for separation and analysis. Gas chromatography has two phases, namely mobile phase and stationary phase. The stationary phase can be a solid adsorbent or a liquid coated on an inert porous support. The mobile phase (also known as carrier gas) is some gases that do not react chemically with the stationary phase and the sample under test. Obviously, the distribution of the sample (or the third component) between the two phases is closely related to the thermodynamic and kinetic properties of the interaction between the sample and the stationary phase.

The sample is driven by the mobile phase through the void of the fixed phase with large specific surface area. The continuous thermodynamic distribution can be carried out and repeated between the gas phase and the fixed phase. Due to the tiny difference in the properties of each component for the sample, the separation purpose can be achieved, and the thermodynamic function of the non-electrolyte solution with infinite dilution can also be measured. After the sample enters the chromatograph through the inlet, part of it is left in the gas phase, and part of it is dissolved in the stationary liquid of the chromatographic column and formed a solution with the stationary liquid. As the carrier gas flows, after a period of time the solute sample will be taken out of the column. Fig. 19-1 is an ideal chromatogram. The peak at t_d indicates that there is "dead space" in the gas path of the chromatograph. The true sample peak appears at t_r. Nitrogen is selected as the mobile phase (carrier gas) in this experiment.

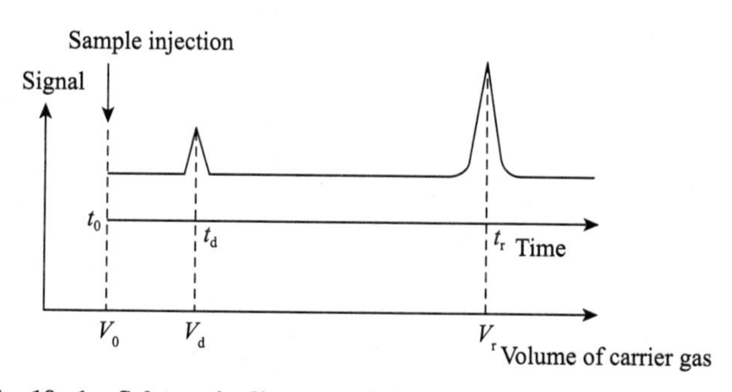

Fig. 19-1　Schematic diagram of pulse injection chromatography

The time t_r from solute injection to the concentration maxima of the detector is called the retention time. The carrier gas flow measured by the soap film flowmeter F Times t_r is the retention volume V_r. The product of F and t_d, V_d, is called the dead volume, which is independent of the dissolution process and is only related to the size of the three parts of the chromatograph: the injector, the column and the detector, so the difference between t_r and t_d characterizes solute dissolution or solution properties. Meanwhile, it is more reasonable to discuss the retention volume with the flow rate F_c of the carrier gas in the column. However, it can be seen from the properties of the column that the pressure in the column is the function of the column length, so the actual flow rate of each part in the column is not constant, which can be corrected by the pressure correction factor j. If the influence of other factors is taken into account, the sample retention volume V_g^0 on the fixed solution per unit mass should be expressed to actually reflect the characteristics of the interaction between the solute and the fixed solution as a solvent

$$V_g^0/\mathrm{dm^3 \cdot kg^{-1}} = (t_r - t_d) \cdot j \cdot \frac{(p_0 - p_w)}{p_0} \cdot \frac{273}{T_r} \cdot \frac{F}{m_1} \tag{1}$$

In the formula

$$j = \frac{3}{2} \cdot \frac{[(p_i/p_0)^2 - 1]}{[(p_i/p_0)^3 - 1]} \tag{2}$$

T_r is the temperature of soap film flowmeter (K, T_r=room temperature+273.15); p_i, p_0, p_w are the column pressure before the column, outlet pressure (atmospheric pressure) and the saturated vapor pressure of water at T_r; m_1 is the mass of fixed solution. If the concentration of solute in the gas phase and the liquid phase are defined by different concepts, then the distribution coefficient of solute in the two phases at 273 K can be expressed as follows

$$K_D = \frac{\text{Solute mass of fixed solution/fixed solution mass}}{\text{Solute mass of mobile phase/volume of mobile phase}} = \frac{m_2^s/m_1}{m_2^g/V_d}$$

In the formula, subscripts 1 and 2 represent fixed solution and solute, respectively. The superscripts s and g represent the fixed liquid and gas phases, respectively.

Under ideal conditions, the chromatographic peak shape should be symmetrical. At t_r, then, exactly half of the solute is carried away from the detector by the carrier gas, while the other half is remained in the column. The two parts have equal mass, and the solute in the column is in the gas phase and the liquid phase, respectively. Therefore

$$V_r^c \cdot \frac{m_2^g}{V_d} = V_d^c \cdot \frac{m_2^g}{V_d} + V_s \cdot \frac{m_2^s}{m_1} \cdot \rho_1 \tag{3}$$

Here, ρ_1 is the density of the fixed solution, and V_r^c and V_d^c represent the retained and dead volumes at column temperature and pressure, respectively. Transfer and make pressure and temperature correction, Formula (4) is obtained

$$(t_r - t_d) \cdot j \cdot \frac{(p_0 - p_w)}{p_0} \cdot \frac{273}{T_r} \cdot F \cdot \frac{m_2^g}{V_d} = V_s \cdot \rho_1 \cdot \frac{m_2^s}{m_1} \tag{4}$$

Since $V_s \cdot \rho_1 = m_1$, and then compare (4) with Equations (1) and (2) respectively, Equation (5) is obtained

$$V_g^0 = \frac{m_2^s/m_1}{m_2^g/V_d} = K_D \tag{5}$$

Since the pulse injection volume is very small, the behavior of the sample in the gas-liquid phase can be approximated by the ideal gas equation and Raoult's law respectively

$$p_2 V_d = nRT_c$$

$$p_2 = \frac{m_2^g RT}{V_d M_2} \tag{6}$$

$$p_2^* = \frac{p_2}{x_2} = p_2\left(\frac{n_1 + n_2}{n_2}\right) \approx p_2 \cdot \frac{n_1}{n_2} = p_2 \cdot \frac{M_2}{M_1} \cdot \frac{m_1}{m_2^s} \tag{7}$$

In the formula, p_2^* and p_2 are the vapor pressure of pure solute and solute in solution, respectively. x_2 is the mole fraction of the solute in solution; M_1 and M_2 are molar mass of fixed solution and solute, respectively. n_1 and n_2 are the amount of substance in fixed solution and solute, respectively. By correcting the vapor pressure from column pressure T_c to 273 K, and substituting Equations (5) and (6) into Equation (7), Equation (8) is obtained

$$p_2^* = p_2 \cdot \frac{273}{T_c} \cdot \frac{M_2}{M_1} \cdot \frac{V_d}{K_D \cdot m_2^g} = \frac{273R}{K_D \cdot M_1} \tag{8}$$

When combined with Equation (5), the following formula is obtained

$$V_g^0 = \frac{273R}{p_2^* \cdot M_1} \tag{9}$$

In fact, the boiling point and vapor pressure of stationary liquid of chromatographic column are high, the vapor pressure is very low, while the molar mass and molar volume are large. However, the physical properties of sample used as solute are far from it, so the solution properties tend to deviate from their Raoult's law. However, in this dilute solution, the actual vapor pressure of solute molecules mainly depends on the interaction between solute and solvent molecules, so Henry's law can be used to deal with it. Therefore, Equation (9) can be expressed as

$$V_g^0/\text{m}^3 \cdot \text{kg}^{-1} = \frac{273R}{\gamma_2^\infty \cdot p_2^* \cdot M_1} \tag{10}$$

$$\gamma_2^\infty = \frac{273R}{V_2^\infty \cdot p_2^* \cdot M_1} \tag{11}$$

The equations above relate the specific retention volume V_g^0, a unique concept of chromatography, with the activity coefficient γ_2^∞ of infinite dilution, an important parameter of solution thermodynamics.

According to Clausius-Clapeyron equation and Henry's law, Equation (12) can be obtained

$$d(\ln p_2^* / \mathrm{kPa}) = \frac{\Delta_{\mathrm{vap}} H_{\mathrm{m}}}{RT^2} dT \tag{12}$$

$$d[\ln(p_2^* x_2 \gamma_2^\infty / \mathrm{kPa})] = \frac{\Delta_{\mathrm{vap}} H_{2,\mathrm{m}}}{RT^2} dT \tag{13}$$

$\Delta_{\mathrm{vap}} H_{2,\mathrm{m}}$ represents the partial molar enthalpy of gasification of a solute from a solution. For an ideal solution, $\gamma_2^\infty = 1$, the partial pressure of the solute can be represented by $p_2^* \cdot x_2$, and the partial molar enthalpy of gasification is equal to the partial molar enthalpy of gasification of the pure solute, and the partial molar enthalpy of dissolution is equal to the enthalpy of gasification, namely $\Delta_{\mathrm{vap}} H_{2,\mathrm{m}} = \Delta_{\mathrm{vap}} H_{\mathrm{m}} = -\Delta_{\mathrm{sol}} H_{2,\mathrm{m}} = -\Delta_{\mathrm{sol}} H_{\mathrm{m}}$. The partial molar dissolution enthalpies of nonideal solutions are also equal to $-\Delta_{\mathrm{vap}} H_{2,\mathrm{m}}$, but they depend on the activity coefficient.

Take the logarithm of Equation (10) and differentiate it by T, then substitute Equation (13) into it, we can get

$$\frac{d(\ln V_g^0 / \mathrm{m}^3 \cdot \mathrm{kg}^{-1})}{dT} = -\frac{d\ln(p_2^* \cdot \gamma_2^\infty / \mathrm{kPa})}{dT} = -\frac{\Delta_{\mathrm{vap}} H_{2,\mathrm{m}}}{RT^2} \tag{14}$$

In a certain temperature range, $\Delta_{\mathrm{vap}} H_{2,\mathrm{m}}$ can be regarded as a constant, and the integral can be obtained

$$d(\ln V_g^0 / \mathrm{m}^3 \cdot \mathrm{kg}^{-1}) = \frac{\Delta_{\mathrm{vap}} H_{2,\mathrm{m}}}{RT} + C \tag{15}$$

By subtracting equations (infinite dilution solution $x_2 \to 0$) from Equations (13) and (12) and replacing them with enthalpy of dissolution, Equation (16) is obtained

$$d(\ln \gamma_2^\infty) = \frac{(\Delta_{\mathrm{vap}} H_{2,\mathrm{m}} - \Delta_{\mathrm{vap}} H_{\mathrm{m}})}{RT^2} dT = -\frac{(\Delta_{\mathrm{sol}} H_{2,\mathrm{m}} - \Delta_{\mathrm{sol}} H_{\mathrm{m}})}{RT^2} dT \tag{16}$$

As Equation (15), integration can be obtained

$$\ln \gamma_2^\infty = -\frac{(\Delta_{\mathrm{vap}} H_{2,\mathrm{m}} - \Delta_{\mathrm{vap}} H_{\mathrm{m}})}{RT} + D = \frac{(\Delta_{\mathrm{sol}} H_{2,\mathrm{m}} - \Delta_{\mathrm{sol}} H_{\mathrm{m}})}{RT} + D = \frac{\Delta_{\mathrm{sol}} H^{\mathrm{E}}}{RT} + D \tag{17}$$

C and D in Equations (15) and (17) are integral constants. $\Delta_{\mathrm{sol}} H^{\mathrm{E}}$ is the difference of enthalpy of dissolution between the non-ideal solution and the solute in the ideal solution, which is called partial molar excess enthalpy of dissolution

$$\Delta_{sol}H^{E} = \Delta_{sol}H_{2,m} - \Delta_{sol}H_{m} = -(\Delta_{vap}H_{2,m} - \Delta_{vap}H_{m}) \quad (18)$$

When $\gamma_2^{\infty} > 1$, the solution has a positive deviation to Raoult's law, and the force between solute and solvent molecules is less than that between solutes, $\Delta_{sol}H^{E} > 0$; otherwise the opposite.

Apparatus and reagents

Gas chromatograph
Chromatographic workstation (including computer)
Infrared heating lamp
Dinonyl phthalate
Acetone (AR)
Nitrogen cylinder
Microsyringe
Stopwatch
Cyclohexane (AR)

Experimental Procedures

1. Preparation before the experiment

A certain amount of dinonyl phthalate is used as fixed solution and accurately weighed in the evaporating pan. Then appropriate amount of acetone is added to the pan to dilute fixed solution. According to the ratio of fixed solution and carrier 1 : 5 to weigh the red carrier, pour into the pan immersion, and use a hair dryer to blow hot air to dry acetone.

The carrier coated with the stationary liquid is carefully loaded into a cleaned and dry chromatographic column. One end of the column is plugged with a small amount of glass wool, connected to a vacuum pump, and the carrier is added from the other end of the column with a small funnel, while constantly vibrating the column tube. After filling, a small amount of glass wool is used to plug it. Accurately calculate the mass of fixed solution remaining in the column.

Check the gas chromatograph according to the operation procedure, make sure the gas circuit and electric circuit of the chromatograph are normal. Open the nitrogen cylinder valve and use the pressure reducing valve and the needle valve of the chromatograph to regulate the gas flow to about 50 $mL \cdot min^{-1}$. Plug the outlet of the carrier gas, the label of the rotor flowmeter in front of the column should be reduced to zero, which indicates good gas tightness. If the flowmeter shows air flow, it means that the system is leaking gas. Usually soapy water can be used to check each joint in sequence, if necessary, the joint should be re-tightened until there is no gas leakage throughout the gas path.

The nitrogen flow rate is maintained, and the column temperature (or chromatography chamber temperature) is adjusted to 130 ℃ for about 4 h to age the stationary phase. Be careful not to exceed 150 ℃.

2. Test

(1) Determination of the retention time

The temperature of the column is adjusted to about 70 ℃, and the pressure before the column is about 2×10^5 Pa (gauge pressure was 1.1 kgf · cm^{-2}). Turn on the heat conduction power and adjust the bridge current to 120 mA, and the temperature of the gasification chamber is about 130 ℃. After the baseline recorded by the chromatographic workstation is stable, the sample can be injected.

Use a 10 μL microsyringe to draw 1 μL cyclohexane for sample injection. Press the start button of the chromatographic workstation before sample injection and press the stopwatch immediately while sample injection is taking place. The computer will automatically record the curve. Note that when the sample peak reaches the highest point and stops the watch, this time is the retention time.

Sample is injected twice at each column temperature, and the average retention time is taken. The column pressure, atmospheric pressure, chromatography chamber temperature (column temperature) and temperature of soap film flowmeter are recorded, and the flow rate of carrier gas is measured with soap film flowmeter.

(2) The relationship between retention time and column temperature

Raise the temperature of the chromatography chamber. Repeat step 2.(1) to measure the retention time and other data at different column temperatures. The temperature increasing can be controlled at 2 ℃ every time, a total of 5 experiments from 70 ℃ to 78 ℃ need to be measured. Two measurements need to be conducted every temperature, and the time error of each measurements is less than 0.5 s.

3. After the experiment, turn off the switch of each part one by one, then turn off the power supply, wait and turn off the gas source until the temperature of the chromatography chamber approaches the room temperature.

Data recording and processing

1. Design reasonable tables which include the original data and calculation results.

(1) Calculation of saturated vapor pressure of a cyclohexane

$$p_2^*/\text{mmHg} = \exp[15.957 - 2879.9/(228.20 + t/℃)]$$

$$p_2^*/\text{Pa} = 133.3 \times \exp[15.957 - 2879.9/(228.20 + t/℃)]$$

(2) Calculation of saturated vapor pressure of water

$$p_w/\text{mmHg} = 4.5829 + 0.33173t/℃ + 1.1113 \times 10^{-2}(t/℃)^2 + 1.6196 \times 10^{-4}(t/℃)^3 + 3.5957 \times 10^{-6}t/℃)^4$$

$$p_w/\text{Pa} = 6.1100 \times 10^2 + 4.4227 \times 10\ t/℃ + 1.4816(t/℃)^2 + 2.1593 \times 10^{-2}(t/℃)^3 + 4.7939 \times 10^{-4}(t/℃)^4$$

Applicable temperature range: 0～40 ℃.

(3) Calculation of specific retention volume and activity coefficient

According to Equation (1), V_g^0 at different column temperatures can be calculated, and γ_2^∞ can be calculated from Equation (11).

2. Calculate the partial molar enthalpy of dissolution and excess molar enthalpy of dissolution

Plot $1/T$ with the $\ln(V_g^0/\text{m}^3 \cdot \text{kg})$ and $\ln\gamma_2^\infty$ of cyclohexane at different column temperatures, respectively, to obtain two lines. The $\Delta_{vap}H_{2,m}$ and $\Delta_{sol}H^E$ of cyclohexane can be calculated from the slope according to Equations (15) and (17), respectively.

3. Calculate the molar enthalpy of gasification

The molar enthalpy of gasification of pure cyclohexane was calculated by Equation (18).

Thinking and discussing

1. What should be paid attention to if hydrogen is used as carrier gas in the experiment?

2. What kind of solution system is suitable for the determination of thermodynamic function by gas chromatography?

3. Try to discuss the reasonable choice of temperature range from the two perspectives of the dependence of thermodynamic function on temperature and experimental measurement error.

Reference materials

1. Li M, Liu Y, Fu W, et al. Selection of experimental conditions for gas chromatography determination of thermodynamic functions of non-electrolyte solutions. Chemical Bulletin, 1988, (4): 54.

2. Cao C, Li X, Nie H. Study on thermodynamics of anion dynamic exchange in tungsten metallurgical system by ion chromatography. Nonferrous Metals Science and Engineering, 2019, 10(04): 1-6.

3. Deng L, Wang Q, Zhang Z, et al. Determination of thermodynamic parameters of ionic liquid 1-hexyl-3-methylimidazolium trifluoromethyl sulfonate by gas chromatography. Chromatography, 2014, 32(02): 169-173.

实验 20　生物酶催化反应动力学常数的测定

实验目的

1. 了解生物酶的结构与性能之间的关系。

2. 用分光光度法测定辣根过氧化物酶的米氏常数和最大反应速率。

3. 通过上述实验理解酶催化动力学的研究方法。

基本原理

酶是存在于生物体内的一些具有专一性和高催化活性的蛋白质，它贯穿于生命活动的全过程，生命系统中的各种生物化学反应绝大多数都是在酶催化下完成的。酶是生物大分子，酶反应体系和一般化学反应体系相比要复杂得多。影响酶催化反应速率的因素有浓度因素（酶浓度、底物浓度），底物数目与结构，产物，抑制剂，激活剂，酶的结构与性质及环境条件（温度、pH、离子强度、压力）等。浓度因素是最基本的影响因素，从浓度因素的有关实验求出速率常数，进一步考虑各种影响因素的相互关系，是酶反应动力学的主要内容。

Michaelis（米哈利）和 Menten（门顿）先后研究了酶催化反应动力学，提出了酶催化反应的历程，即 Michaelis-Menten 机理，指出酶（E）与底物（S）先生成中间化合物（ES），然后中间化合物再进一步分解为产物，并释放出酶（E）。

$$S+E \underset{k_{-1}}{\overset{k_1}{\rightleftharpoons}} ES \xrightarrow{k_2} E+P$$

ES 分解为产物（P）为控速步骤，采用稳态法处理

$$\frac{d[ES]}{dt} = k_1[S][E] - k_{-1}[ES] - k_2[ES] = 0$$

所以

$$[ES] = \frac{k_1[S][E]}{k_{-1} + k_2} = \frac{[E][S]}{K_M} \tag{1}$$

式中 $K_M = \dfrac{k_{-1} + k_2}{k_1}$ 称为米氏常数（Michaelis constant），这个公式也叫作米氏公式。所以，反应速率为

$$v = \frac{d[P]}{dt} = k_2[ES] \tag{2}$$

将[ES]的表示式代入后，得

$$v = k_2[ES] = \frac{k_2[E][S]}{K_M}$$

变形得

$$K_M = \frac{[E][S]}{[ES]}$$

所以米氏常数实际上相当于反应 E + S = ES 的不稳定常数。

若令酶的原始浓度为 $[E_0]$，反应达到稳态后，一部分变为中间化合物 [ES]，另一部分仍处于游离状态，所以 $[E_0] = [E] + [ES]$ 或 $[E] = [E_0] - [ES]$

代入式(1)后得

$$[ES] = \frac{[E_0][S]}{K_M + [S]} \tag{3}$$

代入式(2)得到速率公式

$$v = \frac{d[P]}{dt} = \frac{k_2[E_0][S]}{K_M + [S]} \tag{4}$$

当底物浓度远远大于酶浓度，即 $K_M \ll [S]$，则 $K_M + [S] \approx [S]$，速率趋于最大值 v_{max}，则 $v_{max} = k_2[E_0]$。由此可以得到

$$v = \frac{v_{max}[S]}{K_M + [S]} \tag{5}$$

对(5)进行进一步变形，两边同时取倒数得

$$\frac{1}{v} = \frac{K_M}{[S]v_{max}} + \frac{1}{v_{max}} \tag{6}$$

这是一个直线方程，以 $\frac{1}{v}$ 对 $\frac{1}{[S]}$ 作图，截距为 $\frac{1}{v_{max}}$，斜率为 $\frac{K_M}{v_{max}}$，因而通过作图可得到 v_{max} 和 K_M。

过氧化物酶是一种对氢受体(例如 H_2O_2)底物有特异性，对氢供体底物缺乏特异性的酶，它可催化过氧化氢氧化多种多元酚或多元胺类底物发生显色、荧光或化学发光反应，可用于微量过氧化氢含量测定，也可以和其他酶反应系统偶联用于测定许多与生命相关的物质，如葡萄糖、半乳糖、氨基酸、尿酸及胆甾醇等，亦是免疫分子和核酸分析中常用的标记物。通常使用的过氧化物酶是从植物辣根中提取的，因此也称为辣根过氧化物酶(horseradish peroxidase, HRP)。HRP 属于含血红素蛋白的一种，是由辅基——氯化血红素(铁卟啉)和脱辅基蛋白(糖蛋白)组成，相对分子质量为 44 000。

邻苯二胺能够被 H_2O_2 氧化，辣根过氧化物酶(HRP)对此反应有强的催化作用。在所选择的酶催化反应条件下，实验表明，酶催化反应只生成一种有色物质，即 2,3-二氨基吩嗪。

本实验中反应方程式为

$$2\,C_6H_4(NH_2)_2 + 3H_2O_2 \xrightarrow{HRP} C_{12}H_6N_2(NH_2)_2 + 6H_2O$$

由朗伯-比尔定律

$$A = ad[\mathrm{P}] \tag{7}$$

A 为产物的吸光度，a 为摩尔消光系数，d 为样品池厚度。d = 1 cm 时

$$A = a[\mathrm{P}] \tag{8}$$

则有

$$\frac{\mathrm{d}[\mathrm{P}]}{\mathrm{d}t} = \frac{1}{a}\left(\frac{\mathrm{d}A}{\mathrm{d}t}\right) \tag{9}$$

即

$$v = \frac{\mathrm{d}[\mathrm{P}]}{\mathrm{d}t} = \frac{1}{a}\left(\frac{\mathrm{d}A}{\mathrm{d}t}\right) \tag{10}$$

综合式(6)及(10)，利用分光光度法可测辣根过氧化物酶催化过氧化氢氧化邻苯二胺反应的米氏常数和最大反应速率。

仪器和试剂

721 型分光光度计　　pHS-2C 型酸度计

辣根过氧化物酶　　邻苯二胺（分析纯）

30%过氧化氢　　醋酸-醋酸钠缓冲溶液（pH = 5.00）

实验步骤

1. 分光光度计的光吸收波长选择为 446.3 nm，配制溶液的溶剂为醋酸盐缓冲溶液，pH 选择在 5.00，反应温度为室温（最好不要低于 10 ℃）。

2. 准确称取 2,3-二氨基吩嗪 0.0017 g，用缓冲溶液定容至 50 mL（浓度约为 16×10^{-5} mol · L^{-1}），并逐渐稀释至 8×10^{-5}，4×10^{-5}，2×10^{-5}和 1×10^{-5} mol · L^{-1}。在 446.3 nm 下测定上述溶液的吸光度 A。

3. 准确称取邻苯二胺（底物）0.0011 g，用缓冲溶液定容至 50 mL，得到约 0.2 mmol · L^{-1} 邻苯二胺溶液，随后逐渐稀释得到浓度约为 0.1，0.05，0.025，0.0125 mmol · L^{-1}的邻苯二胺溶液。

4. 取 5 mL 0.0125 mmol · L^{-1}的邻苯二胺溶液于称量瓶中，加入 10 μg · mL^{-1} HRP 溶液5 μL，30% H_2O_2 5 μL，迅速摇匀后加入比色皿中，从加入 HRP 开始计时，记录 446.3 nm 处 A 随 t 的变化，读数时间间隔可为 2 min，20 min 后停止测试，得 A-t 曲线。

5. 只改变底物浓度为 0.025 mmol · L^{-1}，其余物质浓度及用量不变，重复前面的操作，

以此类推，继续增大底物浓度，重复操作，得不同底物浓度下的 A-t 曲线。

数据记录与处理

1. 根据朗伯-比尔定律计算摩尔消光系数 a 。

2. 将所测得的不同浓度邻苯二胺溶液的 A 对 t 作图，其直线斜率为 $\left(\frac{\mathrm{d}A}{\mathrm{d}t}\right)$ ，由 $\frac{1}{a}\left(\frac{\mathrm{d}A}{\mathrm{d}t}\right)$ 可计算出不同浓度邻苯二胺溶液时的反应速率。

3. 以 $\frac{1}{v}$ 对 $\frac{1}{[\mathrm{S}]}$ 作图，通过直线截距和斜率，得到 $v_{\max}$ 和 K_{M} 。

思考与讨论

1. 酶催化反应与一般催化反应相比有何特点？
2. 研究酶催化动力学对绿色化学发展有什么重要意义？
3. 酶催化反应为什么需要在最适温度和最适 pH 下进行？

参考资料

1. 袁勤生.现代酶学.上海：华东理工大学出版社，2001：9.

2. 冯春梁，李红丹，张振彦，等.催化动力学光度法测定碱性磷酸酶的理论分析与条件优化.辽宁师范大学学报：自然科学版，2003，32(2)：195-198.

3. 傅献彩，沈文霞，姚天扬.物理化学（第三版）.北京：高等教育出版社，2004.

Experiment 20

Determination of kinetic constants of reaction catalyzed by bioenzyme

Experimental purpose

1. Understand the relation between the structure and properties of bioenzymes.

2. Determinate the Michaelis constant and maximum reaction rate of the Horseradish peroxidase by spectrophotometer.

3. Comprehend the research methods of enzyme kinetics through the above experiment.

Basical principles

Enzymes are proteins with specificity and high catalytic activity that exist in living organisms, which exist throughout the life activities. Most biochemical reactions in living systems are catalyzed by enzymes.

Enzymes are biological macromolecules and enzymatic reaction systems are much more complex compared to general chemical reaction systems. Factors affecting the rate of enzyme-catalyzed reactions include concentration factors (enzyme concentration, substrate concentration), number and structure of substrates, products, inhibitors, activators, structure and properties of the enzyme and environmental conditions (temperature, pH, ionic strength, pressure) etc. Concentration are the most basic influencing factors. From the relevant experiments on concentration factors to obtain the rate constants, and further considering the relationship among various influencing factors, are the main content of the enzyme kinetics.

Michaelis and Menten studied the kinetics of enzyme-catalyzed reactions and proposed the course of enzyme-catalyzed reactions, namely the Michaelis-Menten mechanism, which states that the enzyme (E) and the substrate (S) first form an intermediate compound (ES), then, ES further decomposed into products and released the enzyme (E).

$$\mathrm{S+E} \underset{k_{-1}}{\overset{k_1}{\rightleftharpoons}} \mathrm{ES} \xrightarrow{k_2} \mathrm{E+P}$$

The decomposition of ES to product (P) is a rate-controlled step, which is treated by the steady-state method

$$\frac{\mathrm{d}[\mathrm{ES}]}{\mathrm{d}t} = k_1[\mathrm{S}][\mathrm{E}] - k_{-1}[\mathrm{ES}] - k_2[\mathrm{ES}] = 0$$

So,

$$[\mathrm{ES}] = \frac{k_1[\mathrm{S}][\mathrm{E}]}{k_{-1} + k_2} = \frac{[\mathrm{E}][\mathrm{S}]}{K_{\mathrm{M}}} \tag{1}$$

In the formula, $K_{\mathrm{M}} = \dfrac{k_{-1} + k_2}{k_1}$ is called the Michaelis constant. This formula is also called the Michaelis formula. Therefore, the reaction rate is

$$v = \frac{\mathrm{d}[\mathrm{P}]}{\mathrm{d}t} = k_2[\mathrm{ES}] \tag{2}$$

After the [ES] expression is substituted in

$$v = k_2[\mathrm{ES}] = \frac{k_2[\mathrm{E}][\mathrm{S}]}{K_{\mathrm{M}}}$$

Changed the form of formula

$$K_{\mathrm{M}} = \frac{[\mathrm{E}][\mathrm{S}]}{[\mathrm{ES}]}$$

So the Michaelis constant is actually equivalent to the instability constant for the reaction E + S = ES .

If the original concentration of the enzyme is $[E_0]$, after the reaction has reached steady state, part of it becomes the intermediate compound [ES], and the other parts remain in the free state. So, $[E_0]=[E]+[ES]$, that is to say $[E]=[E_0]-[ES]$.

After substituting in (1), the following formula is obtained

$$[ES]=\frac{[E_0][S]}{K_M+[S]} \tag{3}$$

When Equation (3) is substituted in to Equation (2), the following equation is obtained

$$v=\frac{d[P]}{dt}=\frac{k_2[E_0][S]}{K_M+[S]} \tag{4}$$

When the substrate concentration is much greater than the enzyme concentration, that is to say $K_M \ll [S]$, then $K_M+[S]\approx[S]$, v tends to a maximum v_{max}, then $v_{max}=k_2[E_0]$. From this we can obtain

$$v=\frac{v_{max}[S]}{K_M+[S]} \tag{5}$$

Further deformation of Equation (5), taking the inverse of both sides simultaneously, Equation (6) is obtained

$$\frac{1}{v}=\frac{K_M}{[S]v_{max}}+\frac{1}{v_{max}} \tag{6}$$

This is a linear equation, plotted as $\frac{1}{v}$ against $\frac{1}{[S]}$, with intercept $\frac{1}{v_{max}}$, and slope $\frac{K_M}{v_{max}}$, thus giving v_{max} and K_M.

Peroxidase is an enzyme with specificity for hydrogen acceptor substrates (for example H_2O_2) and lack of specificity for hydrogen donor substrates. It can catalyse hydrogen peroxide to oxidise many polyphenols or polyamine substrate for chromogenic, fluorescence or chemiluminescence reactions, and can be used for the determination of trace amounts of hydrogen peroxide. It can also be coupled with other enzymatic systems for the determination of many life-related substances such as glucose, galactose, amino acids, uric acid and cholesterol, and it is also a common marker for immunomolecular and nucleic acid analysis. The commonly used peroxidase enzyme is derived from the plant horseradish, for this reason it is also known as horseradish peroxidase (HRP). HRP is a type of haemoglobin-containing protein, consisting of a cofactor, haemoglobin chloride (ferriporphyrin), and an apoprotein (glycoprotein), which has a molecular weight of 44 000.

The *o*-phenylenediamine (OPD) is capable of being oxidized by H_2O_2. HRP has a strong catalytic effect on this reaction. Under the conditions chosen for the enzyme-catalysed reaction, it produces only one coloured substance, 2,3-diaminophenothiazine.

The equation for the reaction in this experiment is

$$2\,C_6H_4(NH_2)_2 + 3H_2O_2 \xrightarrow{\text{HRP}} C_{12}H_6N_2(NH_2)_2 + 6H_2O$$

By Lambert-Beer law

$$A = ad[P] \tag{7}$$

In this equation, A is the absorbance of the product, a is the molar extinction coefficient and d is the thickness of the sample pool.

When $d = 1\text{cm}$,

$$A = a[P] \tag{8}$$

then

$$\frac{d[P]}{dt} = \frac{1}{a}\left(\frac{dA}{dt}\right) \tag{9}$$

that is

$$v = \frac{d[P]}{dt} = \frac{1}{a}\left(\frac{dA}{dt}\right) \tag{10}$$

Combination Equations (6) and (10), the v_{max} and K_M of the OPD oxidation by H_2O_2 catalyzed by HRP can be measured by using the spectrophotometer .

Apparatus and reagents

721 type spectrophotometer	PHS-2 acidity meter
HRP	OPD (AR)
30% H_2O_2	2,3-diaminophenothiazine (AR)
Acetic Acid	Sodium acetate

Experimental procedures

1. The wavelength of light absorption for the spectrophotometer is chosen as 446.3 nm. The solvent used to prepare the solution is an acetate buffer solution (ABS). The pH is set to 5.00. The reaction temperature is set at room temperature (preferably no less than 10 ℃).

2. Accurately weigh 0.0017 g of 2, 3-diaminophenothiazine, and add ABS to constant volume to 50 mL (concentration is about 16×10^{-5} mol · L^{-1}). Then the solutions are gradually diluted to 8×10^{-5}, 4×10^{-5}, 2×10^{-5} and 1×10^{-5} mol · L^{-1}. Measure the absorbance A of these solutions at 446.3 nm.

3. Accurately weigh 0.0011 g of OPD (substrates), and add ABS to constant volume to

50 mL(concentration is about 0.2 mmol · L^{-1}). The solution is then gradually diluted to 0.1, 0.05, 0.025 and 0.0125 mmol · L^{-1}.

4. Take 5 mL 0.0125 mmol · L^{-1} OPD in a weighing bottle, and add 5 μL of 10 μg · mL^{-1} HRP and 5 μL 30% H_2O_2. Shake well and add the solution to the cuvette. The time is started from the addition of H_2O_2. Record the changes of A with time at 446.3 nm. The interval of reading time is 2 min. After 20 min, stop the measurement and the A-t curve is obtained.

5. Only change the concentration of the substrate to 0.025 mmol · L^{-1}, keep the concentration and amount of the remaining substances unchanged, repeat the previous operation and so on, continue to increase the concentration of the substrate and repeat the operation to obtain the A-t curves at different substrate concentrations.

Data recording and processing

1. According to Lambert-Beer law, calculate the molar extinction coefficient a.

2. The A measured at different concentrations of OPD is plotted against t. The slope $\frac{dA}{dt}$ of the line is obtained, and the initial rate v of the reaction for different concentrations of the OPD is calculated from $\frac{1}{a}\left(\frac{dA}{dt}\right)$ 。

3. The obtained $\frac{1}{v}$ is plotted against $\frac{1}{[S]}$. From the intercept and slope of the line, v_{max} and K_M can be calculated.

Thinking and discussing

1. What are the characteristics of enzyme-catalyzed reactions compared with general catalytic reactions?

2. What is the significance of enzyme kinetics research for the development of green chemistry?

3. Why do enzymatic reactions need to take place at optimum temperature and pH?

Reference materials

1. Yuan Q. Modern Enzymology. Shanghai: East China University of Science and Technology Press, 2001.

2. Feng C, Li H, Zhang Z, et al. Theoretical analysis and optimization of catalytic kinetic spectrophotometric determination of alkaline phosphatase. Journal of Liaoning Normal University (Natural Science Edition), 2003, 32(2): 195–198.

3. Fu X, Shen W, Yao T. Physical Chemistry. (3rd ed). Beijing: Higher Education Press, 2004.

实验21 电导法测定水溶性表面活性剂的临界胶束浓度

实验目的

1. 了解表面活性剂的特性及胶束形成原理。
2. 掌握常见电导率仪的使用方法。
3. 用电导法测定十二烷基硫酸钠的临界胶束浓度。

基本原理

表面活性剂是在两个物理相之间的表面或界面上活跃的化学物质。通常，表面活性剂分子包括两个部分或基团。一部分具有喜水性或亲水性；另一部分具有憎水性或疏水性，有时也被称为亲油性或亲脂性。由于分子的两亲性，表面活性剂可溶于有机溶剂和水，表现出非常强烈的迁移到界面或表面的趋势，并在界面定向排布，极性基团位于水中，非极性基团倾向于置于水外。表面活性剂具有降低表面张力的性质，根据其主要用途可分为洗涤剂、润湿剂、分散剂、乳化剂、发泡剂、杀菌剂、缓蚀剂、抗静电剂等。

按照其化学结构，表面活性剂分子可分为离子型表面活性剂和非离子型表面活性剂。离子型的还可按生成的活性基团是阳离子或是阴离子再进行分类。离子型表面活性剂在水中电离后起表面活性作用的部分带正电荷的称为阳离子表面活性剂，阳离子型表面活性剂的亲水基绝大多数为含氮原子的阳离子，少数为含硫或磷原子的阳离子。分子中的阴离子不具有表面活性，通常是单个原子或基团，如氯、溴、醋酸根离子等。例如 $RN^+(CH_3)_3X^-$，$RN^+(CH_3)_3OH^-$。在水中电离后起表面活性作用的部分带负电荷的表面活性剂称为阴离子表面活性剂。阴离子表面活性剂分为羧酸盐、硫酸酯盐、磺酸盐和磷酸酯盐四大类，如 $RSO_4^-Na^+$。非离子型表面活性剂是在水溶液中不产生离子的表面活性剂。如 $R_2N—(C_2H_4O)_nH$，$R(OC_2H_4)_nOH$ 和 $R—CONH(C_2H_4O)_nH$。

当离子型表面活性剂的浓度较低时，以单个分子形式存在，由于其两亲性质，这些分子聚集在水的表面上，使空气和水的接触面减小，水的表面张力显著降低。当溶解浓度逐渐增大时，不但表面上聚集的表面活性剂增多而形成单分子层，而且溶液体相内表面活性剂的分子也三三两两地以憎水基互相依靠，聚集在一起，开始形成胶束（如图21-1所示）。根据活性剂的性质，形成的胶束可以成球形、棒形或层状。形成胶束的最低浓度为临界胶束浓度（critical micelle concentration, CMC）。继续增加表面活性剂的量，超过临界胶束浓

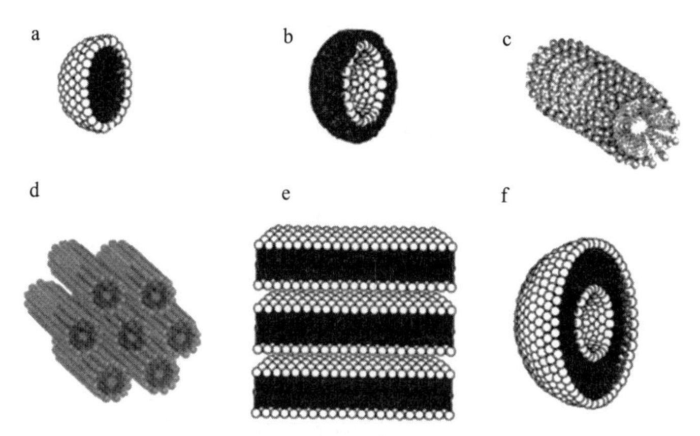

图 21-1　胶束结构示意图。球形(a、b、f),棒形(c、d),层状(e)。白色代表亲水基团,黑色表示憎水基团

度后,由于表面已经占满,只能增加溶液中胶束的数量,即形成表面活性剂憎水基团靠在一起的胶束。由于胶束不具有表面活性,表面张力不再下降,表面张力与表面活性剂浓度的关系曲线上表现为水平线段。

临界胶束浓度可用各种不同方法进行测定,采用的方法不同,测得的 CMC 值也有差别。因此,一般所给的 CMC 值是一个临界胶束浓度的范围,在该浓度范围前后不仅表面张力有显著变化,溶液的其他物理性质如渗透压、电导率、去污能力、浊度、光学性质等也有显著的变化,如图 21-2 所示。CMC 与表面活性剂的结构密切相关,并有一定的规律,此外,温度、电解质、有机物、第二种表面活性剂存在以及水溶性大分子等都对 CMC 值有显著的影响。有研究表明,胶束的形态主要取决于表面活性剂的几何形状,特别是亲水和疏水基在溶液中各自横截面积的相对大小。

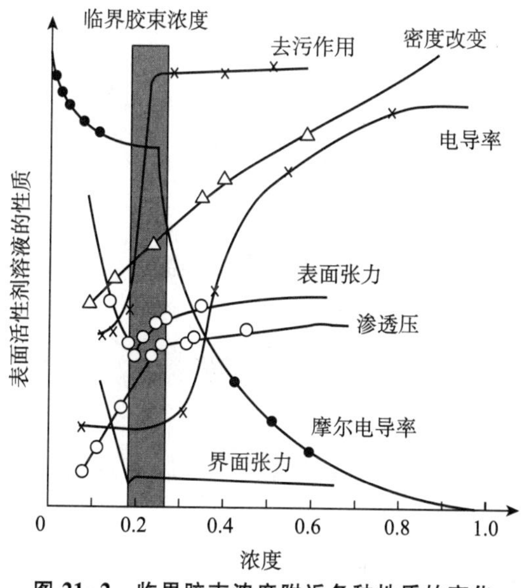

图 21-2　临界胶束浓度附近各种性质的变化

本实验利用电导仪测定不同浓度、不同温度下十二烷基硫酸钠水溶液的电导值（或摩尔电导率），并作电导值（或摩尔电导率）与浓度的关系图，从图中的转折点求得临界胶束浓度，并考察温度对表面活性剂 CMC 的影响。

仪器和试剂

数显电导率仪
铂黑电极
恒温水浴
电子天平
容量瓶（1000 mL、100 mL）
移液管
大试管
十二烷基硫酸钠（分析纯）

实验步骤

1. 开通电导率仪的电源预热 20 min。打开恒温水浴并分别调节水浴温度至 25 ℃、40 ℃。

2. 取烘干好的十二烷基硫酸钠（80 ℃下烘干 3 h），准确配制 0.200 $mol \cdot L^{-1}$的十二烷基硫酸钠溶液。对于这一步，实验台上准备提前配好的母液，学生直接量取稀释即可。

3. 用电导水或重蒸馏水分别准确配制 0.002，0.004，0.006，0.007，0.008，0.009，0.010，0.012，0.014，0.016，0.018，0.020 $mol \cdot L^{-1}$的十二烷基硫酸钠溶液各 100mL。

4. 使用电导率仪，在不同温度下，从稀到浓分别测定上述各溶液的电导率。每次测量前荡洗电极和容器 3 次以上，各溶液测定前必须恒温 10 min，每个溶液的电导率读数 3 次，取平均值。

5. 列表记录各溶液对应的电导率或者摩尔电导率。

6. 实验结束后用蒸馏水洗净试管和电极，并测量所用水的电导率。

数据记录与处理

做出不同温度下电导率（或摩尔电导率）与浓度的关系图，从图中的转折点找出临界胶束浓度，并探讨不同温度对临界胶束浓度的影响。

思考与讨论

1. 溶解的表面活性剂分子与胶束之间的平衡同温度和浓度有关，其关系式可表示为

$$\frac{\mathrm{d}\ln c_{\mathrm{CMC}}}{\mathrm{d}T} = -\frac{\Delta H}{2RT^2}$$

试问如何测出其热效应值？

2. 非离子型表面活性剂能否用本实验方法测定临界胶束浓度？为什么？若不能，则可用何种方法测定？

3. 若要知道所测得的临界胶束浓度是否准确，可用什么实验方法验证？

参考资料

1. 傅献彩,沈文霞,姚天明,等. 物理化学(第五版). 北京:高等教育出版社,2005.

2. 庄继华. 物理化学实验(第三版), 北京:高等教育出版社,2004.

3. http://www.countrywives.co.uk/marvellous-micellar

4. Moradi M, Yamini Y. Surfactant roles in modern sample preparation techniques: a review. Journal of Separation Science, 2012, 35(18): 2319-2340.

Experiment 21

Determination of the critical micelle concentration of water-soluble surfactants by conductivity

Experimental purpose

1. Understand the properties of surfactant and the mechanism of micelles formation.

2. Master the use of common conductivity meters.

3. Determinate critical micelle concentration of sodium dodecyl sulfate by conductivity method.

Basical principles

Surfactants are chemicals that are active at surfaces, or interfaces between two physical phases. Typically, surfactant molecules include two parts or groups. One part is water-liking or hydrophilic. The other part is water-hating or hydrophobic and is sometimes described as oil-liking or lipophilic. Due to the amphiphilic nature of the molecule, surfactants are soluble in both organic solvent and water. This is why amphiphilic molecules exhibit a very strong tendency to migrate to interfaces or surfaces and to orientate so that the polar group lies in water and the nonpolar group is placed out of it. Surfactant has the property of reducing surface tension, which can be classified according to its main uses: detergent, wetting agent, dispersant, emulsifier, foaming agent, bactericide, corrosion inhibitor, antistatic agent, etc.

Surfactant molecules are classified into ionic and non-ionic surfactants, according to their chemical structure. Among these, ionic surfactants can be further classified according to whether the reactive group produced is cationic or anionic. The active part of an ionic surfactant after ionization in water is positively charged, which is called a cationic surfactant. The majority of the hydrophilic groups of cationic surfactants are cations containing nitrogen atoms and a few are cations containing sulphur or phosphorus atoms. The anions in the molecule are not surface active and are usually single atoms or groups such as chloride, bromine, acetate ions etc. For example, $RN^+(CH_3)_3X^-$, $RN^+(CH_3)_3OH^-$.The active part of a surfactant after ionization in water is negatively charged, which is called an anionic surfactant. Anionic surfactants are divided into four categories: carboxylates, sulphates, sulphonates and phosphates, for example, $RSO_4^- Na^+$. Non-ionic surfactant are surfactant that do not produce ions in aqueous solutions, such as $R_2N-(C_2H_4O)_nH$, $R(OC_2H_4)_nOH$, and $R-CONH(C_2H_4O)_nH$.

When ionic surfactants are present in low concentrations, they exist as individual molecules. Due to their amphiphilic nature, these molecules collect on the surface of water, reducing the contact surface between air and water and thus causing a significant reduction in the surface tension of water. As the dissolution concentration gradually increases, not only do more surfactants collect on the surface and form a monomolecular layer, but also the molecules of surfactants in the solution phase depend on each other in twos and threes with water-repellent groups and begin to gather together to form micelles (as shown in Fig. 21-1).

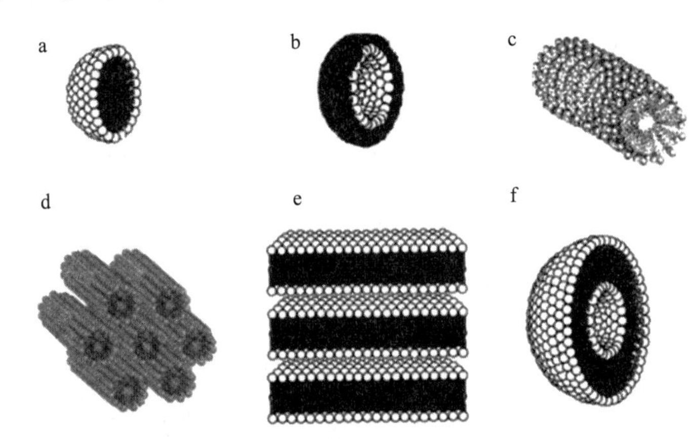

Fig. 21-1 Schematic diagram of micelles structure. Spherical micelles (a、b、f), rod-shaped micelles(c、d), laminar (e).White represents the hydrophilic group, and black represents the hydrophobic group

Depending on the nature of the surfactants, the micelles formed can be spherical, rod-shaped or laminar. The minimum concentration for micelle formation is the critical micelle concentration (CMC). After increasing the amount of surfactant more than the critical micelle concentration, the number of micelles in the solution can only be increased because the surface is already full, i. e., the formation of micelles with surfactant hydrophobic groups. As the micelles are not surface active, the surface tension no longer decreases and the surface tension is shown as a horizontal line segment on the curve of surface tension versus surfactant concentration.

The critical micelle concentration can be determined by a variety of different methods and the CMC values obtained will vary depending on the method used. Therefore, the CMC values generally given are a range of critical micelle concentrations around which are significant changes not only in surface tension but also in other physical properties of the solution, such as osmotic pressure, conductivity, detergency, turbidity, optical properties etc, as shown in Fig. 21-2.

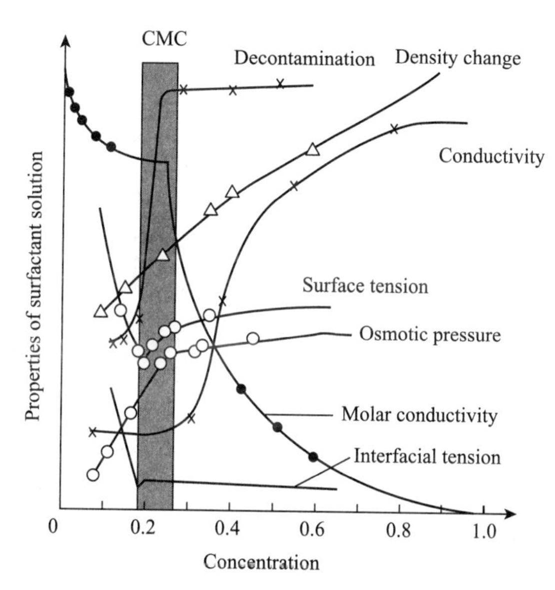

Fig. 21-2 Changes in properties in the vicinity of the critical micelle concentration

CMC is closely related to the structure of the surfactant and has a certain pattern. In addition, temperature, electrolytes, organic matter, a second surfactant and water-soluble macromolecules all have a significant effect on CMC values. It has been shown that the morphology of micelles depends mainly on the geometry of the surfactant, in particular the relative size of the respective cross-sectional areas of the hydrophilic and hydrophobic groups in solution.

In this experiment, the conductivity (or molar conductivity) of aqueous sodium dodecyl sulfate solutions at different concentrations and temperatures are determined using a conductivity meter. By making a graph of conductivity (or molar conductivity) versus concentration, the critical micelle concentration is derived from the turning point in the graph and the effect of temperature on the surfactant CMC is examined.

Apparatus and reagents

Digital conductivity meter

Platinum black electrode

Thermostatic water baths
Volume bottles(1000 mL、100 mL)
Big test-tube
Electronic balance
Pipettes
Sodium dodecyl sulfate (SDS, AR)

Experimental procedures

1. Switch on the conductivity meter and preheat for 20 min.Turn on the thermostatic water bath and adjust the bath temperature to 25 ℃ and 40 ℃ respectively.

2. Take the dried sodium dodecyl sulfate (dried at 80 ℃ for 3h) and accurately configure a 0.200 mol · L^{-1} solution of sodium dodecyl sulfate.

3. Accurately prepare 100mL SDS solution of 0.002, 0.004, 0.006, 0.007, 0.008, 0.009, 0.010, 0.012, 0.014, 0.016, 0.018, 0.020 mol · L^{-1} respectively with conductivity water or redistilled water.

4. The conductivity of the above solutions is measured using a conductivity meter at different temperatures, from dilute to concentrated solution. The electrode and vessel should be washed more than 3 times before each measurement. Every solution should be kept at a constant temperature for 10 min before measurement, and take 3 readings of the conductivity of each solution and take the average.

5. Record the conductivity or molar conductivity corresponding to each solution in a table.

6. At the end of the experiment, wash the tubes and electrodes with distilled water. Measure the conductivity of the water used.

Data recording and processing

1. Make a graph of conductivity (or molar conductivity) versus concentration at different temperatures.

2. Identify the critical micelle concentration from the turning points in the graph.

3. Explore the effect of different temperatures on the critical micelle concentration.

Thinking and discussing

1. The equilibrium between the dissolved surfactant and the micelles depends on temperature and concentration. The equation can be expressed as

$$\frac{\mathrm{d}\ln c_{\mathrm{CMC}}}{\mathrm{d}T} = -\frac{\Delta H}{2RT^2}$$

How do you measure the thermal effect?

2. Can non-ionic surfactant be used to determine critical micelle concentration? Why? If not, what method can be used?

3. What experimental methods can be applied to verify the accuracy of the method used for the measurement of CMC?

Reference materials

1. Fu X, Shen W, Yao T, et al. Physical Chemistry (5th ed). Beijing: Higher Education Press, 2005.

2. Zhuang J, Physical Chemistry Experiment (3rd ed). Beijing: Higher Education Press, 2004.

3. http://www.countrywives.co.uk/marvellous-micellar

4. Moradi M, Yamini Y. Surfactant roles in modern sample preparation techniques: a review. Journal of Separation Science, 2012, 35(18): 2319-2340.

实验 22　溶液法测定极性分子的偶极矩

实验目的

1. 了解偶极矩与分子电性质的关系。

2. 掌握溶液法测定偶极矩的基本原理和实验技术。

3. 用溶液法测定乙酸乙酯的偶极矩。

基本原理

1. 偶极矩与极化度

偶极矩是介质分子极性的量度,不仅对于介质的特性有意义,同时在物理学、化学和生物学等领域都有着重要的应用。如图 22-1 所示,偶极矩的定义为

$$\boldsymbol{\mu} = q \cdot d \tag{1}$$

式中 q 为正、负电荷中心所带的电荷量,d 为正、负电荷中心之间距离。$\boldsymbol{\mu}$ 是一个矢量,其单位为 C · m,$\boldsymbol{\mu}$ 的方向规定为从正电荷到负电荷。

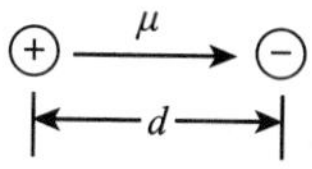

图 22-1　电偶极矩示意图

通过偶极矩的测定可以了解分子结构中有关电子云的分布和分子的对称性等情况,还可以用来判别几何异构体和分子的立体结构等。

极性分子正负电荷中心不重合,具有永久偶极矩。在没有外电场存在时,由于分子的热运动,偶极矩指向各个方向的机会相同,所以偶极矩的统计值等于零。

若将极性分子置于均匀的电场中，则偶极矩在电场的作用下会趋向电场方向排列。这时我们称这些分子被极化了，极化的程度可以用摩尔转向极化度 $\boldsymbol{P}_{转向}$ 来衡量。

$\boldsymbol{P}_{转向}$ 与永久偶极矩平方成正比，与热力学温度 T 成反比，其关系为

$$\boldsymbol{P}_{转向} = \frac{4}{3}\pi N_A \frac{\boldsymbol{\mu}^2}{3kT} = \frac{4}{9}\pi N_A \frac{\boldsymbol{\mu}^2}{kT} \tag{2}$$

式中 k 为玻尔兹曼常数，N_A 为阿伏伽德罗常数。

在外电场作用下，不论极性分子还是非极性分子都会发生电子云对分子骨架的相对移动，分子骨架也会发生变形，这种现象称为诱导极化或变形极化，用摩尔诱导极化度 $\boldsymbol{P}_{诱导}$ 来衡量。显然，$\boldsymbol{P}_{诱导}$ 可分为二项，即电子极化度 $\boldsymbol{P}_{电子}$ 和原子极化度 $\boldsymbol{P}_{原子}$，因此 $\boldsymbol{P}_{诱导} = \boldsymbol{P}_{电子} + \boldsymbol{P}_{原子}$。

如果外电场是交变电场，极性分子的极化情况则与交变电场的频率有关。当处于频率小于 $10^{10}\ s^{-1}$ 的低频电场或静电场中，极性分子所产生的摩尔极化度 $\boldsymbol{P}$ 是转向极化、电子极化和原子极化的总和

$$\boldsymbol{P} = \boldsymbol{P}_{转向} + \boldsymbol{P}_{电子} + \boldsymbol{P}_{原子} \tag{3}$$

当频率增加到 $10^{12} \sim 10^{14} s^{-1}$ 的中频（红外频率）时，电场的交变周期小于分子偶极矩的弛豫时间，极性分子的转向运动跟不上电场的变化，即极性分子来不及沿电场定向，故 $\boldsymbol{P}_{转向} = 0$。此时极性分子的摩尔极化度等于摩尔诱导极化度 $\boldsymbol{P}_{诱导}$。当交变电场的频率进一步增加到大于 $10^{15}\ s^{-1}$ 的高频（可见光和紫外频率）时，极性分子的转向运动和分子骨架变形都跟不上电场的变化，此时极性分子的摩尔极化度等于电子极化度 $\boldsymbol{P}_{电子}$。

因此，原则上只要在低频电场下测得极性分子的摩尔极化度 $\boldsymbol{P}$，在红外频率下测得极性分子的摩尔诱导极化度 $\boldsymbol{P}_{诱导}$，两者相减得到极性分子的摩尔转向极化度 $\boldsymbol{P}_{转向}$，然后代入式（2）就可求出极性分子的永久偶极矩 $\boldsymbol{H}$。

2. 极化度的测定

对于分子间相互作用很小的系统，克劳修斯－莫索提－德拜（Clausius-Mosotti-Debye）从电磁理论推导得到了摩尔极化度 $\boldsymbol{P}$ 与介电常数 ε 之间的关系式

$$\boldsymbol{P} = \frac{\varepsilon - 1}{\varepsilon + 2} \cdot \frac{M}{\rho} \tag{4}$$

式中 M 为被测物质的摩尔质量，ρ 是被测物质的密度，ε 可以通过实验测定。

因式（4）是假定分子与分子间无相互作用而推导得到的，所以它只适用于温度不太低的气相体系。然而测定气相的介电常数和密度，在实验上难度较大，甚至根本无法使某些物质处于稳定的气相状态。因此提出溶液法来解决这一困难。溶液法的基本思路是，在无限稀释的非极性溶剂的溶液中，溶质分子所处的状态与气相时相近，于是无限稀释溶液中溶质的摩尔极化度 $\boldsymbol{P}_2^{\infty}$ 就可以看作式（4）中的 $\boldsymbol{P}$

$$P = P_2^{\infty} = \lim_{x_2 \to 0} P_2 = \frac{3\alpha\varepsilon_1}{(\varepsilon_1 + 2)^2} \cdot \frac{M_1}{\rho_1} + \frac{\varepsilon_1 - 1}{\varepsilon_1 + 2} \cdot \frac{M_2 - \beta M_1}{\rho_1} \tag{5}$$

式中 ε_1、ρ_1 和 M_1 分别是溶剂的介电常数、密度和摩尔质量；M_2 是溶质的摩尔质量；α、β 为常数，其值可利用下列稀溶液近似公式求得

$$\varepsilon_{溶} = \varepsilon_1(1 + \alpha x_2) \tag{6}$$

$$\rho_{溶} = \rho_1(1 + \beta x_2) \tag{7}$$

上述式(6)、(7)中，$\varepsilon_{溶}$、$\rho_{溶}$ 分别是溶液的介电常数和密度，x_2 是溶质的摩尔分数。

已知在红外频率的电场下可以测得极性分子的摩尔诱导极化度。但在实验中由于条件的限制，很难做到这一点。根据光的电磁理论，在同一频率的高频电场作用下，透明物质的介电常数 ε 和折光率 n 的关系为

$$\varepsilon = n^2 \tag{8}$$

习惯上用摩尔折射度 $\boldsymbol{R}_2$ 来表示高频区测得的极化度，因为此时 $\boldsymbol{P}_{转向} = 0$，$\boldsymbol{P}_{原子} = 0$，则

$$\boldsymbol{R}_2 = \boldsymbol{P}_{电子} = \frac{n^2 - 1}{n^2 + 2} \cdot \frac{M}{\rho} \tag{9}$$

同样从式(9)可以推导出无限稀释时溶质的摩尔折射度的公式

$$\boldsymbol{P}_{电子} = \boldsymbol{R}_2^{\infty} = \lim_{x_2 \to 0} \boldsymbol{R}_2 = \frac{n_1^2 - 1}{n_1^2 + 2} \cdot \frac{M_2 - \beta M_1}{\rho_1} + \frac{6n_1^2 M_1 \gamma}{(n_1^2 + 2)^2 \rho_1} \tag{10}$$

上述式(9)、(10)中，n_1 是溶剂的折光率，γ 为常数，其值可利用如下稀溶液近似公式求得

$$n_{溶} = n_1(1 + \gamma x_2) \tag{11}$$

$n_{溶}$是溶液的折光率。

3. 偶极矩的测定

考虑到原子极化度通常只有电子极化度的 5%～10%，而且 $\boldsymbol{P}_{转向}$ 又比 $\boldsymbol{P}_{电子}$ 大得多，故常常忽视原子极化度。

从式(2)、(3)、(5)和(10)可得

$$\boldsymbol{P}_{转向} = \boldsymbol{P}_2^{\infty} - \boldsymbol{R}_2^{\infty} = \frac{4}{9}\pi L \frac{\boldsymbol{\mu}^2}{kT} \tag{12}$$

上式把物质分子的微观性质偶极矩和它的宏观性质介电常数、密度和折光率联系起来了，分子的永久偶极矩可用下面简化式计算

$$\boldsymbol{\mu}/C \cdot m = 0.04274 \times 10^{-30}\sqrt{(\boldsymbol{P}_2^{\infty} - \boldsymbol{R}_2^{\infty})T} \tag{13}$$

在某种情况下，若需要考虑 $\boldsymbol{P}_{原子}$ 影响，只需对 $\boldsymbol{R}_2^{\infty}$ 作部分修正就行了。

上述测量极性分子偶极矩的方法称为溶液法，溶液法测得的溶质偶极矩与气相测得的真实值间存在偏差，造成这种现象的原因是非极性溶剂与极性溶质分子相互间的作

用——溶剂化作用，这种偏差现象称为溶液法测量偶极矩的溶剂效应，感兴趣的读者可阅读有关参考资料[2]。

4. 介电常数与电容

任何物质的介电常数都可借助一个电容器的电容值来表示，即

$$\varepsilon = \frac{C}{C_0} \tag{14}$$

式中 C 为电容器以该物质为介质时的电容值，C_0 为同一电容器在真空时的电容值。通常空气电容的介电常数接近于1，所以介电常数可以近似表示为

$$\varepsilon = \frac{C}{C_{空}} \tag{15}$$

$C_{空}$ 为电容器以空气为介质时的电容值。因此测介电常数就变为测电容了。

实际所测电容 $C'_{样}$ 包括样品的电容 $C_{样}$ 和电容池的分布电容 C_d 两部分，即

$$C'_{样} = C_{样} + C_d \tag{16}$$

对于给定的电容池，必须先测出其分布电容 C_d。可以先测出以空气为介质的电容 $C'_{空}$，再用一种已知介电常数 $\varepsilon_{标}$ 的物质测得其电容 $C'_{标}$。则

$$C'_{空} = C_{空} + C_d \quad （或\ C'_0 = C_0 + C_d） \tag{17}$$

$$C'_{标} = C_{标} + C_d \tag{18}$$

因为

$$\varepsilon_{标} = \frac{C_{标}}{C_0} \approx \frac{C_{标}}{C_{空}} \tag{19}$$

由式(17)～(19)可得

$$C_d = C'_{空} - \frac{C'_{标} - C'_{空}}{\varepsilon_{标} - 1} = \frac{\varepsilon_{标} C'_{空} - C'_{标}}{\varepsilon_{标} - 1} \tag{20}$$

$$C_0 = \frac{C'_{标} - C'_{空}}{\varepsilon_{标} - 1} \tag{21}$$

测出不同浓度溶液的电容 $C'_{样}$，按式(16)计算 $C_{样}$，按式(14)计算出不同浓度溶液的介电常数。

仪器和试剂

阿贝折光仪　　PCM-1A 型精密电容测量仪

电容池　　超级恒温槽

洗耳球　　容量瓶(25 mL，10 mL)

乙酸乙酯(分析纯)　　环己烷(分析纯)

实验步骤

1. 配制溶液

分别量取 0.5,1.0,2.0,3.0 mL 乙酸乙酯于 25 mL 容量瓶中,用称重法配制 4 种不同浓度的乙酸乙酯-环己烷溶液。操作时应注意防止溶质和溶剂的挥发以及吸收极性较大的水气,溶液配好后应迅速盖上瓶盖。

2. 折光率的测定

在(25.0 ± 0.1) ℃条件下用阿贝折光仪测定环己烷及所配制的各溶液的折光率。测定时注意各样品需加样 2 次,每次读取 2 个数据,然后取平均值。

3. 介电常数的测定

本实验采用环己烷作为标准物质,其介电常数的温度公式为

$$\varepsilon_{标} = 2.015 - 0.16 \times 10^{-2}(t/℃ - 25)$$

式中 t 为恒温槽温度。25 ℃时 $\varepsilon_{标}$为 2.015。

4. 溶液密度的测定(用 10 mL 容量瓶)

称 10 mL 小容量瓶空瓶质量(记录),加入环己烷定容,称重(记录)后,溶剂倒入回收瓶中,将容量瓶用洗耳球吹干,加入 1 号溶液定容称重(记录),然后将溶液倒入原 1 号瓶中,再将瓶用洗耳球吹干。以此类推,称出余下的 3 种溶液的质量。

环己烷和各溶液的密度按下式计算

$$\rho^{25\ ℃} = \frac{m_2 - m_0}{m_1 - m_0} \cdot \rho_{水}^{25\ ℃}$$

m_0:空瓶质量(g);m_1:瓶加水的质量(g);m_2:瓶加样品的质量(g);$\rho_{水}^{25\ ℃}$:25 ℃时水的密度(g · cm^{-3})。

数据记录与处理

1. 将数据记录和结果列表表示。

2. 作 $\varepsilon_{溶} - x_2$ 图,由直线斜率求算 α 值;作 $\rho_{溶} - x_2$ 图,由直线斜率求算 β 值;作 $n_{溶} - x_2$ 图,由直线斜率计算 γ 值。

3. 将 ρ_1、ε_1、α 和 β 值代入式(7)计算 $\boldsymbol{P}_2^{\infty}$。

4. 将 ρ_1、n_1、β 和 γ 值代入式(11)计算 $\boldsymbol{R}_2^{\infty}$。

5. 将 $\boldsymbol{P}_2^{\infty}$、$\boldsymbol{R}_2^{\infty}$ 值代入式(13)计算乙酸乙酯分子的偶极矩 $\boldsymbol{\mu}$ 值。

思考与讨论

1. 分析本实验误差的主要来源,如何改进?

2. 试说明溶液法测量极性分子永久偶极矩的要点,有何基本假定,推导公式时做了哪些近似?

3. 如何利用溶液法测量偶极矩的溶剂效应来研究极性溶质分子与非极性溶剂分子的相互作用?

参考资料

1. 项一非,李树家.中级物理化学实验.北京:高等教育出版社,1988: 142.

2. 邢志忠.电子的电偶极矩及其对新物理的可能限.现代物理知识, 2023, 35(03): 35-37.

3. 张岩,周立敏,高先池,等.基于绿色化学理念的偶极矩测定实验综合化设计.山东化工, 2022, 51(02):187-188+203.

附录　文献值

表 22-1　乙酸乙酯分子的偶极矩[3]

μ/D	$\mu\times10^{-30}$/C · m*	状态或溶剂	温度/℃
1.78	5.94	气	30～195
1.83	6.10	液	25
1.76	5.87	CCl_4	25
1.89	6.30	CCl_4	25

* 按 1D = 3.335 64 C · m 换算

Experiment 22

Determination of dipole moments of polar molecules by solution method

Experimental purpose

1. Understand the relationship between dipole moment and molecular electrical properties.

2. Master the basic principles and experimental techniques of dipole moment determined by solution method.

3. Determinate dipole moment of ethyl acetate by solution method.

Basical principles

1. Dipole moment and polarization

The dipole moment is a measure of the polarity of a medium molecule. It is not only significant to the properties of the medium, but also has important applications in physics, chemistry and biology. As shown in Fig. 22-1, the dipole moment is defined as

$$\boldsymbol{\mu} = q \cdot d \tag{1}$$

where q is the amount of charge carried by the positive and negative charge centers, and d is the distance between the positive and negative charge centers. $\boldsymbol{\mu}$ is a vector in units C · m and the direction of the $\boldsymbol{\mu}$ is specified from positive to negative.

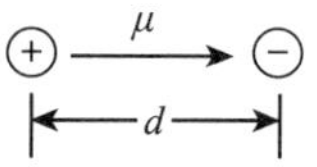

Fig. 22-1 Schematic diagram of electric dipole moment

Through the determination of dipole moment, the distribution of electron clouds and the symmetry of molecules in the molecular structure can be understood, which can also be used to distinguish geometric isomers and three-dimensional structures of molecules.

Polar molecules have permanent dipole moments, which are directed to each due to the thermal motion of the molecules in the absence of an external electric field. The chance of direction is the same, so the statistical value of the dipole moment is equal to zero.

If polar molecules are placed in a uniform electric field, the dipole moment tends to be aligned in the direction of the electric field under the action of the electric field. At present, it can be said that the molecules are polarized, and the degree of polarization can be measured by the molar steering polarization $\boldsymbol{P}_{steer}$.

$\boldsymbol{P}_{steer}$ is proportional to the square of the permanent dipole moment and inversely proportional to the thermodynamic temperature T, and their relationship is

$$P_{steer} = \frac{4}{3}\pi N_A \frac{\mu^2}{3kT} = \frac{4}{9}\pi N_A \frac{\mu^2}{kT} \tag{2}$$

where k is the Boltzmann constant and N_A is the Avogadro constant.

Under the action of the external electric field, both polar and non-polar molecules will undergo the relative movement of the electron cloud to the molecular skeleton, and the molecular skeleton will also undergo deformation, which is called induced polarization or deformation polarization, which is measured by mole-induced polarization $\boldsymbol{P}_{induction}$. Obviously, $\boldsymbol{P}_{induction}$ can be divided into two terms, namely electron polarization $\boldsymbol{P}_{electrons}$ and atomic polarization $\boldsymbol{P}_{atoms}$, so $\boldsymbol{P}_{induction} = \boldsymbol{P}_{electrons} + \boldsymbol{P}_{atoms}$. $\boldsymbol{P}_{induction}$ is proportional to the strength of the external electric field, independent of temperature.

If the external electric field is an alternating electric field, the polarization of the polarmolecules is related to the frequency of the alternating electric field. When in a low-frequency electric or electrostatic field with a frequency less than $10^{10}\ s^{-1}$, the molar polarization

$\boldsymbol{P}$ produced by polar molecules is the sum of steering polarization, electron polarization, and atomic polarization

$$\boldsymbol{P}=\boldsymbol{P}_{\text{steer}}+\boldsymbol{P}_{\text{electron}}+\boldsymbol{P}_{\text{atom}} \tag{3}$$

When the frequency is increased to the intermediate frequency (infrared frequency) of $10^{12}\sim10^{14}\ \text{s}^{-1}$, the alternating period of the electric field is less than the relaxation time of the molecular dipole moment, and the steering motion of polar molecules cannot keep up with the change of the electric field, that is, the polar molecules do not have time to orient along the electric field, so $\boldsymbol{P}_{\text{steer}}=0$. At this time, the molar polarization of the polar molecule is equal to the molar induced polarization $\boldsymbol{P}_{\text{induction}}$. When the frequency of the alternating electric field is further increased to a high frequency (visible light and ultraviolet frequency) greater than $10^{15}\ \text{s}^{-1}$, the steering motion of polar molecules and the deformation of the molecular skeleton cannot keep up with the change of electric field, and the molar polarization of polar molecules is equal to the electron polarization $\boldsymbol{P}_{\text{electron}}$.

Therefore, in principle, as long as the molar polarization $\boldsymbol{P}$ of polar molecules is measured under a low-frequency electric field, and the molar induced polarization $\boldsymbol{P}_{\text{induction}}$ of polar molecules is measured at infrared frequencies, the two are subtracted to obtain the molar steering polarization $\boldsymbol{P}_{\text{steer}}$ of polar molecules, and then the permanent dipole moment of polar molecules $\boldsymbol{\mu}$ can be calculated by substituting Equation (2).

2. Determination of polarization

For systems with minimal molecular interactions, Clausius-Mosotti-Debye derived the relationship between molar polarization $\boldsymbol{P}$ and the $\boldsymbol{\varepsilon}$ of the permittivity from electromagnetic theory

$$\boldsymbol{P}=\frac{\varepsilon-1}{\varepsilon+2}\cdot\frac{M}{\rho} \tag{4}$$

where M is the molar mass of the measured substance, ρ is the density of the measured substance, and ε can be determined by experiment.

Equation (4) is derived from the assumption that there is no interaction between molecules, so it is only suitable for gas-phase systems where the temperature is not too low. However, determining the permittivity and density of the gas phase is experimentally difficult, and some substances cannot even make it in a stable gas phase state. Therefore, a solution method is later proposed to solve this difficulty. The basic idea of the solution method is that in an infinitely diluted solution of a non-polar solvent, the solute molecules are in a state similar to that of the gas phase, so the molar polarization of the solute in an infinitely diluted solution $\boldsymbol{P}_2^{\infty}$ can be seen as $\boldsymbol{P}$ in Equation (4).

$$P = P_2^{\infty} = \lim_{x_2 \to 0} P_2 = \frac{3\alpha\varepsilon_1}{(\varepsilon_1 + 2)^2} \cdot \frac{M_1}{\rho_1} + \frac{\varepsilon_1 - 1}{\varepsilon_1 + 2} \cdot \frac{M_2 - \beta M_1}{\rho_1} \tag{5}$$

where ε_l、$\rho_l$and M_l are the permittivity, density and molar mass of the solvent; M_2 is the molar mass of the solute; α、β is a constant whose value can be obtained using the following approximation formula of dilute solution

$$\varepsilon_{\text{solution}} = \varepsilon_1(1 + \alpha x_2) \tag{6}$$

$$\rho_{\text{solution}} = \rho_1(1 + \beta x_2) \tag{7}$$

In Equations (6) and (7), $\varepsilon_{\text{solution}}$、$\rho_{\text{solution}}$ is the dielectric constant and density of the solution, respectively, and x_2 is the molar fraction of the solute.

Molar-induced polarization of polar molecules is known to be measured under electric fields at infrared frequencies. But in experiments, it is difficult to do this due to the limitations of conditions. According to the electromagnetic theory of light, under the action of a high-frequency electric field of the same frequency, the relationship between the permittivity ε and the refractive index n of a transparent substance is

$$\varepsilon = n^2 \tag{8}$$

It is customary to use molar refractive degree $\boldsymbol{R}_2$ to represent the polarization measured in the high-frequency region, because at this time $\boldsymbol{P}_{\text{steer}} = 0$, $\boldsymbol{P}_{\text{atom}} = 0$, then

$$R_2 = P_{\text{electron}} = \frac{n^2 - 1}{n^2 + 2} \cdot \frac{M}{\rho} \tag{9}$$

Similarly, from Equation (9), the formula for the molar refractive index of the solute on infinite dilution can be derived

$$P_{\text{electron}} = R_2^{\infty} = \lim_{x_2 \to 0} R_2 = \frac{n_1^2 - 1}{n_1^2 + 2} \cdot \frac{M_2 - \beta M_1}{\rho_1} + \frac{6n_1^2 M_1 \gamma}{(n_1^2 + 2)^2 \rho_1} \tag{10}$$

In Equations (9) and (10), n_1 is the refractive index of the solvent, γ is constant, and its value can be obtained using the following dilute solution approximation formula

$$n_{\text{solution}} = n_1(1 + \gamma x_2) \tag{11}$$

n_{solution} is the refractive index of the solution.

3. Determination of dipole moment

Considering that the polarization of atoms is usually only 5% ~ 10% of the electrons polarization, and the $\boldsymbol{P}_{\text{steer}}$ is much larger than that of $\boldsymbol{P}_{\text{electron}}$, the atomic polarization is often ignored.

From Equations (2), (3), (5) and (10)

$$P_{steer} = P_2^\infty - R_2^\infty = \frac{4}{9}\pi L \frac{\mu^2}{kT} \tag{12}$$

Equation (12) relates the microscopic property dipole moment of a substance molecule to its macroscopic properties of the dielectric constant, density and refractive index, and the permanent dipole moment of the molecule can be calculated using the following simplified formula

$$\boldsymbol{\mu}/C \cdot m = 0.04274 \times 10^{-30}\sqrt{(\boldsymbol{P}_2^\infty - \boldsymbol{R}_2^\infty)T} \tag{13}$$

In some cases, if the effect of $\boldsymbol{P}_{atom}$ is to be considered, only partial corrections of $\boldsymbol{R}_2^\infty$ are required.

The above method of measuring the dipole moment of polar molecules is called the solution method. There is a deviation between the solute dipole moment measured by the solution method and the true value measured by the gas phase, and the reason for this phenomenon is the interaction between non-polar solvents and polar solute molecules-solvation, which is called the solvent effect, interested readers can read the relevant reference materials[2].

4. Dielectric constant and capacitance

The permittivity of any substance can be expressed by the capacitance value of a capacitance

$$\varepsilon = \frac{C}{C_0} \tag{14}$$

where C is the capacitance value of the capacitor when the substance is the medium, and C_0 is the capacitance value of the same capacitor in vacuum. Usually, the permittivity of an air capacitor is close to 1, so the dielectric constant can be approximated as

$$\varepsilon = \frac{C}{C_{air}} \tag{15}$$

C_{air} is the capacitance value of the capacitor when air is the medium. Therefore, the dielectric constant becomes the capacitance.

The actual measured capacitance C'_{sample} includes two parts: the capacitance of the sample C_{sample} and the distributed capacitance of the capacitance cell C_d, namely

$$C'_{sample} = C_{sample} + C_d \tag{16}$$

For a given capacitance cell, its distributed capacitance C_d should be firstly measured. The capacitance of the air medium C'_{air} can be measured first, and the capacitance $\varepsilon_{substance}$ can be measured by a substance with a known dielectric constant $C'_{substance}$, i.e.

$$C'_{air} = C_{air} + C_d \quad (\text{or}: C'_0 = C_0 + C_d) \tag{17}$$

$$C'_{substance} = C_{substance} + C_d \tag{18}$$

Because

$$\varepsilon_{\text{substance}} = \frac{C_{\text{substance}}}{C_0} \approx \frac{C_{\text{substance}}}{C_{\text{air}}} \tag{19}$$

From Equations (17) ~ (19), we get

$$C_d = C'_{\text{air}} - \frac{C'_{\text{substance}} - C'_{\text{air}}}{\varepsilon_{\text{substance}} - 1} = \frac{\varepsilon_{\text{substance}} C'_{\text{air}} - C'_{\text{substance}}}{\varepsilon_{\text{substance}} - 1} \tag{20}$$

$$C_0 = \frac{C'_{\text{substance}} - C'_{\text{air}}}{\varepsilon_{\text{substance}} - 1} \tag{21}$$

Measure the capacitance of different concentrations of solutions C'_{sample}, calculate C_{sample} according to Equation (16), and the dielectric constant of different concentrations of solution could be calculated according to Equation (14).

Apparatus and reagents

Abbe refractometer
Capacitor cell
Ear wash bulb
Ethyl acetate (AR)
PCM-1A precision capacitance measuring
Super thermostatic bath
Instrument volume flask (25 mL, 10 mL)
Cyclohexane (AR)

Experimental procedures

1. Solution preparation

Four different concentrations of ethyl acetate-cyclohexane solutions are prepared by weighing method and placed in 25 mL volumetric flasks. During operation, attention should be paid to preventing the volatilization of solutes, solvents and absorbing water vapor with large polarity, and the bottle should be quickly closed after the solution is prepared.

2. Refractive index determination

The refractive index of cyclohexane and the prepared solutions is determined by an Abbe refractometer at (25.0 ± 0.1) ℃. When measuring, note that each sample needs to be sampled twice, read 2 data each time, and then take the average. Read Experiment 4 for the usage of the Abbe refractometer.

3. Determination of dielectric constant

In this experiment, cyclohexane is used as the standard material, and the temperature formula of the dielectric constant is

$$\varepsilon_{\text{substance}} = 2.015 - 0.16 \times 10^{-2}(t/℃ - 25)$$

where t is the temperature of the thermostatic bath. $\varepsilon_{\text{substance}}$ is 2.015 at 25 ℃.

4. Determination of solution density (density measurement with 10 mL volumetric flask)

Weigh the mass of 10 mL small volumetric flask empty bottles (recorded), add cyclohexane volumetric, weigh (record), pour the solvent into the recovery bottle, dry the volumetric flask with ear wash balls, add No. 1 solution to weigh (record), then pour the solution into the original No. 1 bottle, and dry the bottle with ear wash balls. The masses of the remaining three solutions are deduced in turn.

The density of cyclohexane and each solution is calculated as follows

$$\rho^{25\ ℃} = \frac{m_2 - m_0}{m_1 - m_0} \cdot \rho^{25\ ℃}_{\text{water}}$$

m_0: weight of empty bottle (g); m_1: weight of the bottle with water (g); m_2: weight of bottle plus sample; $\rho^{25\ ℃}_{\text{water}}$: density of water at 25 ℃ ($g \cdot cm^{-3}$).

Data recording and processing

1. Representation of data records and result lists.

2. Make a $\varepsilon_{\text{solution}} - x_2$ plot, and calculate the α value from the slope of the line.

3. Make a $\rho_{\text{solution}} - x_2$ plot and calculate the β value from the slope of the straight line. The density of water is shown in Table 1 of Appendix.

4. Make a $n_{\text{solution}} - x_2$ plot and calculate the γ value from the slope of the line.

5. Substitute the value of ρ_1, ε_1, α and β into Equation (7) to calculate $\boldsymbol{P}_2^{\infty}$.

6. Substitute the value of ρ_1, n_1, β and γ into Equation (11) to calculate $\boldsymbol{R}_2^{\infty}$.

7. Substitute the value of $\boldsymbol{P}_2^{\infty}$, $\boldsymbol{R}_2^{\infty}$ into Equation (13) to calculate the dipole moment value μ of the ethyl acetate molecule.

Thinking and discussing

1. The main source of error in analyzing this experiment, how to improve it?

2. Try to explain the key points of measuring the permanent dipole moment of polarized molecules by solution method, what are the basic assumptions, and what approximations are made when deriving the formula?

3. How to use the solution method to measure the "solvent effect" of the dipole moment to study the interaction between polar solute molecules and non-polar solvent molecules?

Reference materials

1. Xiang Y, Li S. Intermediate Physical Chemistry. Beijing: Higher Education Press, 1988: 142.

2. Xing Z. The electric dipole moment of an electron and its possible limit to new physics. Knowledge of Modern Physics,2023, 35(03): 35-37.

3. Zhang Y, Zhou L, Gao X, et al. The integrated design of dipole moment measurement experiment based on the concept of green chemistry. Shandong Chemical Industry, 2022, 51(02): 187-188+203.

Appendix　Bibliographic values

Table 22-1　Dipole moment of ethyl acetate molecule[3]

μ/D	$\mu\times10^{-30}$/C · m*	State or solvent	Temperature/℃
1.78	5.94	gas	30～195
1.83	6.10	liquid	25
1.76	5.87	CCl_4	25
1.89	6.30	CCl_4	25

* Converted by 1D=3.335 64 C · m

实验 23　常压顺-丁烯二酸催化氢化

实验目的

1. 了解烯酸催化氢化基本原理。

2. 通过顺-丁烯二酸催化氢化，掌握常压液相催化氢化操作。

基本原理

催化氢化反应是指在催化剂的作用下，氢分子加成到有机化合物的不饱和基团上的反应。可以使烯键、炔键直接加氢，也可以使许多不饱和官能团得到还原。催化氢化适用于大规模和连续化生产，在工业上有重要用途。例如，石油裂解气中的乙炔和丙炔等通过钯催化部分氢化，可生产高纯度的乙烯和丙烯；在油脂工业中将液态油氢化为固态或半固态的脂肪，可生产人造奶油或肥皂工业用的硬化油。

催化氢化的机理通常被认为是氢和有机分子中的不饱和键首先被吸附在催化剂的表面上，氢和不饱和键被催化剂的活化中心活化形成中间产物，再进一步与活化了的氢作用，生成饱和有机分子，从催化剂表面脱附。氢化用的催化剂种类繁多，常用的有镍、铂和钯等。按氢化的方法不同，催化氢化又可分为常压液相催化氢化、加压液相催化氢化和气相催化氢化。

本实验是顺-丁烯二酸在 Adams 催化剂存在下,在常温下于乙醇溶剂中进行的常压液相催化氢化

$$\begin{matrix} \text{CHCOOH} \\ \| \\ \text{CHCOOH} \end{matrix} + H_2 \xrightarrow[\text{乙醇}]{\text{Adams 催化剂}} \begin{matrix} \text{CH}_2\text{COOH} \\ | \\ \text{CH}_2\text{COOH} \end{matrix}$$

Adams 催化剂($PtO_2 \cdot H_2O$)是铂催化剂的一种,由氯铂酸与硝酸钠熔融分解制得。氧化铂在反应过程中首先吸收氢,迅速转变成活性铂。

催化剂的活性影响催化反应的速度,可以用半氢化时间来量度。所谓半氢化时间是氢化过程进行到被反应物吸收的氢气量达到它全部氢化所需氢气量的一半时所用去的时间,它可以从吸收氢气的体积-时间图求得。

分子中所含不饱和键的数目可根据氢化时所消耗的氢气量计算得到。计算时应将实验条件下消耗的氢气体积换算成标准状态下的体积,再减去催化剂本身消耗的氢气体积。设氢化时温度为 t,大气压力为 p,吸收氢气的总体积为 V,t 时的氢气分压、水和乙醇的蒸气压分别为 p_H,p_W,p_E。则有

$$\frac{p_0 V_0}{T_0} = \frac{p_H V}{T}, \quad p_H = p - p_W - p_E, \quad \frac{101\ 325 \times V_0}{273} = \frac{(p - p_W - p_E) \times V}{273 + t}$$

$$V_0 = \frac{273 \times (p - p_W - p_E) \times V}{101\ 325 \times (273 + t)} \tag{1}$$

而催化剂消耗的氢气体积 V_c 为

$$V_c/\text{mL} = \frac{m_c \times 2}{M_c} \times 22.415 \tag{2}$$

式中:m_c 和 M_c 分别为催化剂的用量(单位:mg)和摩尔质量。所以,反应物实际消耗的氢气体积 V_e 应为

$$V_e = V_0 - V_c$$

故分子含双键数 n 为

$$n = \frac{\text{反应物消耗的氢气物质的量}}{\text{反应物的物质的量}} = \frac{V_e}{22415} \times \frac{M}{m} \tag{3}$$

式中 M 和 m 为不饱和反应物的摩尔质量和称取的量。

由式(3)可求算不饱和化合物中的双键数。

仪器和试剂

常压催化氢化装置　　熔点测定仪
薄膜旋转蒸发器　　真空干燥器

氢气钢瓶
布氏漏斗(4 cm)
瓷蒸发皿
硝酸钠(化学纯)
乙醇(95%)(化学纯)
磁力搅拌器
吸滤瓶(250 mL)
顺-丁烯二酸(化学纯)
氯铂酸(化学纯)

实验步骤

1. Adams 催化剂的制备

在瓷蒸发皿中加入 0.4 g 氯铂酸,用 4 mL H_2O 溶解,再加入 4 g 硝酸钠,反应混合物在搅拌下用小火缓缓加热蒸发至干,在不断搅动下,约在 10 min 内升温至 350～370 ℃。在这一过程中,反应物先变稠发黏,而且发泡放出棕色的二氧化氮气体,并逐渐熔融成液体。再经 5 min 左右温度升至 400 ℃左右,气体逸出量大大减少。再升温到 500～550 ℃,维持此温度 30 min。冷却瓷蒸发皿到室温,加入少量蒸馏水溶解融块,棕色沉淀用蒸馏水倾析洗涤 2 次,然后抽气过滤,沉淀物用蒸馏水洗涤 6～7 次,抽干后置于真空干燥器中干燥备用。

2. 催化氢化操作

在氢化瓶中加入 100 mg Adams 催化剂和 70 mL 乙醇,再加入约 2.0 g 顺-丁烯二酸,加搅拌子,盖上通气瓶塞,置于磁力搅拌器上,如图 23-1 安装好仪器。

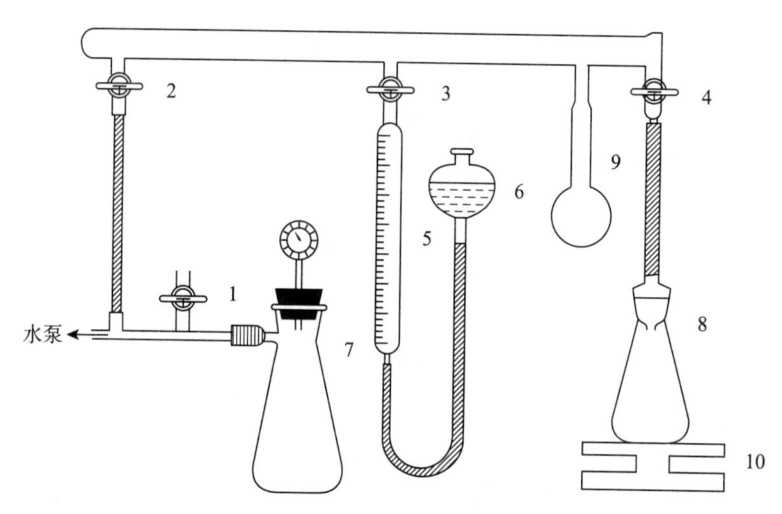

图 23-1 常压催化氢化装置图

1,2,3,4-活塞;5-量气管;6-水准瓶;7-安全瓶;8-氢化瓶;9-阱;10-磁力搅拌器

(1) 排除量气管余气

打开活塞 1 和 3,转动三通活塞 2 使量气管 5 经安全瓶通大气,慢慢升高水准瓶 6 使

量气管内液面上升，液面接近活塞 3 时立即将 3 关闭。然后放低水准瓶，置于架上。

（2）抽空氢化系统并用充氢稀释法排除空气

关活塞 1，开活塞 4，然后用水泵抽空氢化瓶（真空度不宜过高，以免减少乙醇的挥发）。小心转动三通活塞 2 使氢气通入。充满氢气后再小心转动 2 使氢化瓶再次抽空。如此反复 2～3 次，使系统中的空气排尽。最后通入氢气并打开 1，关水泵停止抽气。

（3）量气管充气

打开三通活塞 2 使氢气进入，放低水准瓶位置，使氢气通入量气管。然后关 2，调整水准瓶的高度，使水准瓶内的液面和量气管中液面相平。记录量气管体积、室温和气压。

（4）氢化

保持水准瓶的液面高于量气管的液面，开动搅拌，同时计时。每隔 1 min 记录体积（读取量气管刻度时要使水准瓶和它的水面相平），直至不再吸收氢气为止。关活塞 3 和 4，停止搅拌，取下氢化瓶，记录温度和大气压。

反应物经过滤，滤出催化剂（催化剂连同滤纸放入回收瓶中），滤液蒸去乙醇，得产物，产物经干燥后称重，测熔点和红外谱图。

数据记录与处理

1. 半氢化时间的确定

记录相关数据，根据吸氢体积-时间关系图，由图吸氢一半时所对应的时间确定半氢化时间。

2. 分子中双键数目的确定

将实验测得的吸氢体积换算成为标准状态下的吸氢体积 V_o，再减去催化剂消耗的氢气体积 V_c 得反应物消耗的氢气体积 V_e，再根据下式计算反应物分子中的双键数 n

$$n = \frac{V_e}{22415} \times \frac{M}{m}$$

3. 测定产物质量及熔点

对产品的进行红外表征，并对红外谱图进行解析。

思考与讨论

试讨论影响催化氢化反应的因素。

参考资料

1. 林斯台德，等.有机化学近代技术. 上海科学技术出版社，1960.

2. 霍宁. 有机合成. 北京：科学出版社，1981.

Experiment 23

Catalytic hydrogenation of maleic acid at normal pressure

Experimental purpose

1. Understand the basic principles of catalytic hydrogenation of acrylic acid.

2. Master the operation of liquid phase catalytic hydrogenation at normal pressure by maleic acid catalytic hydrogenation.

Basical principles

Hydrogenation is the reaction in which hydrogen molecules are added to the unsaturated group of an organic compound in the presence of catalyst. It can be directly hydrogenated to the olefin and alkyne bonds, and many unsaturated functional groups can be reduced. Catalytic hydrogenation is suitable for large-scale and continuous production, and has important applications in industry. For example, the partial hydrogenation of acetylene and proparyne in petroleum cracking gas with palladium catalyst can produce high purity ethylene and propylene. The hydrogenation of liquid oil to solid or semi-solid fats in the oil and grease industry, which can produce margarine or hardened oil for soap industry.

The mechanism of catalytic hydrogenation is generally considered that the hydrogen and unsaturated bonds of organic molecules are first adsorbed on the surface of the catalyst, then the hydrogen and unsaturated bonds are activated by the activation center of the catalyst to form intermediate products, furthermore, it reacts with the activated hydrogen to form saturated organic molecules, and desorbs from the surface of the catalyst.There are many kinds of catalysts for hydrogenation, including nickel, platinum and palladium. According to different methods of hydrogenation, catalytic hydrogenation can be divided into liquid phase catalytic hydrogenation at normal pressure, pressurized liquid phase catalytic hydrogenation and gas phase catalytic hydrogenation.

This experiment is liquid phase catalytic hydrogenation of maleic acid in ethanol solvent at room temperature and normal pressure in the presence of Adams catalyst

$$\begin{array}{l}CHCOOH\\ \| \\ CHCOOH\end{array} + H_2 \xrightarrow[\text{Ethanol}]{\text{Adams catalyst}} \begin{array}{l}CH_2COOH\\ | \\ CH_2COOH\end{array}$$

Adams catalyst ($PtO_2 \cdot H_2O$) is a kind of platinum catalyst, which is prepared by melting and decomposing chloroplatinic acid and sodium nitrate. In the reaction process, platinum oxide first absorbs hydrogen and rapidly converts into active platinum.

The activity of catalyst affects the speed of catalytic reaction, which can be measured by half-hydrogenation time. The so-called half-hydrogenation time is the time that takes from the hydrogenation process to the time when the amount of hydrogen absorbed by the reactant reaches half of the amount of hydrogen required for its full hydrogenation. It can be obtained from the volume time diagram of hydrogen absorption.

The number of unsaturated bonds contained in the molecule can be calculated according to the amount of hydrogen consumed during hydrogenation. During calculation, the volume of hydrogen consumed under experimental conditions should be converted into the volume under standard conditions, and then the volume of hydrogen consumed by the catalyst itself should be deducted. Assume that the temperature during hydrogenation is t, the atmospheric pressure is p, the total volume of absorbed hydrogen is V, and the partial pressure of hydrogen, the vapor pressure of water and ethanol at t are p_H, p_W, p_E, respectively. There is

$$\frac{p_0 V_0}{T_0} = \frac{p_H V}{T}, \quad p_H = p - p_W - p_E, \quad \frac{101\,325 \times V_0}{273} = \frac{(p - p_W - p_E) \times V}{273 + t}$$

$$V_0 = \frac{273 \times (p - p_W - p_E) \times V}{101\,325 \times (273 + t)} \tag{1}$$

The volume of hydrogen consumed by the catalyst, V_c, is

$$V_c/\text{mL} = \frac{m_c \times 2}{M_c} \times 22.415 \tag{2}$$

where, m_c and M_c are the amount of catalyst (mg) and molar mass, respectively. Therefore, the volume of hydrogen actually consumed by the reactant, V_e, should be

$$V_e = V_0 - V_c$$

So the double bonds contained in the molecule

$$n = \frac{\text{Molar number of hydrogen consumed by reactants}}{\text{Mole number of reactants}} = \frac{V_e}{22 \cdot 415} \times \frac{M}{m} \tag{3}$$

In the formula, M and m are the molar mass and the weight of the unsaturated reactants.

The number of double bonds in unsaturated compounds can be calculated from Formula (3).

Apparatus and reagents

Atmospheric pressure catalytic hydrogenation unit

Film rotary evaporator

Hydrogen cylinder

Buchner funnel (4 cm)

Melting point tester

Vacuum dryer

Magnetic stirrer

Suction bottle (250 mL)

Porcelain evaporating dish
Sodium nitrate(CP)
Ethanol (95%)
Maleic acid (CP)
Chloroplatinic acid(CP)

Experimental procedures

1. Preparation of Adams catalyst

Add 0.4 g chloroplatinic acid to the porcelain evaporating dish and dissolve it with 4 mL H_2O, then add 4 g sodium nitrate. The reaction mixture is slowly heated and evaporated to dry under stirring, and then the temperature rises to 350～370 ℃ in about 10 min under constant stirring. In this process, the reactants become thick and sticky, and the foaming releases brown nitrogen dioxide gas, which gradually melts into a liquid. After about 5 min, the temperature rises to about 400 ℃, and the gas escape is greatly reduced, and then the temperature rises to 500～550 ℃, and is maintained for 30 min. Cool the porcelain evaporating dish to room temperature, add a small amount of distilled water to dissolve the melting block. Brown precipitates are decanted and washed with distilled water twice, and then filtered by air extraction. The precipitates are washed with distilled water for 6～7 times, and then dried in a vacuum dryer for later use.

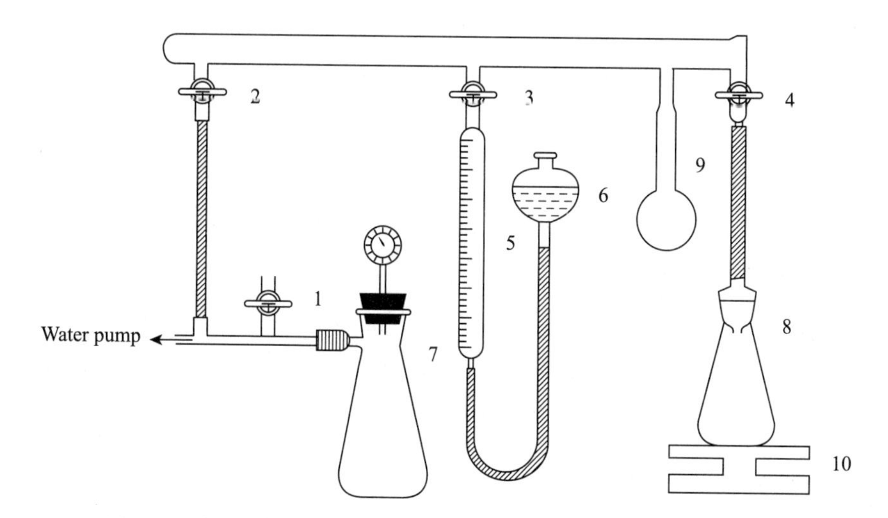

Fig. 23-1 Diagram of catalytic hydrogenation unit at normal pressure.

1,2,3,4-Piston; 5-Measuring trachea; 6-Level bottle; 7-Safety bottle; 8-Hydrogenation bottle; 9-Well; 10-Magnetic stirrer

2. The operation of catalytic hydrogenation

Add 100 mg Adams catalyst and 70 mL ethanol into the hydrogenation bottle. Then about 2.0 g maleic acid and stirrer are added into the bottle, cover the breather stopper, and place the hydrogenation bottle on the magnetic stirrer, as shown in Fig. 23-1. Install the instrument.

(1) Remove the residual gas from the measuring trachea

Open pistons 1 and 3, turn the three-way piston 2 to make the measuring trachea 5 pass through the safety bottle to the atmosphere, slowly raise the level bottle 6 to make the liquid rise in the measuring trachea, and immediately close piston 3 when the liquid level is close to the piston 3. Then lower the level bottle and put it on the rack.

(2) Evacuate the hydrogenation system and exhaust the airby hydrogen dilution method

Close the piston 1, open the piston 4, and then use the water pump to empty the hydrogenation bottle (the vacuum should not be too high so as not to reduce the evaporation of ethanol). Carefully turn the three-way piston 2 to allow hydrogen to enter. After being filled with hydrogen, carefully turn 2 to evacuate the hydrogenation bottle again. Repeat 2 ~ 3 times to exhaust the air in the system. Finally, inject hydrogen and open piston 1, shut down the water pump and stop air extraction.

(3) Fill gas into the measuring trachea

Open the three-way piston 2 to allow hydrogen to enter, lower the position of the level bottle, and allow hydrogen to enter the measuring trachea. Then turn off piston 2 and adjust the height of the level bottle so that the liquid in the level bottle is align horizontally with the liquid in the measuring tube. Record the volume, room temperature and air pressure of the measuring trachea.

(4) Hydrogenation

Keep the liquid of the level bottle higher than that of the measuring trachea, start stirring while timing. Record the volume every 1 min (make level bottle the same with its water surface when reading the scale of the measuring trachea). Until hydrogen is no longer absorbed, close pistons 3 and 4, stop stirring, remove the hydrogenation bottle, and record the temperature and pressure.

The reactant is filtered, and the catalyst is filtered out (the catalyst and filter paper are put into the recovery bottle). Ethanol is evaporated from the filtrate, and the product is obtained. After drying, the product is weighed, and the melting point and infrared spectrum are measured.

Data recording and processing

1. Determination of semi-hydrogenation time

Record the related data. According to the diagram of hydrogen absorption volume-time relationship, determine the half-hydrogenation time from the diagram of half hydrogen absorption corresponding time.

2. Determination of the double bonds number in molecules

The volume of hydrogen absorption measured in the experiment is converted to V_0 in the standard state, and then the volume of hydrogen consumed by catalyst V_c is subtracted to get the volume of hydrogen consumed by reactant V_e, the number of double bond n in the reactant molecule is

$$n = \frac{V_e}{22415} \times \frac{M}{m}$$

3. Determination of the product mass and melting point

The product is characterized by IR and the IR spectrum is analyzed.

Thinking and discussing

Try to discuss the factors affecting the catalytic hydrogenation reaction.

Reference materials

1. Linstad, et al. Modern Technology of Organic Chemistry. Shanghai Science and Technology Press, 1960.

2. Hornin. Organic Synthesis. Beijing: Science Press, 1981.

实验 24 氨分子与水分子的二聚体稳定结构及其氢键强度预测

实验目的

1. 掌握如何使用 Gaussian 09 程序优化分子结构和进行能量计算。

2. 通过对氨分子与水分子的二聚体氢键预测，加深对计算化学的理解。

基本原理

氢键是一种特殊的化学键合形式，在很多物质体系中广泛存在着氢键。它不仅影响着许多物质的构象及物理、化学性质，通过影响生物系统如蛋白质、核酸、生物膜等的结构，还会对人类及动植物的生理、生化过程产生深刻影响。在无机化学、有机化学、分析化学、生物化学、高分子材料等领域中，氢键都具有极其特殊的地位。

氢键是一种被广泛关注的重要的弱相互作用，通常以 X—H…Y 表示，其中 X—H 为质子供体，Y 为质子受体(可以是原子或原子团)。研究中一般用相互作用能描述氢键相互作用的强弱，并定义为氢键键能。据此，由分子 A 和 B 形成的氢键复合物 AB 的氢键强度计算公式如下

$$IE = E_{AB}(\vec{R}_A + \vec{R}_B) - E_A(\vec{R}_A) - E_B(\vec{R}_B)$$

其中 E_{AB} 是复合物 AB 的能量，E_A 和 E_B 分别代表分子 A 和 B 的单体能量，$\vec{R}_A$ 和 $\vec{R}_B$ 是复合物中分子 A 和分子 B 的位置坐标。

本实验中，由氨分子和水分子通过氢键作用形成二聚体的氢键强度，可视为如下广义化学反应的反应热。可通过理论计算获得反应物和产物的能量，进而求得氢键强度。

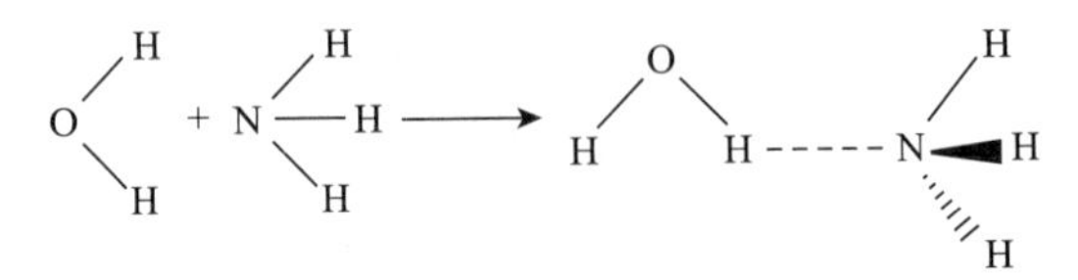

图 24-1 氨分子和水分子形成氢键示意图

仪器和软件

PC 计算机（windows 操作系统）

Gaussian 09 软件

GaussView 5.0 软件

实验步骤

1. 构建分子初始结构，生成输入文件

① 打开 GaussView 软件，选择"Element Fragments"按钮，从中选取 N 的"Nitrogen Tetravalent"和 O 的"Oxygen Tetravalent"，初步构建氨分子与水分子形成的二聚体。而后结合使用 ALT 键和鼠标将二聚体初始结构调整如图 24-2 所示。

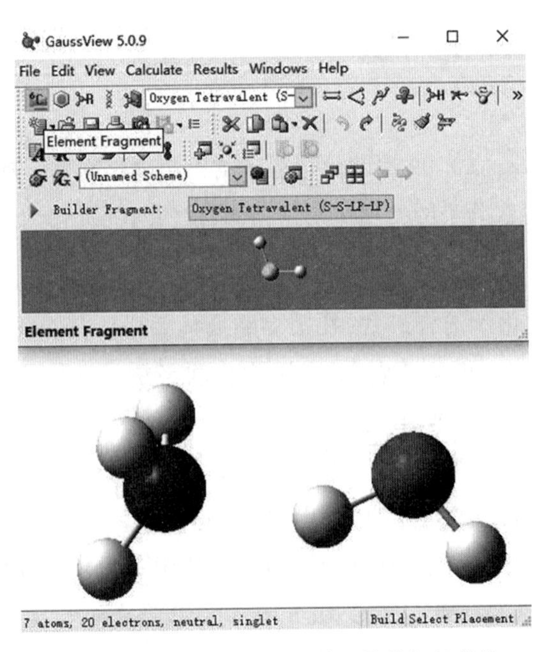

图 24-2 氨分子和水分子的二聚体初始结构

② 鼠标左击:“File—Save”,将文件命名为 NH3-H2O. gjf 并保存成笛卡儿坐标的形式至文件夹中。然后对输入文件进行修改,得到图 24-3 形式。

```
%chk=F:\Lab\NH3-H2O.chk
#p b3lyp/6-31+g** opt freq

b3lyp/6-31+g**

0 1
 N                 -0.37193375    -0.64499701    -0.83031229
 H                 -1.23435416    -0.21781892    -0.55874289
 H                 -0.38665511    -1.61337117    -0.58124451
 H                 -0.25325308    -0.55695954    -1.81933418
 O                  0.69467724     0.89195829     0.87683047
 H                  1.66451360     0.92326549     0.85691286
 H                  0.43601880     0.32985957     0.12904957
```

图 24-3 NH3-H2O. gjf 文件

③ 以同样的方式构建氨分子和水分子单体的初始结构,并生成单体结构优化的输入文件,可命名为 NH3. gjf 和 H2O. gjf。

2. 作业提交和运行

启动 G09W 程序,打开并运行 NH3-H2O. gjf 文件,将运行结果保存在 NH3-H2O.out 文件中,并确认程序顺利结束。而后对单体氨分子和水分子重复上述操作。

3. 数据查找和结果确认

作业计算完成后,须确认优化好的结构为稳定的几何构型:输出文件中是否出现四个“YES”和无虚频。

① 在优化程序末尾出现“Stationary point found”字样,并且有四个“YES”,表明结构已经优化完毕,如图 24-4 所示。

```
         Item               Value     Threshold  Converged?
Maximum Force            0.000049     0.000450     YES
RMS     Force            0.000018     0.000300     YES
Maximum Displacement     0.000459     0.001800     YES
RMS     Displacement     0.000257     0.001200     YES
Predicted change in Energy=-2.249855D-08
Optimization completed.
   -- Stationary point found.
```

图 24-4 结构优化完毕

② 在结果的频率分析部分出现“Frequencies”,对应的第一行第一列的数,即为该结构的最小振动频率,应为正值,代表无虚频,如图 24-5。以上两点均满足即为稳定的几何构型。

```
                             1                      2                      3
                             A                      A                      A
 Frequencies --          24.4217               194.3069               219.3471
 Red. masses --           1.0925                 2.3408                 1.0704
 Frc consts  --           0.0004                 0.0521                 0.0303
 IR Inten    --          98.0828                14.0136                44.6230
  Atom  AN      X      Y      Z        X      Y      Z        X      Y      Z
     1   7     0.00   0.00  -0.03     0.21  -0.04   0.00     0.02   0.01   0.06
     2   1    -0.03  -0.04  -0.24     0.75   0.10   0.01     0.04  -0.03  -0.21
     3   1     0.20   0.24   0.15     0.08   0.19   0.01    -0.56   0.07  -0.19
     4   1    -0.17  -0.20   0.22    -0.11   0.22   0.01     0.57  -0.10  -0.20
     5   8     0.00   0.00   0.07    -0.23   0.01   0.00    -0.02   0.00  -0.02
     6   1     0.01   0.00  -0.83     0.16   0.19   0.01    -0.05  -0.02  -0.31
     7   1     0.00   0.00  -0.07    -0.18  -0.37  -0.01    -0.02   0.03   0.34
```

图 24-5 频率分析

③ 优化后电子能（E_{elec}）的查找：在"Stationary point found"词条后，查找词条"HF"，如图 24-6 所示，对应的能量即为电子能。

```
Leave Link  601 at Fri Apr 21 16:28:59 2023, MaxMem=   33554432 cpu:        0.0
(Enter C:\G09W\l9999.exe)
1|1|UNPC-DESKTOP-2PQPC9R|FOpt|RB3LYP|6-31+G(d,p)|H5N1O1|PCLWANG|21-Apr
-2023|0||#p b3lyp/6-31+g** opt freq||b3lyp/6-31+g**||0,1|N,-0.48860005
82,-0.6946962645,-0.8853245001|H,-1.452646582,-0.3975948333,-0.7556505
739|H,-0.4385587087,-1.6831153044,-0.6527312848|H,-0.2677235131,-0.596
5614244,-1.872777565|O,0.8698062572,1.0541299886,0.994852652|H,1.82853
44231,0.9907259247,0.918220684|H,0.4982017218,0.4390486233,0.326569617
8||Version=IA32W-G09RevA.02|State=1-A|HF=-133.0135065|RMSD=3.666e-009|
RMSF=2.105e-005|Dipole=-0.1647889,-1.0441768,-1.1260171|Quadrupole=2.8
803685,-1.380395,-1.4999736,0.8447156,0.8884238,-0.6758026|PG=C01 [X(H
5N1O1)]||@
```

图 24-6 电子能的查找

数据记录与处理

1. 记录氨分子和水分子二聚体和各自单体的优化后的构型和能量。

2. 计算氨分子和水分子二聚体的相互作用能，即氢键键能。

要求：键长保留 3 位小数；键角和二面角保留 1 位小数；能量保留 6 位小数，如 −133.013507 hartree，相互作用能换算为 $kJ \cdot mol^{-1}$后保留 2 位小数。

思考与讨论

1. Gaussian 09 软件输出的分子电子能是否只包括电子的能量，为什么？

2. 本实验预测的氢键强度是否准确？怎样做才能够提高计算精度？

参考资料

1. 华彤文，王颖霞，等. 普通化学原理（第四版）. 北京：北京大学出版社，2013.

2. 周公度，段连运. 结构化学基础（第五版）. 北京：北京大学出版社，2017.

3. 胡红智，马思渝. 计算化学实验. 北京：北京师范大学出版社，2008.

4. Jeffrey G A. An Introduction to Hydrogen Bonding. New York：Oxford University Press，1997.

5. https://www.gaussian.com/

Experiment 24

Prediction of stable structure and hydrogen bond strength for dimer of ammonia and water molecules

Experimental purpose

1. Learn to use Gaussian 09 to build the structure of molecule.

2. Learn to use Gaussian program to do structure optimization and calculate energy of the system.

Basical principles

Hydrogen bond is an important weak interaction which attracts attention from researchers of different fields. Usually, it is represented by X—H⋯Y, where X—H is the proton donor and Y is the proton acceptor (it can be an atom or a cluster). The interaction energy is usually used to describe the strength of hydrogen bond, which is defined as the hydrogen bond energy. As a result, the formula for calculating the hydrogen bond energy of dimer AB is as follows

$$IE = E_{AB}(\vec{R}_A + \vec{R}_B) - E_A(\vec{R}_A) - E_B(\vec{R}_B)$$

where E_{AB} is the energy of dimer AB, E_A and E_B represents the energy of molecule A and B, $\vec{R}_A$ and $\vec{R}_B$ is the position of molecule A and B.

In this experiment, the hydrogen bond energy of dimers formed by ammonia and water molecules can be regarded as the reaction heat of the following chemical reaction, where the energy of reactants and products can be obtained through theoretical calculations.

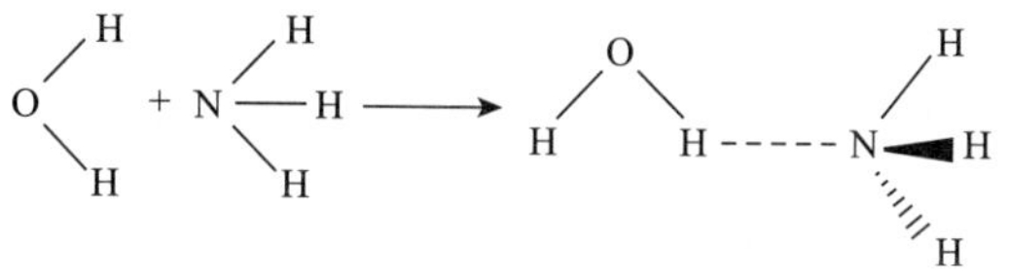

Fig. 24-1 Schematic diagram of hydrogen bond between ammonia and water molecules

Apparatus and software

PC with Windows

Gaussian 09

GaussView 5.0

Experimental procedures

1. Construct the initial structure of molecules and generate the input files.

① Open the GaussView software, click the "Element Fragments" button, select N's

"Nitrogen Tetravalant" and O's "Oxygen Tetravalant" to construct a dimer formed by ammonia and water molecules. Then use the ALT key and mouse to adjust the initial structure of the dimer to the position shown in Fig. 24-2.

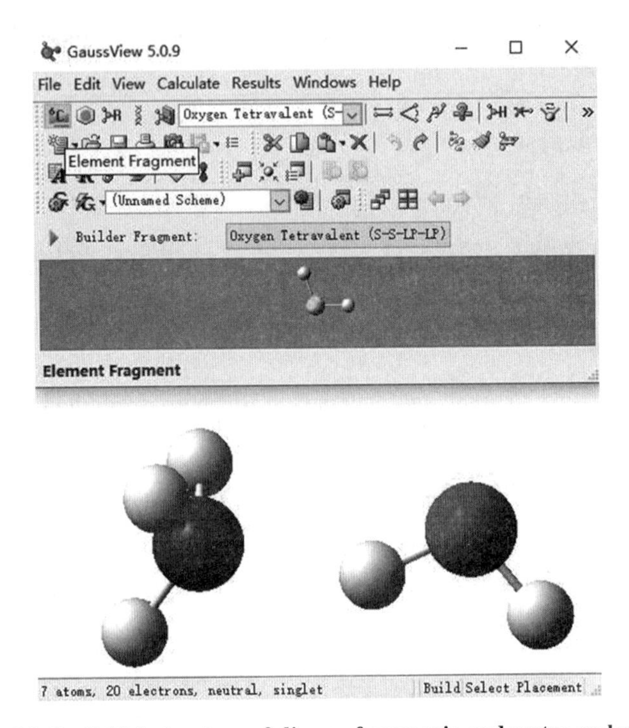

Fig. 24-2 Initial structure of dimer of ammonia and water molecules

② Left click: "File—Save", name the file as NH3-H2O. gjf, and save it in the form of Cartesian coordinates to the folder. Then modify the input file as the following form of Fig. 24-3.

```
%chk=F:\Lab\NH3-H2O.chk
#p b3lyp/6-31+g** opt freq

b3lyp/6-31+g**

0 1
 N                 -0.37193375    -0.64499701    -0.83031229
 H                 -1.23435416    -0.21781892    -0.55874289
 H                 -0.38665511    -1.61337117    -0.58124451
 H                 -0.25325308    -0.55695954    -1.81933418
 O                  0.69467724     0.89195829     0.87683047
 H                  1.66451360     0.92326549     0.85691286
 H                  0.43601880     0.32985957     0.12904957
```

Fig. 24-3 Input file of NH3-H2O. gjf

③ Construct the initial structures of ammonia and water monomers in the same way, and generate input files for structure optimization, which can be named as NH3. gjf and H2O. gjf.

2. Job submission

Start the G09W program, open and run the NH3-H2O. gjf file, save the results in the NH3-H2O.out file, and confirm that the program ends normally. Then repeat the above operation for monomer ammonia and water molecules.

3. Data and results searching

After the job finishes, it is necessary to confirm that the optimized structure is a stable geometric configuration: there are four "YES" and no imaginary frequencies in the output file.

① At the end of the * . out file, the words "Stationary point found" appear with four "YES", indicating that the structure is optimized, as shown in Fig. 24-4.

```
         Item               Value     Threshold  Converged?
Maximum Force            0.000049     0.000450     YES
RMS     Force            0.000018     0.000300     YES
Maximum Displacement     0.000459     0.001800     YES
RMS     Displacement     0.000257     0.001200     YES
Predicted change in Energy=-2.249855D-08
Optimization completed.
   -- Stationary point found.
```

Fig. 24-4 Completed structure optimization

② In the frequency analysis section, the number in the first row and first column which stands for the minimum vibrational frequency of the structure must be positive, indicating no imaginary frequency, as shown in Fig. 24-5. The stable geometric configuration is found when both of the above requirements are met.

```
                         1                      2                      3
                         A                      A                      A
Frequencies --       24.4217               194.3069               219.3471
Red. masses --        1.0925                 2.3408                 1.0704
Frc consts  --        0.0004                 0.0521                 0.0303
IR Inten    --       98.0828                14.0136                44.6230
 Atom  AN      X      Y      Z        X      Y      Z        X      Y      Z
   1    7     0.00   0.00  -0.03     0.21  -0.04   0.00     0.02   0.01   0.06
   2    1    -0.03  -0.04  -0.24     0.75   0.10   0.01     0.04  -0.03  -0.21
   3    1     0.20   0.24   0.15     0.08   0.19   0.01    -0.56   0.07  -0.19
   4    1    -0.17  -0.20   0.22    -0.11   0.22   0.01     0.57  -0.10  -0.20
   5    8     0.00   0.00   0.07    -0.23   0.01   0.00    -0.02   0.00  -0.02
   6    1     0.01   0.00  -0.83     0.16   0.19   0.01    -0.05  -0.02  -0.31
   7    1     0.00   0.00  -0.07    -0.18  -0.37  -0.01    -0.02   0.03   0.34
```

Fig. 24-5 Frequency analysis

③ Search for electronic energy (E_{opt}): After words of "Stationary point found", search for the keyword "HF", as shown in Fig. 24-6. The corresponding energy is the electronic energy.

```
Leave Link  601 at Fri Apr 21 16:28:59 2023, MaxMem=   33554432 cpu:       0.0
(Enter C:\G09W\l9999.exe)
1|1|UNPC-DESKTOP-2PQPC9R|FOpt|RB3LYP|6-31+G(d,p)|H5N1O1|PCLWANG|21-Apr
-2023|0||#p b3lyp/6-31+g** opt freq||b3lyp/6-31+g**||0,1|N,-0.48860005
82,-0.6946962645,-0.8853245001|H,-1.452646582,-0.3975948333,-0.7556505
739|H,-0.4385587087,-1.6831153044,-0.6527312848|H,-0.2677235131,-0.596
5614244,-1.872777565|O,0.8698062572,1.0541299886,0.994852652|H,1.82853
44231,0.9907259247,0.918220684|H,0.4982017218,0.4390486233,0.326569617
8||Version=IA32W-G09RevA.02|State=1-A|HF=-133.0135065|RMSD=3.666e-009|
RMSF=2.105e-005|Dipole=-0.1647889,-1.0441768,-1.1260171|Quadrupole=2.8
803685,-1.380395,-1.4999736,0.8447156,0.8884238,-0.6758026|PG=C01 [X(H
5N1O1)]||@
```

Fig. 24-6 Searching for electronic energy

Data recording and processing

1. Record the energy of dimers and monomers of ammonia and water molecules.

2. Calculate the interaction energy of dimer of ammonia and water molecules.

Requirement: Bond length, keep 3 decimal places; bond angle and dihedral angle, keep 1 decimal place; Energy, keep 6 decimal places, such as −133.013507 hartree; keep 2 decimal places after interaction energy is converted to $kJ \cdot mol^{-1}$; Vibrational frequency, keep 2 decimal places, such as 4.56 cm^{-1}.

Thinking and discussing

1. If the calculated energy only includes the electronic energy of a molecule? Why?

2. What is the hydrogen bond energy for dimer of ammonia and water molecules obtained in the experiment? Why are the predicted results from simulation different from the experimental values? How can we improve the accuracy of simulation to match the experimental values?

Reference materials

1. Hua T, Wang Y, et al. Principles of General Chemistry (4th ed). Beijing: Peking University Press, 2013.

2. Zhou G, Duan L. Fundamentals of Structural Chemistry (5th ed). Beijing: Peking University Press, 2017.

3. Hu H, Ma S. Experiment of Computational Chemistry. Beijing: Beijing Normal University Press, 2008.

4. Jeffrey G A. An Introduction to Hydrogen Bonding. New York: Oxford University Press, 1997.

5. https://www.gaussian.com/

第四章

研究设计型实验

实验 25　氧氟沙星在固体电极表面的电化学行为

实验目的

1. 了解不可逆循环伏安曲线、示差脉冲伏安曲线特点，掌握相关的实验技术。
2. 利用循环伏安法筛选能催化氧氟沙星的电极材料。
3. 利用选择的电极及示差脉冲伏安法测定实际样品中氧氟沙星含量。

基本原理（围绕下面主题，撰写完成）

1. 不可逆循环伏安曲线特点。
2. 示差脉冲伏安法的原理及应用。
3. 氧氟沙星的结构及性能。
4. 氧氟沙星在电极表面的催化氧化机理。

仪器和试剂

CHI660E 电化学工作站
氧氟沙星（分析纯）
对氨基苯磺酸（分析纯）
电极（锌、铜、玻碳、饱和甘汞、铂丝电极各一支）
磷酸盐缓冲溶液（pH 6.5，含 0.1 mol/L 氯化钾）

实验步骤

1. 制备聚对氨基苯磺酸修饰玻碳电极

使用玻碳电极作为工作电极，饱和甘汞电极作为参比电极，铂丝电极作为辅助电极，在 0.02 $mol \cdot L^{-1}$的对氨基苯磺酸溶液中利用循环伏安法制备聚对氨基苯磺酸修饰玻碳电极。设置参数：电位范围为$-1.5 \sim 2.5$ V，扫速为 0.1 $V \cdot s^{-1}$，灵敏度为 1×10^{-4} $A \cdot V^{-1}$。

2. 利用循环伏安法筛选出能催化氧氟沙星的电极材料

(1) 用磷酸盐缓冲溶液配制 2×10^{-3} $mol \cdot L^{-1}$氧氟沙星溶液。

（2）筛选催化氧氟沙星效果最好的电极材料

分别使用锌电极、铜电极、裸玻碳电极、聚对氨基苯磺酸修饰玻碳电极作为工作电极，饱和甘汞电极作为参比电极，铂丝电极作为辅助电极，在 2×10^{-3} mol · L^{-1} 氧氟沙星溶液中扫循环伏安图，根据峰电流高低选择合适的催化电极材料。设置参数：电位范围为 0.4～1.6 V，扫速为 0.15 V · s^{-1}，采样间隔为 0.001 V，灵敏度为 1×10^{-4} A · V^{-1}。

3. 绘制检测系列浓度氧氟沙星的标准曲线

（1）配制系列浓度氧氟沙星溶液

使用磷酸盐缓冲溶液配制浓度分别为 6.5×10^{-4}，3.5×10^{-4}，2.0×10^{-4}，1×10^{-4}，5×10^{-5} mol · L^{-1} 氧氟沙星标准溶液。

（2）使用聚对氨基苯磺酸修饰玻碳电极作为工作电极，饱和甘汞电极作为参比电极，铂丝电极作为辅助电极，在上述浓度的氧氟沙星溶液中分别做示差脉冲伏安扫描，记录峰电流。设置参数：电位范围为 0.5～1.2 V，扫速为 0.1 V · s^{-1}，采样间隔为 0.004 V，灵敏度为 1×10^{-4} A · V^{-1}。

（3）利用所得到的峰电流值对浓度作图，绘制氧氟沙星溶液的标准曲线。

4. 测定实际样品的氧氟沙星含量

（1）用 pH 为 6.5 的磷酸盐缓冲溶液稀释市售氧氟沙星滴耳液，至上述工作曲线的浓度范围内。

（2）使用聚对氨基苯磺酸修饰玻碳电极作为工作电极，饱和甘汞电极作为参比电极，铂丝电极作为辅助电极，在上述浓度氧氟沙星溶液中扫描示差脉冲伏安曲线，记录峰电流值，代入标准曲线的回归方程中，根据标准曲线即可求得氧氟沙星滴耳液中氧氟沙星含量。

（3）利用所得数据求制备的聚对氨基苯磺酸修饰玻碳电极检测氧氟沙星的误差。

数据记录与处理

1. 筛选对氧氟沙星具有较好催化氧化作用的电极材料。

2. 绘制电极对氧氟沙星检测的标准曲线。

3. 利用示差脉冲伏安法测定实际样品中氧氟沙星的含量，并计算误差。

思考与讨论

1. 与循环伏安方法相比，示差脉冲伏安法有哪些优点和特点？

2. 本实验中，选择催化氧氟沙星电极材料时，除了可以使用循环伏安法，还可以用什么方法？具体如何操作？

参考资料（根据前面查阅的资料完成）

Experiment 25

Electrochemical behaviour of ofloxacin on the surface of solid electrode

Experimental purpose

1. Understand the characteristics of irreversible cyclic voltammetric curves and differential pulse voltammetric(DPV) curves, and master related experimental techniques.

2. Select of electrode materials which can catalyze ofloxacin by cyclic voltammetry.

3. Determinate ofloxacin content in real samples by selective electrode via DPV.

Basical principles (Around the following topics, complete the writing)

1. Characteristics of irreversible cyclic voltammetric curves.

2. Mechanism and application of DPV.

3. Structure and properties of ofloxacin.

4. Catalytic oxidation mechanism of ofloxacin on the electrode surface.

Apparatus and reagents

Electrochemical workstation

Electrodes (zinc, copper, glass carbon, saturated calomel, platinum wire electrode each one)

Ovofloxacin (AR)

Phosphate buffer solution(pH = 6.5, containing 0.1 mol · L^{-1} KCl)

4-aminobenzenesulfonic acid(ABSA, AR)

Experimental procedures

1. Preparation of glassy carbon electrode (GCE) modified with poly(ABSA) by CV

The glassy carbon electrode is used as working electrode, saturated calomel electrode (SCE) as reference electrode, and platinum wire as auxiliary electrode. The GCE modified with poly(ABSA) is prepared by CV in 0.02 mol · L^{-1} ABSA using the following parameters: the potential range is −1.5 to 2.5 V, scan rate is 0.1 V · s^{-1}, the sampling interval is 0.001 V, and the sensitivity is 1×10^{-4} A · V^{-1}. The obtained electrode is expressed with PABSA/GCE.

2. Selection of electrode materials which can catalyze ofloxacin by cyclic voltammetry

(1) Preparation of 2×10^{-3} mol · L^{-1} oxifloxacin solution by phosphate buffer solution.

(2) Screening the best electrode material for the catalysis of of loxacin.

The zinc, copper, bare GCE, GCE modified with poly (ABSA) are used as working electrode, respectively. SCE is as reference electrode, and platinum wire is as auxiliary electrode. The electrode material is selected according to the peak current by scanning cyclic voltammetry in 2×10^{-3} mol · L^{-1} ofloxacin solution using the following parameters: the potential range is 0.4 to 1.6 V, scan rate is 0.15 V · s^{-1}, the sampling interval is 0.001 V, and the sensitivity is 1×10^{-4} A · V^{-1}.

3. Drawing a standard curve by measuring a series of ofloxacin

(1) Preparation of a series of ofloxacin solutions

Ofloxacin standard solutions of 6.5×10^{-4}、3.5×10^{-4}、2.0×10^{-4}、1×10^{-4}、5×10^{-5} mol · L^{-1} are prepared by phosphate buffer solutions.

(2) The GCE modified with poly (ABSA) are used as working electrode. SCE is as reference electrode, and platinum wire is as auxiliary electrode. DPV is conducted in above concentration of ofloxacin solution and the peak current is recorded. The following parameters is used: the potential range is 0.5 to 1.2 V · s^{-1}, scan rate is 0.1 V · s^{-1}, the sampling interval is 0.004 V, and the sensitivity is 1×10^{-4} A · V^{-1}.

(3) The peak current obtained is plotted with the concentration to draw the standard curve of the ofloxacin solution.

4. Determination of the ofloxacin in the actual sample

(1) The commercially available ofloxacin ear drops are diluted with phosphate buffer solution of pH 6.5 to the concentration range of the above working curve.

(2) Scan DPV in the above ofloxacin solutions using GCE modified with poly(ABSA) as the working electrode, SCE as the reference electrode, and a platinum wire electrode as the auxiliary electrode. The peak current is recorded and substituted into the regression equation of the standard curve. The concentration of ofloxacin in ear drops could be calculated.

Data recording and processing

1. Screen electrode materials for catalytic oxidation of ofloxacin.

2. Draw a standard curve of ofloxacin detection via electrode.

3. Determine the concentration of ofloxacin in actual samples by DPV, and calculate the error.

Thinking and discussing

1. Compared with cyclic voltammetry, what are the advantages and characteristics of differential pulse voltammetry?

2. In this experiment, in addition to the CV, what other methods can be used to select the electrode material for catalytic ofloxacin? How does it work?

Reference materials (complete on the basis of the information consulted earlier)

实验26　稀土金属直接加氢制备纳米稀土金属氢化物

实验目的

1. 熟悉催化加氢方法。
2. 了解稀土金属加氢反应的基本原理。
3. 了解纳米稀土金属氢化物表征方法。

基本原理(围绕下面主题，撰写完成)

1. 纳米稀土氢化物简介。
2. 催化加氢方法。
3. 稀土金属加氢反应机理。
4. 纳米稀土金属氢化物的表征。

重点介绍如何利用透射电镜、X射线衍射和BET方法对材料进行表征。

仪器和试剂

无水无氧装置
反应器
量气管
磁力搅拌器
缓冲瓶
恒温水浴
真空泵
氢气发生器
镧系金属(La, Dy, Yb,分析纯)
C_2H_5I(分析纯)
四氢呋喃(分析纯)
高纯氩

实验步骤

1. 溶剂纯化

将一定量经550 ℃焙烧干燥的5A分子筛放入四氢呋喃中,浸泡24小时,初步脱水。然后在金属钠和苯甲酮溶液中回流至溶液为深蓝色,蒸出四氢呋喃备用。

2. 气体纯化

氢气和氩气经过5A分子筛脱水,220 ℃铜柱脱氧后使用。

3. 合成纳米稀土金属氢化物

以金属镧为例。取 3.5 g 金属镧（纯度大于 99.95%），并将其分割为小块状（20～30 目）。将干燥的反应瓶抽真空，再充入氩气，反复 3 次。在氩气保护下，加入块状金属镧。利用氩气，吹扫金属镧 10 min。将反应瓶与恒压氢气量气管相连后，依次注入 15 mL 四氢呋喃、0.03 mL 溴代乙烷。油浴加热，反应温度 40 ℃，磁力搅拌。氢气压力 0.1 MPa，吸氢停止后，认为反应结束。

4. 纯化纳米稀土金属氢化物

将合成液转移到离心瓶中，离心后，倾倒出液体。用四氢呋喃洗涤固体 3 次。将所得固体 80 ℃ 真空干燥 1 h 后，得到产品。

5. 采用 BET 方法测量该固体比表面积，X 射线衍射方法测量晶格结构，透射电镜测量纳米粒子大小及分布。

数据记录与处理

1. 做出吸氢量随时间变化的动力学曲线图。

2. 利用吸氢量计算纳米稀土金属氢化物的计量式。

3. 利用 BET、X 射线衍射和透射电镜表征的数据，给出纳米稀土金属氢化物结构。

思考与讨论

1. 该实验成败的关键是什么？

2. 为什么纳米稀土金属氢化物的计量式不是 MH_3？

参考资料（根据前面查阅的资料完成）

Experiment 26

Preparation of nano rare earth metal hydride by direct hydrogenation of rare earth metal

Experimental purpose

1. Be familiar with catalytic hydrogenation method.

2. Understand the principle of hydrogenation reaction for rare earth metal.

3. Understand the characterization method of rare earth metal hydride with nano size.

Basical principles (Around the following topics, complete the writing)

1. Brief introduction of nano-rare earth hydride.

2. Catalytic hydrogenation method.

3. Hydrogenation mechanism of rare earth metals.

4. Characterization of nano rare earth hydride.

The characterization of materials by transmission electron microscopy (TEM), X-ray diffraction (XRD) and BET are highlight introduced.

Apparatus and reagents

Anhydrous and oxygen-free device	Reactor
Measuring pipe	Magnetic stirrer
Buffer bottle	Thermostatic water bath
Vacuum pump	Hydrogen generator
Lanthanide metals (La, Dy, Yb, AR)	C_2H_5I (AR)
Tetrahydrofuran (AR)	High purity argon

Experimental procedures

1. Solvent purification

Put a certain amount of 5A molecular sieve calcined and dried at 550 ℃ into tetrahydrofuran, soak for 24 hours, and preliminarily dehydrate. Then reflux in sodium metal and benzophenone solution until the solution is dark blue, and steam out tetrahydrofuran for later use.

2. Gas purification

Hydrogen and argon are dehydrated by 5A molecular sieve and deoxidized by 220 ℃ copper column before use.

3. Synthesis of rare earth metal hydride with nano size

Use metal lanthanum as an example. Take 3.5 g metallic lanthanum (purity greater than 99.95%) and divide it into small pieces (20～30 mesh). The dry reaction bottle is vacuumed, then filled with argon gas, and repeat it three times. Add bulk metal lanthanum under the protection of argon. Use argon to purge the metal lanthanum for 10 minutes. After connecting the reaction bottle with the constant pressure measuring tube of hydrogen, inject 15 mL tetrahydrofuran and 0.03 mL bromoethane successively. Take oil bath heating, reaction temperature 40 ℃, and magnetic stirring. When the hydrogen pressure is 0.1 MPa and the hydrogen absorption stops, the reaction is considered to be over.

4. Purification of rare earth metal hydride with nano size

Transfer the synthetic solution into a centrifuge bottle, and pour out the liquid after centrifugation. Wash the solid with tetrahydrofuran three times. The product is obtained by vacuum drying the solid at 80 ℃ for 1 hour.

5. BET method is used to measure the specific surface area of the solid, XRD is used to measure the lattice structure, and TEM is used to measure the size and distribution of nanoparticles.

Data recording and processing

1. Make the kinetic curve diagram of hydrogen absorption amount changing with time.

2. Calculate the metering formula of nano rare earth metal hydride by hydrogen absorption.

3. Using the data measured by BET, XRD, and TEM, give the structure of nano rare earth metal hydride.

Thinking and discussing

1. What is the key to the success of the experiment?

2. Why is the metering formula of nano rare earth metal hydride not an integer 3, that is, MH_3?

Reference materials (complete on the basis of the information consulted earlier)

实验 27 低温等离子体直接分解 NO

实验目的

1. 了解物质的第四态——等离子体的概念及存在形式。

2. 掌握等离子体直接分解 NO 方法。

基本原理(围绕下面主题，撰写完成)

1. 等离子体介绍。

2. 常用的等离子体的产生方式,重点介绍介质阻挡放电。

3. 等离子体直接分解 NO 机理。

仪器和试剂

气相色谱仪

等离子体电源

放电反应器

NO 气体(高纯)

氢气发生器

气体流量控制仪

控温仪

高纯氦(高纯)

实验步骤

1. 安装反应器

实验用反应器分别为 A、B、C 和 D 四种类型,其中 A、B、C 为一段式反应器;D 为两段式反应器。具体为,反应器 A:内径 15 mm、厚 2 mm 的刚玉管,其中心为一直径2 mm 的不锈钢管,与交流高压电源输出端相连。管外紧密缠绕以不锈钢网(不锈钢网长度控制放电空间大小),与交流高压电源的接地端相连。反应器 B:内径 10 mm、厚 1 mm 的石英管,放电电极部分同上述刚玉管。反应器 C:内径 8 mm、厚 1 mm 的石英管,放电电极部分同上述刚玉管。反应器 D:内径 10 mm、厚 1 mm 的石英管,将放电部分与催化剂分开,两部分相距 10 mm,在电炉中可同时被加热到相同温度。一段放电反应部分添加无催化活性及表面积很小的石英砂作为空白。

2. 测温装置调试

反应器用电炉加热,反应温度的测量采用如下方法:一个测温热电偶插入反应器中心不锈钢管中,置于催化剂床层高度中点;另一个测温热电偶紧靠反应器外壁放置,高度与里面的热电偶一致。未放电时,反应气流使中心热电偶指示的温度低于外面热电偶指示的温度,当施加介质阻挡放电时,中心热电偶指示的温度略有增加,将内外温度的平均值作为反应温度。

3. 等离子体直接分解 NO

实验用气体为高纯 NO(纯度×99.99%)。将不同流速的 NO 气体通入反应器,控制一定温度,施加不同功率的等离子体,利用气相色谱对反应后气体在线分析。用 NO 转化百分数评价脱除 NO 的活性,其定义为

$$NO\% = \frac{[NO]_0 - [NO]_{eq}}{[NO]_0} \times 100\%$$

数据记录与处理

1. 计算不同实验条件下 NO 转化百分数。
2. 作出 NO 转化百分数随等离子体功率变化、气体流速变化、反应温度变化曲线图。
3. 由前面各曲线图,优化等离子体直接分解 NO 的实验条件。

思考与讨论

1. 等离子体分解 NO 的效率受哪些因素影响?
2. 如何理解反应器的直径对等离子体分解 NO 的影响?

参考资料(根据前面查阅的资料完成)

Experiment 27

Direct decomposition of NO by low-temperature plasma

Experimental purpose

1. Understand the concept and existing form of plasma, the fourth state of matter.

2. Master the method of direct decomposition of NO by plasma.

Basical principles (Around the following topics, complete the writing)

1. Introduction to plasma.

2. Plasma generation in common use, with emphasis on dielectric barrier discharge.

3. Mechanism of NO decomposition by plasma.

Apparatus and reagents

Gas chromatography
Plasma power supply
Discharge reactor
NO gas (high purity)
Hydrogen generator
Gas flow controller
Temperature controller
Helium gas (high purity)

Experimental procedures

1. Install the reactor

There are four types of reactors used in the experiment: A, B, C and D, among which A, B and C are "one-stage" reactors; D is a "two-stage" reactor. Specifically, reactor A is a corundum tube with an inner diameter of 15 mm and a thickness of 2 mm, and its center is a stainless steel tube with a diameter of 2 mm, which is connected to the output end of high-voltage power supply with alternating current (AC). The stainless steel mesh (the length of the stainless steel mesh controls the size of the discharge space) is tightly wound outside the pipe and connected to the grounding terminal of the AC high-voltage power supply. Reactor B: quartz tube with inner diameter of 10 mm and thickness of 1 mm. The discharge electrode part is the same as the above corundum tube. Reactor C: quartz tube with inner diameter of 8 mm and thickness of 1 mm. The discharge electrode part is the same as the above corundum tube. Reactor D: a quartz tube with an inner diameter of 10 mm and a thickness of 1 mm. It separates the discharge part from the catalyst, and the two parts are 10 mm apart, which can be heated to the same temperature at the same time in the electric furnace. In the first discharge reaction part, quartz sand with no catalytic activity and small surface area is added as blank.

2. Commissioning of temperature measuring device

The reactor is heated by electric furnace, and the reaction temperature is measured by the following methods: one temperature thermocouple is inserted into the stainless steel pipe in the center of the reactor and placed at the midpoint of the height of the catalyst bed. The other temperature thermocouple is placed close to the outer wall of the reactor, with the height consistent of that of the thermocouple inside. When not discharged, the temperature indicated by the central thermocouple is lower than that indicated by the external thermocouple due to the reaction gas flow. When the dielectric barrier discharge is applied, the temperature indicated by the central thermocouple increases slightly, and the average value of the internal and external temperatures is taken as the reaction temperature.

3. Direct decomposition of NO by plasma

The experimental gas is high purity NO (purity>99.99%). The NO gas with different flow rates is introduced into the reactor, the temperature is controlled, and the plasma with different power is applied. The gas after reaction is analyzed online by gas chromatography. The activity of NO removal is evaluated by the percentage of NO conversion, which is defined as

$$NO\% = \frac{[NO]_0 - [NO]_{eq}}{[NO]_0} \times 100\%$$

Data recording and processing

1. Calculate the percentage of NO conversion under different experimental conditions.

2. Draw the curve of NO conversion percentage changing with plasma power, gas flow rate and reaction temperature.

3. Optimize the experimental conditions for the direct decomposition of NO by plasma by the above curves.

Thinking and discussing

1. What factors affect the efficiency of NO decomposition by plasma?

2. How to understand the effect of reactor diameter on NO decomposition by plasma?

Reference materials (complete on the basis of the information consulted earlier)

实验 28　氢质子交换膜燃料电池的组装与性能测试

实验目的

1. 了解质子交换膜燃料电池的结构和工作原理。
2. 掌握硼氢化钠水解制氢原理和方法。
3. 掌握膜电极的制备和单燃料电池的组装技术。
4. 了解纳米催化剂在燃料电池中的应用。

基本原理(围绕下面主题，撰写完成)

1. 燃料电池简介。
2. 质子交换膜燃料电池简介。
3. 质子交换膜燃料电池的结构及工作原理。

仪器和试剂

质子交换膜燃料电池配件一套	电流表
电压表	烘箱
恒温水浴锅	电吹风
丙酮(分析纯)	浓硫酸
无水乙醇(分析纯)	30%的 H_2O_2
NaOH(分析纯)	硼氢化钠(分析纯)
Pt 催化剂浆料	去离子水

实验步骤

1. 查阅文献

明确硼氢化钠水解制氢原理和方法,设计实验方案,制备氢气。

2. 质子交换膜及碳纸的处理

将质子交换膜(4 cm×4 cm)放入温度为 80 ℃、3%的 H_2O_2 水溶液中浸泡 1 h,取出用去离子水洗净;在去离子水中煮沸 1 h 后,再在 1 mol · L^{-1}的硫酸溶液中浸泡 1 h,取出用去离子水洗净;再在去离子水中煮沸 1 h 并洗至中性,最后浸泡保存在去离子水中备用。

将 2 片(2.5 cm×2.5 cm)碳纸置于丙酮溶液中浸泡 0.5 h 后,用去离子水清洗,置于烘箱内烘干备用。

3. 催化剂的固定

清洗涂膜夹具及垫圈,并用无水乙醇擦拭干净。在底座上放上一块密封垫,然后放上质子交换膜,再放上一块密封垫,将夹具面板盖上,扭紧螺丝将膜固定。将配好的催化剂浆料均匀涂在质子交换膜上，用电吹风器吹干。再将质子交换膜从夹具上取下，在膜的反面用同样方法涂覆催化剂。

4. 燃料电池组装

首先将4个螺丝装在有机玻璃的氧气侧端板上，依次安装集流板、密封垫、碳纸、涂覆催化剂的质子交换膜、密封垫、碳纸、集流板、密封垫和氢气端板，用螺丝将电池锁紧，装上电极接头。

5. 燃料电池的应用

将组装好的燃料电池与小风扇连接，将氢气导管连到燃料电池的阳极。当硼氢化钠水解产生的氢气经导管进入燃料电池，燃料电池开始工作，可观察小风扇转动情况。

6. 燃料电池极化曲线的测定

将电压表并联于外电路，将电流表串联于外电路，测试燃料电池的电流随电压的变化。

7. 电极材料的回收

测试结束后，将燃料电池与小风扇分离，拆卸燃料电池，将使用过的碳纸置于丙酮溶液中浸泡0.5 h，然后用去离子水清洗，置于烘箱内烘干备用。将涂覆催化剂的膜浸入乙醇中，催化剂层会溶解脱落，将收集的催化剂溶液蒸发至一定浓度以重复使用。将去除催化剂的膜在去离子水中煮沸1 h，然后浸入去离子水中重复使用。将其他配件清洗干净。

数据记录与处理

1. 归纳总结硼氢化钠水解制氢的实验条件，阐明影响产氢效率的关键因素。
2. 将电流、电压数据列表，绘制电流、电压关系曲线（极化曲线）。

思考与讨论

1. 以甲醇或乙醇代替氢燃料，阳极发生什么反应？
2. 举例说明其他制氢方法。
3. 燃料电池作为一种新的能源技术，有什么发展优势和应用前景？

参考资料（根据前面查阅的资料完成）

注意事项

1. 本实验中有氢气产生，实验室严禁明火，并保持良好通风。
2. 防止质子交换膜在操作过程中被戳破。如果膜有破损则须更换。
3. 必须将催化剂浆料均匀涂覆于膜上（此时膜会发生卷曲），并及时吹干。
4. 安放碳纸时注意将碳纸准确放置在密封垫中空处，避免漏气。
5. 燃料电池装配时，螺丝应均匀、交叉拧紧，以达最佳密封。
6. 利用一体化燃料电池系统时，要严格禁止制氢装置中的水流入燃料电池，以免损坏燃料电池。

Experiment 28

Fabrication and properties determination of hydrogen proton exchange membrane fuel cells

Experimental purpose

1. Understand the structure and working principles of proton exchange membrane fuel cell.

2. Master the principles and methods of hydrogen preparation by sodium borohydride hydrolysis.

3. Master the fabrication method of membrane electrodes and the assembly technology of single fuel cell.

4. Understand the application of nano-catalysts in fuel cells.

Basical principles (Around the following topics, complete the writing)

1. Introduction to fuel cells.

2. Introduction to proton exchange membrane fuel cell.

3. The structure and working principle of the proton exchange membrane fuel cell.

Apparatus and reagents

Proton exchange membrane fuel cell accessories set	Ammeter
Voltmeter	Ovens
Constant temperature water bath	Electric hairdryer
Acetone (AR)	Concentrated sulfuric acid
Anhydrous ethanol(AR)	30% H_2O_2
NaOH (AR)	Sodium borohydride (AR)
Pt catalyst slurry	Deionized water

Experimental procedures

1. Check the literature, clarify the principle and method of hydrogen preparation by hydrolysis of sodium borohydride, and design an experimental scheme to prepare hydrogen gas.

2. Handling of proton exchange membrane and carbon paper

The proton exchange membrane (4 cm×4 cm) is soaked in 3% aqueous H_2O_2 solution at 80 ℃ for 1 h, and washed with deionized water. After boiling in deionized water for 1 h, the proton exchange membrane is soaked in 1 mol · L^{-1} sulfuric acid solution for 1 h. Wash it and boil it again in deionized water for 1 h. Wash it with deionized water to neutral, finally soak and save in deionized water for later use.

Two pieces (2.5 cm×2.5 cm) of carbon paper are soaked in acetone solution for 0.5 h, then washed with deionized water and placed in an oven for drying.

3. Immobilization of catalyst

Clean the film coating fixture and gasket, and wipe them clean with anhydrous ethanol. Put a gasket on the base, a proton exchange membrane, and another gasket to cover the jig panel, and twist the screws to immobilize the membrane. Apply the prepared catalyst slurry evenly on the proton exchange membrane and blow it dry with an electric hair dryer. Remove the proton exchange membrane from the fixture and coat the reverse side of the membrane with the catalyst in the same way.

4. Fabrication of fuel cell

Firstly, four screws are installed on the oxygen side end plate of plexiglass. Install the collector plate, gasket, carbon paper, proton exchange membrane coated with catalyst, gasket, carbon paper, collector plate, gasket and hydrogen end plate in order. The obtained cell is locked with screws and installed with electrode connector.

5. Application of fuel cells

Connect the obtained fuel cell to the small fan, and connect the hydrogen conduit to the anode of the fuel cell. When the hydrogen gas produced by the hydrolysis of sodium borohydride enters the fuel cell through the conduit, the fuel cell starts to work and the rotation of the small fan can be observed.

6. Determination of fuel cell polarization curve

The voltmeter is connected in parallel and an ammeter is connected in series to the external circuit to test the variation of the fuel cell current with voltage.

7. Recycling of the electrode materials

After the test, the fuel cell is separated from the small fan, and the fuel cell is disassembled. The carbon paper used is soaked in acetone solution for 0.5 h and then washed with deionized water, and finally is dried in an oven for later use. The membrane coated with catalyst is immersed in ethanol, and the catalyst layer will dissolve and fall off. The collected catalyst solution is evaporated to a certain concentration for reuse. Boil the membrane which has been removed catalyst in deionized water for 1 h and then immerse it in deionized water for reuse. Clean the other accessories.

Data recording and processing

1. Summarize the experimental conditions of hydrogen preparation by hydrolysis of sodium

borohydride and clarify the key factors affecting the efficiency of hydrogen preparation.

2. Tabulate the current and voltage data. Plot the current and voltage relationship curve (polarization curve).

Thinking and discussing

1. What reaction occurs at the anode when methanol or ethanol is used instead of hydrogen fuel?

2. Give examples of other methods of hydrogen preparation.

3. What are the development advantages and application prospects of fuel cells as a new energy technology?

Reference materials (complete on the basis of the information consulted earlier)

Cautions

1. Hydrogen gas is generated in this experiment, open fire should be strictly prohibited in the laboratory and good ventilation is maintained.

2. Prevent the proton exchange membrane from being punctured during operation. If the membrane is damaged, it should be replaced again.

3. The catalyst slurry must be evenly applied to the membrane (the membrane will curl at this time) and blown dry in time.

4. When place the carbon paper, pay attention to place the carbon paper accurately in the hollow of the gasket to avoid air leakage.

5. When assemble the fuel cell, the screws should be evenly and cross-tightened to achieve the best seal.

6. When using the integrated fuel cell system, strictly prohibit the flow of water from the hydrogen production device into the fuel cell to avoid damage to the fuel cell.

实验 29　甲烷生成焓和燃烧焓的理论计算

实验目的

1. 通过计算机辅助量子化学计算软件，学习创建和修改分子结构模型。

2. 利用理想气体模型，通过统计热力学方法计算甲烷生成焓和燃烧焓等热力学量变化值。

3. 了解理论计算在研究甲烷生成和燃烧过程中的重要意义。

基本原理(请围绕下面主题，撰写完成)

1. 生成焓和燃烧焓简介。

2. 量子化学软件创建分子结构模型的常用方法。

3. 量子化学常用计算方法。

仪器和软件

计算机　　Gaussian 程序

GaussView 程序

实验步骤

1. 使用 GaussView 等软件分别创建甲烷(CH_4)和氢气(H_2)的分子结构,并设置正确的电荷和自旋多重度等参数。使用 Gaussian 程序,在 B3LYP/6-311G(d,p)水平上对上述分子的结构进行优化,并计算振动频率。

以甲烷分子为例,参考操作步骤如下。

(1) 构建分子初始构型

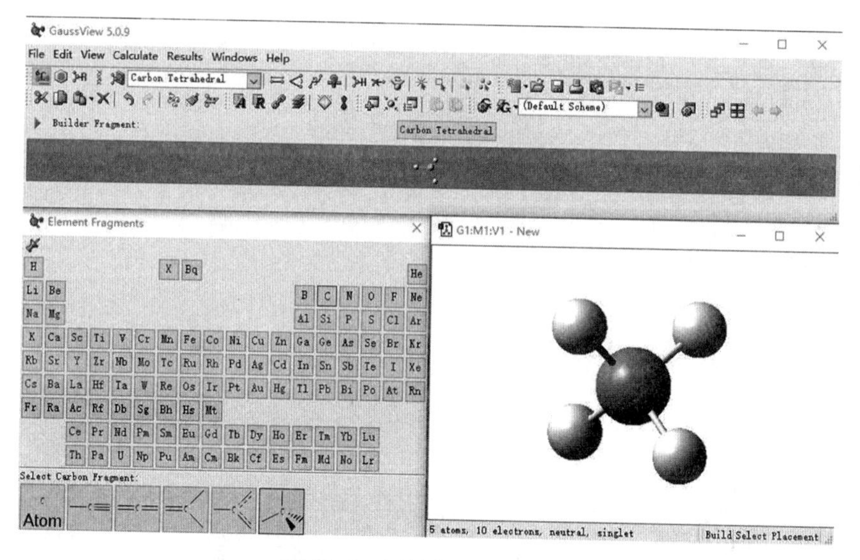

图 29-1　甲烷分子结构构建

如图 29-1 所示,在 GaussView 软件中,打开“Element Fragments”工具,选取合适的原子类型。依次点击“C—Carbon Tetrahedral”来选取四面体构型的 C 原子,然后再用鼠标左键在显示窗口中单击一次,以获得相应构型的原子或原子团。在计算中需要保持分子所属点群不变,可点击“Edit—Point Group”并在弹出窗口中选择“Enable Point Group Symmetry”选项。

（2）设置计算参数

点击“Calculate—Gaussian Calculation Setup”，并在弹出窗口中设置计算参数。

① 选择“Job Type”选项卡，设置计算任务为“Opt+Freq”。

② 选择“Method”选项卡，如图 29-2 所示，设置计算方法、基组、电荷和自旋多重度信息。对于甲烷分子，计算方法选择“Ground State—DFT—Default Spin—B3LYP”；基组选择“6-311G”“ ”“d”“p”；电荷输入“0”；自旋多重度选择“Singlet”。

（3）保存输入文件

将输入文件保存到指定文件夹中。

（4）执行计算任务

使用 Gaussian 程序打开输入文件，运行计算任务。

2. 使用 GaussView 等软件创建碳原子（C），使用 Gaussian 程序，在 B3LYP/6-311G(d, p) 水平上计算碳原子在三重态的能量。

3. 使用 GaussView 等软件分别创建甲烷（CH_4）、氧气（O_2）、二氧化碳（CO_2）、水（H_2O）的分子结构，并设置正确的电荷和自旋多重度等参数。使用 Gaussian 程序，在 B3LYP/6-311G(d,p) 水平上对上面的分子的结构进行优化，并计算振动频率。

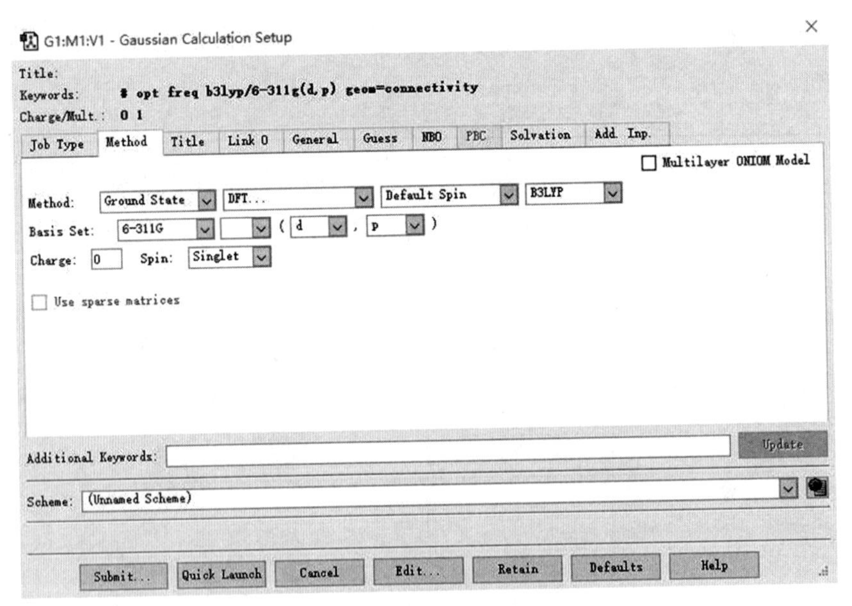

图 29-2 甲烷分子计算方法设置

数据记录与处理

1. 记录计算获得的各种物质的热力学函数据（U、H、G、S 等），验证热力学参数之间的关系。

2. 根据理论计算得到的相关数据计算 298.15 K、100 kPa 下甲烷的标准摩尔生成焓和标准摩尔燃烧焓。

3. 将理论计算得到的 $\Delta_f H_m^{\circ}(CH_4)$ 和 $\Delta_c H_m^{\circ}(CH_4)$ 与文献中的实验值进行比较,讨论理论计算值与实验值之间的差,并提出合理的提高理论计算结果准确度的建议。

思考与讨论

1. 分析分子结构修改对生成热和燃烧热的影响。
2. 探讨不同计算方法在热力学计算中的优劣。
3. 针对实验结果,讨论理论计算在研究甲烷生成和燃烧中的应用前景。

参考资料

1. Levine I N. Quantum Chemistry. Pearson, 2009.
2. Szabo A, Ostlund N S. Modern Quantum Chemistry: Introduction to Advanced Electronic Structure Theory. Dover Publications, 1996.

Experiment 29

Theoretical calculation of enthalpy of formation and combustion enthalpy of methane

Experimental purpose

1. Learn to create and modify molecular structure models using computer-aided quantum chemistry software.

2. Use the ideal gas model and statistical thermodynamics methods to calculate the thermodynamic quantities, such as the enthalpy of formation and combustion of methane.

3. Understand the significant importance of theoretical calculations in studying the processes of methane formation and combustion.

Basical principles (Around the following topics, complete the writing)

1. A brief introduction to the formation enthalpy and the combustion enthalpy.

2. Common methods for creating molecular structure models using quantum chemistry software.

3. Calculation methods of quantum chemistry.

Apparatus and software

Computer

Gaussian program

GaussView program

Experimental procedures

Molecular structure model creation, optimization, and frequency calculation.

1. Use software like GaussView to create the molecular structures of methane (CH_4) and hydrogen (H_2), set the correct parameters such as charge and spin multiplicity. Optimize the structure using the B3LYP/6-311G(d,p) method and calculate vibrational frequencies.

Taking methane molecules as an example, the procedural steps are outlined as follows.

(1) Constructing the initial molecular configuration

Referencing Fig. 29-1, in the GaussView software, open the "Element Fragments" tool and select appropriate atomic types. Sequentially click "C—Carbon Tetrahedral" to choose the C atom with tetrahedral configuration. Then, left-click once in the view window to obtain the corresponding configuration of atoms or atomic groups. To maintain the molecular point group during calculations, click "Edit—Point Group" and choose the "Enable Point Group Symmetry" option in the pop-up window.

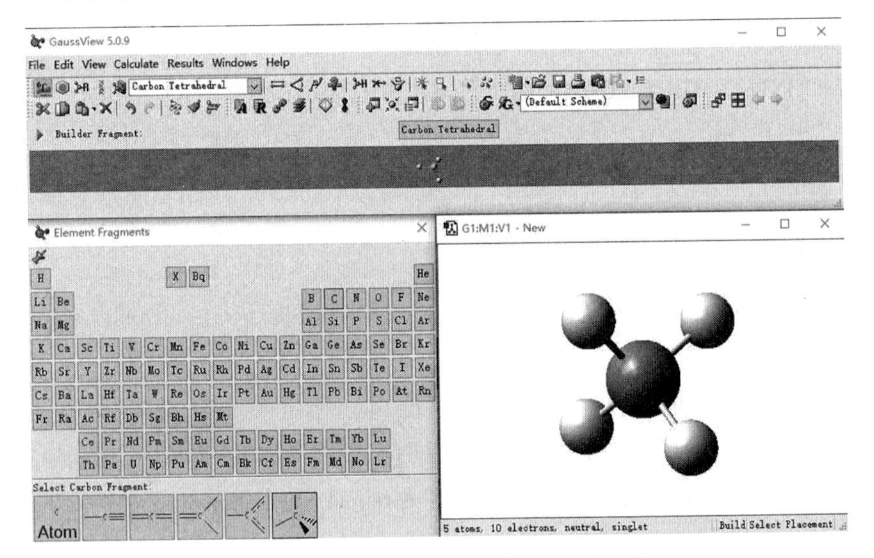

Fig. 29-1 Structure of methane molecule

(2) Setting calculation parameters

Click "Calculate—Gaussian Calculation Setup" and set the calculation parameters in the pop-up window.

① Select the "Job Type" tab and set the calculation task to "Opt+Freq".

② Select the "Method" tab (Fig. 29-2) and set the calculation method, basis set, charge, and spin multiplicity information. For methane molecules, choose "Ground State—DFT—Default Spin—B3LYP" as the calculation method; select "6-311G" " " "d" "p" as the basis set; input "0" for charge; choose "Singlet" for spin multiplicity.

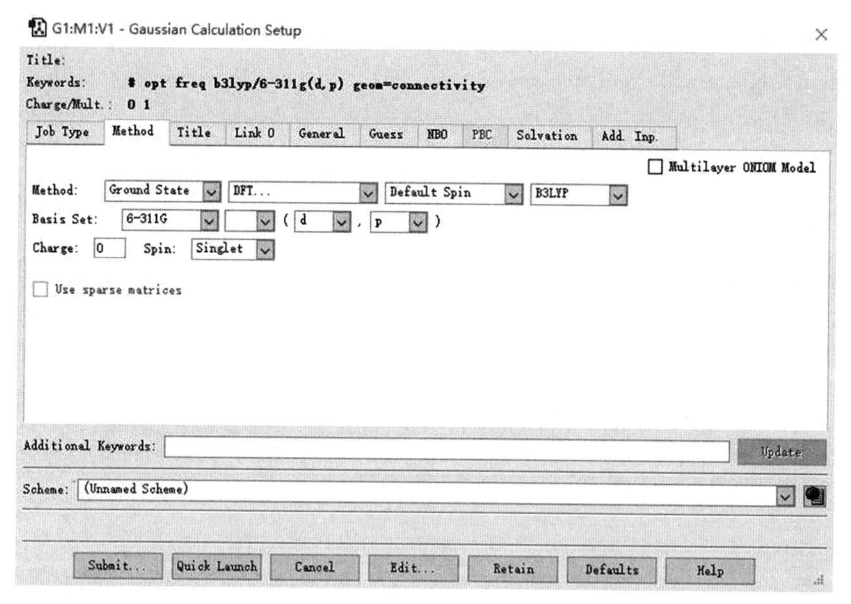

Fig. 29-2　Calculation setup for methane

(3) Save input file

Save the input file to the specified folder.

(4) Execute calculation task

Use the Gaussian program to open the input file and run the calculation task.

2. Create structure for carbon atoms (C) using software like GaussView, calculate the energy of the carbon atom in triplet states using the B3LYP/6-311G(d,p) method.

3. Create molecular structures for methane (CH_4), oxygen (O_2), carbon dioxide (CO_2), and water (H_2O) using software like GaussView, set the correct parameters, and optimize the structure. Calculate vibrational frequencies.

Data recording and processing

1. Record the calculated thermodynamic functions (U, H, G, S, etc.) obtained for various substances, verify the relationships between thermodynamic parameters.

2. Based on the relevant data obtained from theoretical calculations, compute the standard molar enthalpy of formation and standard molar combustion enthalpy for methane at 298.15 K and 100 kPa.

3. Compare the theoretically calculated values $\Delta_f H_m^\ominus(CH_4)$ and $\Delta_c H_m^\ominus(CH_4)$ with experimental values in literature, discuss the differences between theoretical and experimental results, and propose reasonable suggestions to improve the accuracy of theoretical calculations.

Thinking and discussing

1. Analyze the impact of structural modifications on the enthalpy of formation and

combustion enthalpy.

2. Discuss the strengths and weaknesses of different computational methods in thermodynamic calculations.

3. Based on experimental results, discuss the potential application prospects of theoretical calculations in studying methane formation and combustion.

Reference materials (complete on the basis of the information consulted earlier)

实验 30　分子中原子电荷的理论计算

实验目的

1. 进一步熟悉 ChemOffice 软件的使用。

2. 利用 Gaussian 程序进行分子结构优化和原子电荷计算。

基本原理(请围绕下面主题，撰写完成)

1. 如何利用量子化学方法计算原子上电荷分布。

2. Mulliken 电荷布局介绍。

仪器和软件

计算机　　ChemOffice 软件

Gaussian 程序　　GaussView 软件

实验步骤

1. 构建分子

① 双击打开 ChemOffice 软件中的 Chem3D，单击文本工具按钮 A 。

② 将鼠标移至模型窗口，单击鼠标，出现文本输入框，在输入框中输入“CH2CHCHCH2”如图 30-1，并按回车键，得到丁二烯分子的结构，如图 30-2。

图 30-1　利用文本工具建立模型

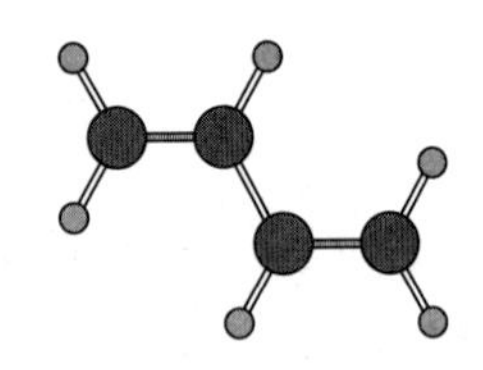

图 30-2　丁二烯的 3D 结构

③ 保存成笛卡儿坐标的形式。执行 file 菜单下的 save 命令，保存类型选择为 Gaussian Input(∗. GJC)形式，命名为 C4H6。

2. 编写输入文件并提交作业

以记事本的方式打开文件 C4H6.GJC，并以图 30-3 的方式编写文件。

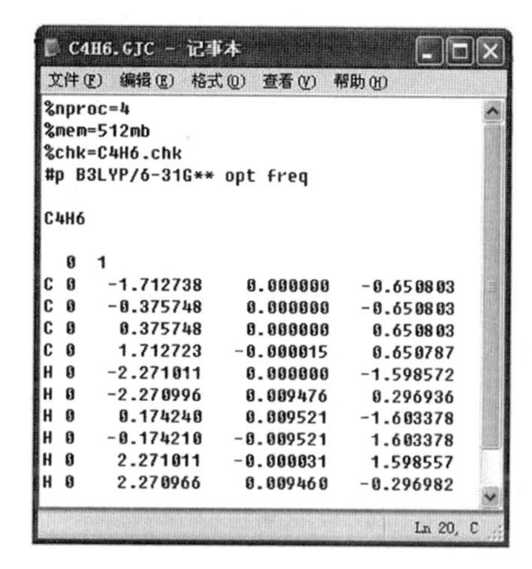

```
%nproc=4
%mem=512mb
%chk=C4H6.chk
#p B3LYP/6-31G** opt freq

C4H6

 0 1
C 0   -1.712738    0.000000   -0.650803
C 0   -0.375748    0.000000   -0.650803
C 0    0.375748    0.000000    0.650803
C 0    1.712723   -0.000015    0.650787
H 0   -2.271011    0.000000   -1.598572
H 0   -2.270996    0.009476    0.296936
H 0    0.174240    0.009521   -1.603378
H 0   -0.174210   -0.009521    1.603378
H 0    2.271011   -0.000031    1.598557
H 0    2.270966    0.009460   -0.296982
```

图 30-3 C4H6.GJC 文件

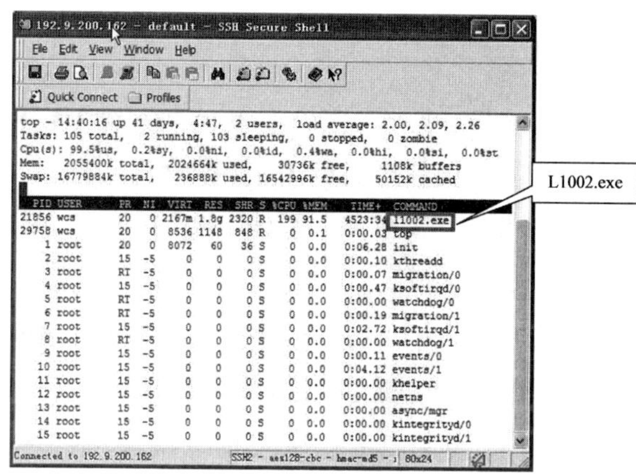

图 30-4 作业的实时监控

通过 SSH Secure Shell Client 软件连接到计算机，并提交作业(详见实验 24)。输入文件命名为 C4H6.dat，输出文件命名为 C4H6.out。

3. 结果的筛选

(1) 作业正常结束后，首先判断优化得到的结构是不是稳定构型(详见实验 24)。

(2) Mulliken 电荷的查找

在“Stationary point found”词条后，查找词条“Mulliken”，如图 30-5，即为对应原子的 Mulliken 电荷。

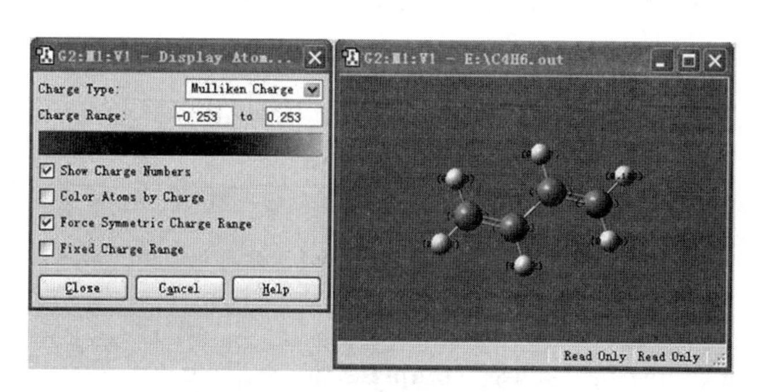

图 30-5 Mulliken 电荷的查找

```
Mulliken atomic charges:
             1
   1  C   -0.252806
   2  C   -0.028707
   3  C   -0.028695
   4  C   -0.252807
   5  H    0.097691
   6  H    0.101751
   7  H    0.082066
   8  H    0.082066
   9  H    0.097689
  10  H    0.101752
Sum of Mulliken charges=   0.00000
```

图 30-6 Mulliken 电荷布局图

(3) 输出文件的导出

① 单击 SSH Secure Shell Client 界面上的,进入 SSH Secure File Transfer 界面。

② 在 remote name 下,双击 homework 进入文件夹 homework,选中 C4H6.out 文件拖拽到 local name 下的目标文件夹中,即完成了文件的导出。

4. 结果的表达

用 GaussView 软件来获得 Mulliken 电荷布局图。

① 双击桌面上 GaussView 图标打开软件。

② 执行 File 菜单下的 Open 命令,选择文件类型为 Gaussian Output Files(* . out, * . log),打开输出文件 C4H6.out。

③ 执行 Result 菜单下的 Charges 命令。

④ 选中"Show Charge Numbers"和"Force Symmetric Charge Rang"两项,如图 30-5,即可得到 Mulliken 电荷布局图,如图 30-6。

⑤ 执行 File 菜单下的 save image 命令,即可将 Mulliken 电荷布局图保存。

⑥ 整理实验结果。

数据记录与处理

表达实验所得的丁二烯的稳定几何构型和 Mulliken 电荷布局。

思考与讨论

1. 理论上计算原子电荷的方案都有哪些? 各自有什么优点和不足?
2. 能否通过实验手段获得分子中各个原子上的电荷? 为什么?

参考资料(根据前面查阅的资料完成)

Experiment 30

Theoretical calculation of atomic charge in molecule

Experimental purpose

1. Be further familiar with the use of ChemOffice software.

2. Use Gaussian program to optimize the molecular structure and calculate the atomic charge.

Basical principles (Around the following topics, complete the writing)

1. How to calculate the charge distribution on an atom by using quantum chemical method?

2. Introduction to Mulliken charge distribution.

Apparatus and software

Computer ChemOffice

Gaussian09 program GaussView

Experimental procedures

1. Construction of molecule

① Double-click to open Chem3D in ChemOffice software, and click the text tool button A .

② Move the cursor to the model window, click, and the text input box will appear. Enter "CH2CHCH2" in the input box, as shown in Fig. 30-1, and press Enter key to get the structure of butadiene molecule, as shown in Fig. 30-2.

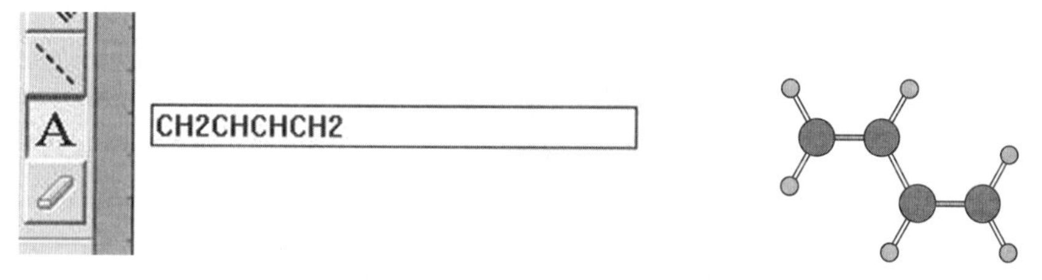

Fig. 30-1 Modeling with text tool **Fig. 30-2 3D structure of butadiene**

③ Save in the form of Descartes coordinates. Execute the "Save" command under the "File" menu. The save type is Gaussian Input (*. GJC), named C4H6.

2. Write input files and submit assignments

Open the file C4H6.GJC in the form of notepad and write the file in the form of Fig. 30-3.

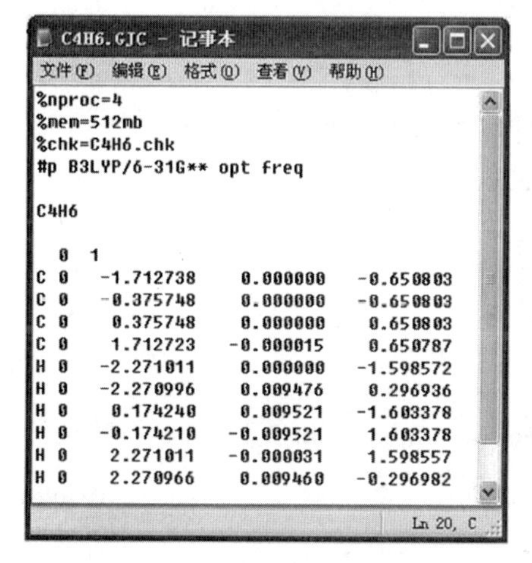

```
C4H6.GJC - 记事本
文件(F) 编辑(E) 格式(O) 查看(V) 帮助(H)
%nproc=4
%mem=512mb
%chk=C4H6.chk
#p B3LYP/6-31G** opt freq

C4H6

 0  1
C 0   -1.712738    0.000000   -0.650803
C 0   -0.375748    0.000000   -0.650803
C 0    0.375748    0.000000    0.650803
C 0    1.712723   -0.000015    0.650787
H 0   -2.271011    0.000000   -1.598572
H 0   -2.270996    0.009476    0.296936
H 0    0.174240    0.009521   -1.603378
H 0   -0.174210   -0.009521    1.603378
H 0    2.271011   -0.000031    1.598557
H 0    2.270966    0.009460   -0.296982
```

Fig. 30-3 C4H6.GJC document

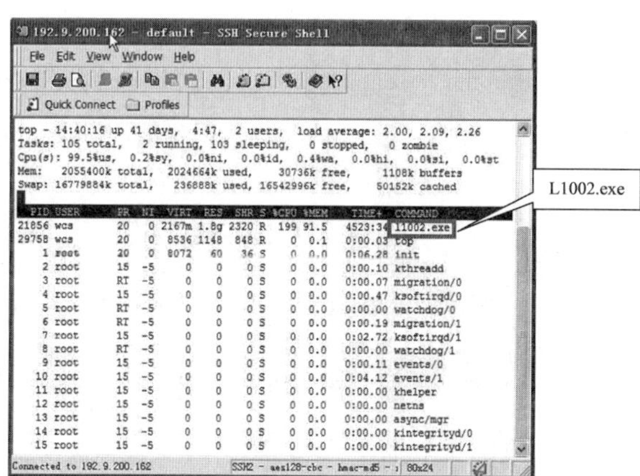

Fig. 30-4 Real-time monitoring of the job

Connect to the computer through SSH Secure Shell Client software and submit the job (see Experiment 24 for details). The input file is named C4H6.dat, and the output file is named C4H6.out.

3. Filtering results L1002.exe

(1) After the operation is completed normally, first judge whether the optimized structure is stable (see Experiment 24 for details).

(2) The search of mulliken charge

After the entry of Stationary point found, find the entry Mulliken, as shown in Fig. 30-5, which is the Mulliken charge of the corresponding atom.

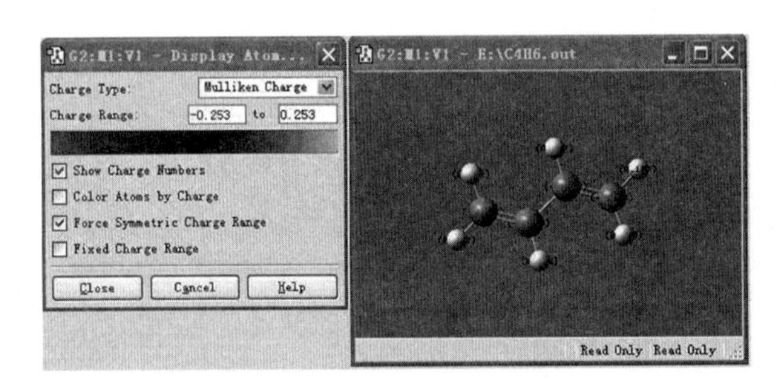

Fig. 30-5　Mulliken charge search

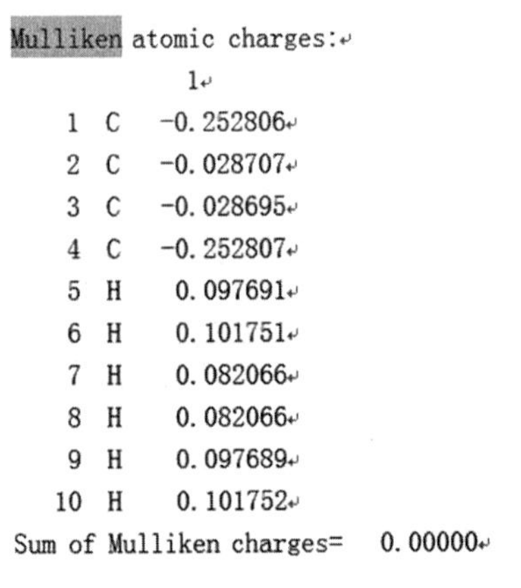

Mulliken atomic charges:

		1
1	C	-0.252806
2	C	-0.028707
3	C	-0.028695
4	C	-0.252807
5	H	0.097691
6	H	0.101751
7	H	0.082066
8	H	0.082066
9	H	0.097689
10	H	0.101752

Sum of Mulliken charges=　0.00000

Fig. 30-6　Mulliken charge layout

(3) Export of output file

① Click on the SSH Secure Shell Client interface to enter the SSH Secure File Transfer interface.

② Under the "Remote name", double-click "Home" to enter the folder home, select the C4H6.out file and drag it to the target folder under the local name to complete the export of the file.

4. Expression of results

Use GaussView software to obtain Mulliken charge layout.

① Double-click the GaussView icon on the desktop to open the software.

② Execute the open command under the file menu, select the file type as Gaussian Output Files (*.out, *.log), and open the output file C4H6.out.

③ Execute the "Charges" command under the result menu.

④ Select "Show Charge Numbers" and "Force Symmetric Charge Range", as shown in

Fig. 30-5, to get the Mulliken charge layout.

⑤ Execute the save image command under the file menu to save the Mulliken charge layout.

⑥ Organize the experimental results.

Data recording and processing

Express the stable geometry and Mulliken charge distribution of butadiene obtained from the experiment.

Thinking and discussing

1. What are the theoretical schemes for calculating atomic charge? What are their advantages and disadvantages?

2. Can we obtain the charge on each atom in the molecule by experimental method? Why?

Reference materials (Complete on the basis of the information consulted earlier)

第五章

课程思政建议

随着经济全球化的快速发展,各种文化思潮和价值观念大量涌入,极大地冲击了我国高校学生的思想。在这种时代背景下,仅靠传统的几门思政课程,很难做好高校学生的思想政治工作。2016 年 12 月全国高校思想政治工作会议的召开,掀起了“课程思政”建设的高潮。会议指出,高校立身之本在于立德树人,而课程思政就是把立德树人作为教育的根本任务,通过各类课程与思想政治理论结合,培育德才兼备、全面发展的高素质人才。

化学是一门以实验为基础的自然科学,物理化学实验是研究物质化学行为的原理、规律的实验课程,它在化学教学中十分必要。若能深入挖掘物理化学实验课程所蕴含的德育元素,对于实现“各门课都要守好一段渠、种好责任田,使各类课程与思想政治理论课同向而行,形成协同育人效应”具有重要意义。

表 1　物理化学实验中的思政元素

序号	思政元素
①	爱国主义精神、民族自尊心和民族自信心
②	为国家做贡献的社会责任感
③	严谨细致的科学态度和精益求精的工匠精神
④	安全责任意识
⑤	实践是检验真理的唯一标准
⑥	热爱科学、孜孜不倦的精神
⑦	理论联系实际的能力
⑧	量变与质变的关系原理
⑨	透过现象看本质的哲学思想
⑩	不畏困难、刻苦钻研的精神
⑪	用发展的眼光看问题的哲学思想
⑫	世界是普遍联系的哲学思想
⑬	绿色发展理念、生态文明建设意识

（续表）

序号	思政元素
⑭	专业自豪感和责任意识
⑮	科学家勇攀高峰、敢为人先的拼搏精神
⑯	培养科学思维、创新精神以及团队合作意识
⑰	学院老教师勇于担当、甘于奉献的精神

笔者深入挖掘本课程中所蕴含的思政元素，结果如表1所示。同时，笔者对本实验各项目所蕴含的思政要点和承载的德育功能进行分析，结合表1所列出的思政元素，给出思政建议，结果如表2所示。从表2中可以看出，每个实验项目都承载了5～8个思政元素，但受于学时限制，以及考虑不同实验题目采用相同的思政教学案例时，学生容易失去兴趣，因此，实际教学中可仅挑选较为突出的思政元素开展教学，见表2中的思政教学思路。

表2　不同实验项目中的思政元素及思政教学思路

序号	实验内容	实验类型	思政元素	思政教学思路
1	燃烧焓的测定	基础	①②③④⑦⑮	燃烧热的测定过程中，尝试引入清洁能源——可燃冰，引导学生思考我国目前面临的能源问题，培养学生为国家做贡献的社会责任感。
2	氨基甲酸铵分解反应平衡常数的测定	基础	②③④⑫⑮	通过真空泵的使用、真空实验技术的比较等，培养学生的安全责任意识。
3	凝固点降低法测定葡萄糖的摩尔质量	基础	③④⑤⑦⑩⑪⑬⑰	利用凝固点降低法测定物质的摩尔质量这一实验，对传统版和改良版的研究体系进行比较，培养学生“绿水青山就是金山银山”的环保理念。
4	双液系的气-液平衡相图	基础	③④⑩⑫⑭⑮	通过引导学生精确读取相关数据及准确绘制组成图，培养学生严谨细致的科学态度和精益求精的工匠精神。
5	$KCl-HCl-H_2O$ 三组分系统相图的绘制	基础	②③④⑦⑩⑭	通过介绍中国科学院院士、国际知名相图专家金展鹏同志的事迹，引导学生体会老一辈科学工作者在新中国成立初期艰苦奋斗、不屈不挠、永不放弃的科学精神，加深对社会主义制度优越性的认识。
6	蔗糖水解反应速率常数的测定	基础	②③④⑦⑨⑪⑮	利用旋光法测定蔗糖水解反应的速率常数，探究一级反应的特点，引导学生将实验结果与理论课所学知识进行对比，培养学生理论联系实际的能力。

（续表）

序号	实验内容	实验类型	思政元素	思政教学思路
7	乙酸乙酯皂化反应速率常数的测定	基础	②③④⑤⑦⑮	利用电导法测定乙酸乙酯皂化反应的速率常数，探究二级反应的特点，引导学生将实验结果与理论课所学知识进行对比，培养学生理论联系实际的能力。
8	原电池电动势的测定和相关热力学函数的计算	基础	③④⑤⑥⑨⑮	利用对消法测定电池电动势以及一些电极的电极电势，与所学理论进行联系比较，使学生深入理解“实践是检验真理的唯一标准”的哲学思想。
9	电势-pH 曲线的测定与应用	基础	③④⑤⑩⑪⑫⑬⑮	通过向学生讲授实验背后科学家（布拜）的故事，引导学生学习科学家的精神。
10	循环伏安法研究铁氰化钾的电化学行为	基础	③④⑨⑩⑮	引导学生思考电极材料、扫描速率、电位范围和浓度等对循环伏安图的影响，加深学生对透过现象看本质哲学思想的理解。
11	最大泡压法测定溶液的表面张力	基础	③④⑤⑧⑨⑩	通过测定乙醇分子的横截面积并与文献值进行对比，使学生深入理解“实践是检验真理的唯一标准”的哲学思想。
12	溶液吸附法测量固体物质的比表面	基础	③④⑤⑧⑨⑩⑭	能源是世界各国的战略聚焦点，页岩气作为新兴的非常规天然气，正在改变世界能源结构与政治格局。通过页岩气吸附问题的介绍，培养学生居安思危、立志报国的精神，同时增强学生的专业自豪感和责任意识。
13	溶胶的制备和性质研究	基础	①②③⑦⑧⑩⑫⑭	引入我国古代四大发明之一的造纸术，及风靡全球的美食——豆腐的制作过程中涉及的胶体相关知识，激发学生的民族自豪感、专业自豪感，培养学生世界是普遍联系的哲学思想。
14	黏度法测定水溶性高聚物的平均摩尔质量	基础	③④⑥⑫⑭⑮	利用乌氏黏度计，通过引导学生多次实验，准确读取溶剂及不同浓度高聚物溶液的流出时间，培养学生严谨细致的科学态度和精益求精的工匠精神。
15	配合物的磁化率测定	基础	③④⑨⑩⑪⑫⑮	由配合物磁化率与不成对电子数之间的内在关系，加深学生对透过现象看本质哲学思想的理解。
16	甲醛分子的结构和性质的计算化学研究	基础	②③④⑦⑨⑪⑫⑮	唐敖庆院士是我国理论化学学科的奠基人，同时也是我国科学基金事业的创始人和高等教育事业的领军人物之一。通过介绍我国化学家在分子领域的贡献，培养学生正确的人生价值观、家国情怀和责任意识。

（续表）

序号	实验内容	实验类型	思政元素	思政教学思路
17	甲烷分子的结构和性质的计算化学研究	基础	②③⑦⑨⑪⑫⑮⑯	通过理论计算化学前辈邓从豪院士推导出“邓势”函数的故事，引导学生体会前辈科学家求实创新的科学态度和用发展的眼光看问题的哲学思想，提高学生学习分子模拟的兴趣，培养学生的科学素养和家国情怀。
18	富勒烯在不同工作介质中的电化学行为研究（虚拟仿真实验）	基础	⑥⑨⑩⑪⑫⑭⑮⑯	通过富勒烯的发现到结构被确定的相关知识介绍，引导学生理解“科研没有坦途，更没有捷径”的思想。同样，人生也不是一帆风顺的，只有面对困难不妥协，才能克服人生道路上的巨大困难。
19	气相色谱法测定非电解质溶液的热力学函数	综合	③④⑥⑫⑭⑮	利用气相色谱法测定非电解质溶液的热力学函数实验过程，体会学科交叉在科研中的应用，强化世界是普遍联系的哲学思想。
20	生物酶催化反应动力学常数的测定	综合	③⑥⑪⑫⑬⑮	由酶催化研究进展引出我国著名科学家——单分子酶学的创始人谢晓亮院士，通过讲述谢晓亮院士的故事，引导学生学习他勇攀高峰、敢为人先的拼搏精神。
21	电导法测定水溶性表面活性剂的临界胶束浓度	综合	③④⑧⑫⑮	利用电导法测定表面活性剂溶液的临界胶束浓度实验过程中如实记录实验数据，感受事物是由量变到质变的转变过程。
22	溶液法测定极性分子的偶极矩	综合	③⑥⑦⑩⑮⑯	通过讲解计算偶极矩的复杂公式，引导学生学习科学家不畏困难、刻苦钻研的精神。
23	常压顺-丁烯二酸催化氢化	综合	②③④⑦⑭	由“氢化油”引出食品安全问题，培养学生健康意识、责任意识；同时，实验涉及高温及易爆气体——氢气的使用，可引导学生树立安全意识和自我保护意识。
24	氨分子与水分子的二聚体稳定结构及其氢键强度预测	综合	③⑦⑨⑪⑫⑮⑯	通过分享2013年诺贝尔化学奖三位得主的科研经历，引导学生除了向书本和老师学习之外，还要重视向同伴学习，重视合作与交流。
25	氧氟沙星在固体电极表面的电化学行为	研究设计	②③⑤⑦⑮⑯	通过对实际药物样品的预处理和国家标准分析方法介绍，培养学生理论联系实际的意识，加强学生运用所学知识分析解决实际问题的能力。
26	稀土金属直接加氢制备纳米稀土金属氢化物	研究设计	①②③⑥⑦⑭⑮⑯	通过“中国稀土之父”徐光宪院士的爱国主义事迹，激发、培养学生的爱国主义精神，对祖国的忠诚心、民族自尊心和民族自信心，激励和培养学生学以报国，建设国家的崇高理想。

（续表）

序号	实验内容	实验类型	思政元素	思政教学思路
27	低温等离子体直接分解 NO	研究设计	②③ ④⑦ ⑨⑩ ⑪⑭	将中国现代科技发展及现实的国情教育等思政元素与等离子体的专业知识有机融合，培养学生严谨、认真的科研态度和大国重器的担当精神。
28	氢质子交换膜燃料电池的组装与性能测试	研究设计	②③ ④⑭ ⑮⑯	在引导学生设计、组装燃料电池的过程中，培养学生的科学思维、创新精神以及团队合作意识。
29	甲烷生成焓和燃烧焓的理论计算	研究设计	①② ④⑥ ⑪⑬ ⑭⑯	通过量子化学相关进展的讨论，使学生了解化学不再是纯实验科学了，从而强化用发展的眼光看问题的哲学思想。通过我国自主研发的超级计算机（神威太湖之光）的介绍，激发学生的爱国意识以及民族自豪感。
30	分子中原子电荷的理论计算	研究设计	②③ ④⑥ ⑪⑭ ⑮⑯	在引导学生设计合理的计算模拟方案时，使学生思考理论计算对实验的作用，培养学生用发展的眼光看问题，不要局限于眼前。

从表 2 内容可以看出，物理化学实验课程教学主要从严谨求实的职业素养、与时俱进的求学态度、强国有我的爱国情怀等方面着手，深入挖掘思政元素，培养“有能力、有担当、有情怀”的化学专业高素质人才。具体做法为：

1. 深挖实验哲学思想，提升学生的政治素养

教学中有意识地挖掘马列主义哲学思想，可成为物理化学实验课程思政的主要做法。例如，在实验 6“原电池电动势的测定和相关热力学函数的计算”中，实验前可对学生进行预习提问，如“化学热力学与电化学联系的桥梁是什么?”“什么情况下测得的电池电动势可用于相关热力学函数的计算?”“实验中如何实现这一条件?”等，通过一系列提问促进学生将理论知识与实验操作结合起来，实验后通过实验测定值与文献值进行对比，可使学生深入理解“实践是检验真理的唯一标准”的哲学思想。又如，在实验 9“蔗糖水解反应速率常数的测定”中，可通过让学生观察旋光度与时间的变化关系、蔗糖浓度对体系旋光度变化的影响等，引导学生总结一级反应的特点，再与理论课上所学到的知识进行对照，培养学生理论联系实际的能力。类似地，实验 10“乙酸乙酯皂化反应速率常数的测定”则可通过理论课与实验课的关于二级反应的特点，培养学生理论联系实际的能力。实验 15“配合物的磁化率测定”和实验 7“循环伏安法研究铁氰化钾的电化学行为”，则可引导学生从实验机理出发，思考产生实验现象的本质原因，加深学生对透过现象看本质哲学思想的理解。而在实验 17“甲烷分子的结构和性质的计算化学研究”中，可引导学生开展量子化学

相关进展的讨论，结合1998年及2013年诺贝尔化学奖都颁给了理论化学家，使学生了解化学不再是纯实验科学，现阶段化学的两大支柱是实验和形式理论，从而强化学生"用发展的眼光看问题"的哲学思想。

2. 关注实验教学内容，培育学生的社会责任感

培育学生的社会责任感，可成为物理化学实验课程思政的另一重要方面。例如，在实验1"燃烧焓的测定"中，可适时引导学生对当前世界的能源问题进行简单讨论，并引入清洁能源——可燃冰，从我国可燃冰储量世界第一出发，激发学生的爱国主义情感，培养学生为国家做贡献的社会责任感。环境问题是人类生存面临的一个主要问题，在实验教学中有意识地培养学生的环保意识，是化学实验课程必须承担的思政内容。除了贯穿于所有实验的"三废"处理问题，也可以针对一些典型实验项目，着重强调环境保护的重要性。例如，凝固点降低法测定摩尔质量是一个经典的基础物理化学实验，传统教材中一般使用"苯-萘"体系进行测试，苯是一种致癌物质，人暴露在大量的苯中会引起白血病。本教程选择水为溶剂测定葡萄糖的摩尔质量。通过现行实验项目的"水-葡萄糖"体系与传统的"苯-萘"体系做对比，可很好地培养学生的环保理念。

同时，安全责任意识的培养也不能忽视，笔者学院经过教学实践的逐步探索，已经将实验安全教育列为实验教育的重要环节。在实验开始之前，教师通过课前提问促进学生主动思考实验中可能遇到的安全问题，并通过引导、讨论，启发学生解决问题。实验完成后，学生要通过小组讨论、文献查阅等方式对实验中涉及的安全主题进行总结，并以汇报的形式在班级内展示。例如，在实验2"氨基甲酸铵分解反应平衡常数的测定"中，学生通过"真空实验技术的比较""真空实验的操作安全"等讨论、汇报环节，强化安全责任意识的同时，也通过安全隐患等看到了实验课程的本质特性，以便之后实验中更好地发挥主观能动性，减少化学事故的发生。

3. 立足实验教学环节，培养学生的科研态度

培养求真务实的科学态度一直是理科实验教学的重要目标，也是课程思政的主要内容之一。事实上，所有的物化实验项目都应注意科学态度的培养，只不过学时有限，有的实验承载了其他的思政元素，科学态度就不再单独拿出来强调了。这里仅举几个典型实验项目，用于介绍笔者学院在培养严谨求实的科学态度方面的做法。例如，在实验4"双液系的气-液平衡相图"过程中，所需记录的数据量较大，消耗时间长，仪器操作和数据处理较为复杂，教师通过引导学生耐心进行实验操作，精确读取相关数据及准确绘制组成图，可培养学生严谨细致的科学态度。另外，实验14"黏度法测定水溶性高聚物的平均摩尔质量"实验中，也可通过引导学生多次实验，准确读取溶剂及不同浓度高聚物溶液的流出时间，培养学生严谨细致的科学态度和精益求精的工匠精神。

4. 聆听实验背后科学家的故事，学习科学家的精神

聆听实验背后科学家的故事，学习科学家的精神，是比较简单的思政手段，但往往会有意想不到的教学效果。例如，在实验9“电势-pH曲线的测定与应用”中，可引入比利时科学家布拜的故事。布拜和同事在20世纪60年代初一起完成了所有元素的电势-pH图，当时布拜不到20岁，跟做实验的同学们年龄差不多。布拜的最后一部著作是《气相存在下化学和电化学平衡图集(固-气平衡)》一书，他出版这部专著时已经92岁高龄。笔者课上每每讲到这两个故事，同学们都会一片沸腾，既感动于布拜的努力及对科学的酷爱，也折服于他对科学孜孜不倦、为科学奋斗终生的精神。

往常一讲到科学家的故事，教师往往介绍的都是国外科学家，容易让学生误认为国内科学家比不上国外科学家，于是，在实验20“生物酶催化反应动力学常数的测定”中，笔者引入我国著名科学家——单分子酶学的创始人谢晓亮院士的故事，讲述谢晓亮院士全职在北京大学工作，以及他在基础研究、技术开发和医学研究三个领域的成就，引导学生学习他淡泊名利、潜心研究的奉献精神，以及勇攀高峰、敢为人先的拼搏精神。

除了国内外著名科学家，学校学院老教师的事迹也是思政育人的落脚点之一。例如，笔者学院老教师在过去市场经济很不发达的年代，甚至计算机都没有几台的时代，潜心开发出实验数据处理软件用以满足教学，十分不易；还比如，二十年前，为了完善实验教学内容，笔者学院老教师采用旧电线、大玻璃缸等组成了简易的凝固点降低法实验装置用于教学，这种勇于担当、甘于奉献的精神同样值得学习。

5. 重视研究设计型实验，强化创新精神以及团队意识

研究设计型实验难度一般比较大，在实验过程中常存在着挫折和失败，尽管如此，这类实验对培养学生的科研态度、创新精神以及团队意识仍然很重要，一定要重视它们的思政作用。例如，“质子交换膜燃料电池组装活化及性能测试”是学院常开设的研究设计型实验，对于这类实验项目，教师一般会给出实验课题，每组学生任选一个并完成。学生在实验中发挥主体作用，通过查阅中英文文献，自主设计实验方案，并对实验中可能出现的问题进行分析、解决。同时，可要求学生按科学论文的格式完成实验报告，论文主体要说清楚三个问题：一是为什么要做实验，即实验目的是什么；二是怎么做的实验，即具体的实验方案是什么；三是得到了什么结论，即实验结果和讨论，并指出若想实验结果更完善，还需要做哪些努力。实验报告不仅在内容上要与科学论文看齐，在形式上也要与科学论文一致。通过进行实验方案交流，互为补充或互为验证的合作性实验，论文报告的讨论交流等必要环节，可大大提高学生的创新精神、思维能力、归纳总结能力以及团队合作意识等，有效促进学生多样化成才。

参考资料

1. 孙越，杨钻.新时代背景下物理化学实验的课程思政教学初探.大学化学，2021，36(08)：25-31.

2. 张恒，刘刚，马莹，等.拔尖计划特色课分子模拟实验中的思政案例设计.大学化学，2019，34(11)：827.

3. 邓留，黄健涵，陈立妙，等.分析化学实验教学中的思政教育——以络合滴定为例.大学化学，2021，36(09)：111-117.

4. 罗青枝，安静，崔敏，等.胶体与界面化学课程中的思政内涵与融入.广东化工，2021，48(15)：239-240.

附 录

附表 1　不同温度下水的密度*

t/℃	ρ/(kg · m^{-3})	t/℃	ρ/(kg · m^{-3})	t/℃	ρ/(kg · m^{-3})	t/℃	ρ/(kg · m^{-3})
10	999.6996	20	998.2041	30	995.6473	40	992.2158
11	999.6051	21	997.9925	31	995.3410	41	991.8298
12	999.4974	22	997.7705	32	995.0262	42	991.4364
13	999.3771	23	997.5385	33	994.2030	43	991.0358
14	999.2444	24	997.2965	34	994.3715	44	990.6280
15	999.0996	25	997.0449	35	994.0319	45	990.2132
16	998.9430	26	996.7837	36	993.6842	46	989.7914
17	998.7749	27	996.5132	37	993.3287	47	989.3628
18	998.5956	28	996.2335	38	992.9653	48	988.9273
19	998.4052	29	995.9445	39	992.5943	49	988.4851

*也可用以下方程计算：

$$\rho/\mathrm{kg}\cdot\mathrm{m}^{-3} = (999.83952 + 16.945176t/℃ - 7.9870401\times10^{-3}t/℃ - 46.170461\times10^{-6}t/℃ + 105.56302\times10^{-9}t/℃ - 280.5425\times10^{-12}t/℃)/(1 + 16.879850\times10^{-3}t/℃)$$

资料参见:附录末参考资料[2]F-4。

附表 2　一些有机化合物的密度与温度的关系

化合物	ρ_0/(g · cm^{-3})	α	β	γ	误差范围	温度范围/℃
乙醇* C_2H_6O	0.78506	−0.8591	−0.56	−5		10～40
乙酸乙酯 $C_4H_8O_2$	0.92454	−1.168	−1.95	+20	0.00005	0～40
乙醚 $C_4H_{10}O$	0.73629	−1.1138	−1.237		0.0001	0～70
苯 C_6H_6	0.90005	−1.0638	−0.0376	−2.213	0.0002	11～72
酚 C_6H_6O	1.03893	−0.8188	−0.670		0.001	40～150

表中所列有机化合物之密度可用下列方程计算：

$$\rho_t = [\rho_0 + 10^{-3}\alpha t + 10^{-6}\beta t^2 + 10^{-9}\gamma t^3] \pm 10^{-4}\Delta$$

式中 ρ_0 为 0 ℃时的密度，ρ_t 为 t℃时的密度。

*0.78506 为 25 ℃时的密度，利用上述方程式计算时，温度项应该用（t-25）代入。

资料参见：The International Critical Tables of Numerical Data, Physics, Chemistry and Technology. Vol Ⅱ:27.

附表 3　某些溶剂的凝固点降低常数

溶剂	凝固点 t_f/℃	K_f/(℃ · kg · mol^{-1})
醋酸 $C_2H_4O_2$	16.66	3.9
四氯化碳 CCl_4	-22.95	29.8
苯 C_6H_6	5.533	5.12
环己烷 C_6H_{12}	6.54	20.0
萘 $C_{10}H_8$	80.290	6.94
水 H_2O	0	1.86

资料参见：[1]157，[2]D-186。

附表 4　常用参比电极的电极电势及其温度系数

名称	体系	E/V*	$\frac{dE}{dT}$/(mV · K^{-1})
饱和甘汞电极	Hg, Hg_2Cl_2 | 饱和 KCl	0.2415	-0.761
标准甘汞电极	Hg, Hg_2Cl_2 | 1mol · dm^{-3} KCl	0.2800	-0.275
银-氯化银电极	Ag, AgCl | 0.1mol · dm^{-3} KCl	0.290	-0.3
硫酸亚汞电极	Hg, Hg_2SO_4 | 1mol · dm^{-3} Hg_2SO_4	0.6758	

*25 ℃；相对于标准氢电极（NHE）。

附表 5　饱和标准电池的温度校正值*

t/℃	ΔE_t/μV	t/℃	ΔE_t/μV	t/℃	ΔE_t/μV
10	+296.90	21.5	-61.97	36	-843.93
11	+277.26	22.0	-83.53	37	-908.25
12	+255.21	23	-127.94	38	-973.73
13	+230.83	24	-174.06	39	-1014.32
14	+204.18	25	-221.84	40	-1108.00

* 相对于 20.0 ℃时 $E_{20}=1.018\ 45\text{V}$。

也可按下式计算：

$$\Delta E_t/\mu\text{V} = -39.94(t/℃-20) - 0.929(t/℃-20)^2 + 0.0090(t/℃-20)^3 - 0.00006(t/℃-20)^4$$

式中 t 为摄氏温度。

附表 6 KCl 溶液的电导率*

t/℃ \ κ \ c/(mol·L⁻¹)**	1.0000	0.1000	0.0200	0.0100
10	0.08319	0.00933	0.001994	0.001020
15	0.09252	0.01048	0.002243	0.001147
16	0.09441	0.01072	0.002294	0.001173
17	0.09631	0.01095	0.002345	0.001199
18	0.09822	0.01119	0.002397	0.001225
19	0.10014	0.01143	0.002449	0.001251
20	0.10207	0.01167	0.002501	0.001278
21	0.10400	0.01191	0.002553	0.001305
22	0.10594	0.01215	0.002606	0.001332
23	0.10789	0.01239	0.002659	0.001359
24	0.10984	0.01264	0.002712	0.001386
25	0.11180	0.01288	0.002765	0.001413
26	0.11377	0.01313	0.002819	0.001441
27	0.11574	0.01337	0.002873	0.001468
28		0.01362	0.002927	0.001496
29		0.01387	0.002981	0.001524
30		0.01412	0.003036	0.001552
35		0.01539	0.003312	

* κ 单位 $\text{S}\cdot\text{cm}^{-1}$。

** 在空气中称取 74.56 g KCl，溶于 18 ℃水中，稀释到 1 L，其浓度为 1.000 $\text{mol}\cdot\text{L}^{-1}$（密度 1.0449 $\text{g}\cdot\text{cm}^{-3}$），再稀释得其他浓度溶液。

附表 7　一些电解质水溶液的摩尔电导率[*]

Λ_m \ cm/(mol·L^{-1}) 化合物	无限稀	0.0005	0.001	0.005	0.01	0.02	0.05	0.1
KCl	149.79	147.74	146.88	143.48	141.20	138.27	133.30	128.90
NH_4Cl	149.6	—	146.7	134.4	141.21	138.25	133.22	128.69
NaCl	126.39	124.44	123.68	120.59	118.45	115.70	111.01	106.69
$NaOOCCH_3$	91.0	89.2	88.5	85.68	83.72	81.20	76.88	72.76
NaOH	247.7	245.5	244.6	240.7	237.9	—	—	—

[*] Λ_m 单位：S·cm^2·mol^{-1}，25 ℃。

资料参见：[2]D-167。

附表 8　水溶液中离子的极限摩尔电导率[*]

Λ_m^∞ \ t/℃ 离子	0	18	25	50
Na^+	26.5	42.8	50.1	82
$1/2Ba^{2+}$	34.0	54.6	63.6	104
$1/2Ca^{2+}$	31.2	50.7	59.8	96.2
OH^-	105	171	198.3	(284)
Cl^-	41.0	66.0	76.3	(116)
NO_3^-	40.0	62.3	71.5	(104)
CH_2COO^-	20.0	32.5	40.9	(67)
$1/2SO_4^{2-}$	41	68.4	80.0	(125)
$1/4[Fe(CN)_6]^{4-}$	58	95	110.5	(173)

[*] 单位：S·cm^2·mol^{-1}

资料主要参见：[1]191，194。

附表 9　一些电解质的活度系数（25 ℃）

r \ m/(mol·kg^{-1}) 电解质	0.1	0.2	0.3	0.4	0.5	0.6	0.7	0.8	0.9	1.0
$CuSO_4$	0.150	0.104	0.082	0.070	0.062	0.056	0.051	0.048	0.045	0.042
KCl	0.770	0.718	0.688	0.666	0.649	0.637	0.626	0.618	0.610	0.604
$KClO_3$	0.749	0.681	0.635	0.599	0.568	0.541	0.518	—	—	—
$K_4Fe(CN)_6$	0.139	0.099	0.081	0.069	0.061	0.056	0.051	0.048	0.045	—

（续表）

$m/(mol \cdot kg^{-1})$ / r / 电解质	0.1	0.2	0.3	0.4	0.5	0.6	0.7	0.8	0.9	1.0
KH_2PO_4	0.731	0.653	0.602	0.561	0.529	0.501	0.477	0.456	0.438	0.421
NH_4NO_3	0.740	0.677	0.636	0.606	0.582	0.562	0.545	0.530	0.516	0.504
NaH_2PO_4	0.744	0.675	0.629	0.593	0.563	0.539	0.517	0.499	0.483	0.468
$Zn(NO_3)_2$	0.531	0.489	0.474	0.469	0.473	0.480	0.489	0.501	0.518	0.535
$ZnSO_4$	0.150	0.140	0.084	0.071	0.063	0.057	0.052	0.049	0.046	0.044

资料参见:[2]D-169。

附表 10　IUPAC 推荐的五种标准缓冲溶液的 pH

溶液 / t/℃	①	②	③	④	⑤
10		3.998	6.923	7.472	9.332
15		3.999	6.900	7.448	9.276
20		4.002	6.881	7.429	9.225
25	3.557	4.008	6.865	7.413	9.180
30	3.552	4.015	6.853	7.400	9.139
35	3.549	4.024	6.844	7.389	9.102
38	3.548	4.030	6.840	7.384	9.081
40	3.547	4.035	6.838	7.380	9.068

① 25 ℃下的饱和酒石酸氢钾溶液(0.0341 $mol \cdot L^{-1}$)。

② 0.05 $mol \cdot L^{-1}$ 的邻苯二甲酸氢钾溶液。

③ 0.025 $mol \cdot L^{-1}$ 的 KH_2PO_4 和 0.025 $mol \cdot L^{-1}$ 的 Na_2HPO_4 溶液。

④ 0.008695 $mol \cdot L^{-1}$ 的 KH_2PO_4 和 0.03043 $mol \cdot L^{-1}$ 的 Na_2HPO_4 溶液。

⑤ 0.01 $mol \cdot L^{-1}$ 的 $Na_2B_4O_7$ 溶液。

资料参见:[1]231,[2]D-146.

附表 11　不同温度下水的表面张力 σ

t/℃	$\sigma/(10^{-3}\ N \cdot m^{-1})$	t/℃	$\sigma/(10^{-3}\ N \cdot m^{-1})$	t/℃	$\sigma/(10^{-3}\ N \cdot m^{-1})$
5	74.92	17	73.19	25	71.97
10	74.22	18	73.05	26	71.82
11	74.07	19	72.90	27	71.66
12	73.93	20	72.75	28	71.50
13	73.78	21	72.59	29	71.35
14	73.64	22	72.44	30	71.18
15	73.49	23	72.28	35	70.38
16	73.34	24	72.13	40	69.56

资料参见:[1]310,[2]F-32。

附表 12　一些有机化合物的折光率及温度系数

化合物	n_D^{15}	n_D^{20}	n_D^{25}	$10^5 \times \frac{dn}{dt}$
四氯化碳 CCl_4	1.4631	1.4603	1.459	-55
乙醇 C_2H_6O	1.3633	1.3613	1.359	-40
丙酮 C_3H_6O	1.3616	1.3591	1.357	-49
溴苯 C_6H_5Br	1.5625	1.5601	1.557	-48
苯 C_6H_6	1.5044	1.5011	1.498	-66
正丁酸乙酯 $C_6H_{12}O_2$		1.4000		
甲苯 C_7H_8	1.4999	1.4969	1.4941	-57
二硫化碳 CS_2	1.6319	1.6280		-78

资料参见：[1]48，[2]C-42，E-367。

附表 13　一些化合物的磁化率

化合物	T/K	质量磁化率		摩尔磁化率	
		①	②	③	④
$Cu(NO_3)_2 \cdot 3H_2O$	293	6.5	81.7	1570.0	19.73
$CuSO_4 \cdot 5H_2O$	293	5.85	73.5	1460.0	18.35
			74.4		
$FeSO_4 \cdot 7H_2O$	293.5	40.28	506.2	11200.0	140.7
$K_3Fe(CN)_6$	297	6.96	87.5	2290.0	28.78
$K_4Fe(CN)_6$	室温	-0.3739	4.699	-130.0	-1.634
$K_4Fe(CN)_6 \cdot 3H_2O$	室温	-0.3739		-172.3	-2.165
$NH_4Fe(SO_4)_2 \cdot 12H_2O$	293	30.1	378	14500	182.2
$(NH_4)_2Fe(SO_4)_2 \cdot 6H_2O$	293	31.6	397	12400	155.8

① χ_m 单位（CGSM 制）：$10^{-6} cm^3 \cdot g^{-1}$；资料参见：[1]447～455。

② $1\ cm^3 \cdot g^{-1}$（SI 质量磁化率）=（$10^3/4\pi$）$cm^3 \cdot g^{-1}$（CGSM 制质量磁化率），本栏数据由①按此式换算而得，χ_m 的 SI 单位为 $10^{-9} m^3 \cdot kg^{-1}$。

③ χ_M 单位（CGSM 制）：$10^{-6} cm^3 \cdot mol^{-1}$；资料参见：[2]E-116～128。

④ 本栏数据参照注②由③换算而得，χ_M 的单位为 $10^{-9} cm^3 \cdot mol^{-1}$。

附表 14 一些液体的介电常数

化合物	介电常数①		温度系数	适用温度范围
	20 ℃	25 ℃	a 或 α	℃
四氯化碳 CCl_4	2.238	2.228	0.200②	−20～+60
三氯甲烷 $CHCl_3$	4.806		0.160③	0～50
乙醇 C_2H_6O		24.35	0.270③	−5—+70
乙酸甲酯 $C_3H_6O_2$		6.68	2.2②	25—40
乙酸乙酯 $C_4H_8O_2$		6.02	1.5②	25
1,4-二氧六环 $C_4H_8O_2$		2.209	0.170②	20～50
环己烷 C_6H_{12}	2.023	2.015	0.160②	10～60
正己烷 C_6H_{14}	1.890		1.55②	−10～+50
正己醇 $C_6H_{14}O$		13.3	0.35③	15—35
水 H_2O	80.37	78.54	0.200③	15—30

① 常压;真空介电常数为 1。

② $a = -10^2 \cdot \frac{d\varepsilon}{dt}$。

③ $\alpha = -10^2 \cdot \frac{d(\lg\varepsilon)}{dt}$。

资料参见:[2]E-49,E-50。

附表 15 气相中分子的偶极矩

化合物	偶极矩 μ		化合物	偶极矩 μ	
	CGS	SI**		CGS	SI**
四氯化碳 CCl_4	0*	0**	甲酸乙酯 $C_3H_6O_2$	1.93	6.44
三氯甲烷 $CHCl_3$	1.01	3.37	乙酸乙酯 $C_4H_8O_2$	1.78	5.94
甲醇 CH_4O	1.70	5.67	溴苯 C_6H_5Br	1.70	5.67
乙醛 C_2H_4O	2.69	8.97	氯苯 C_6H_5Cl	1.69	5.64
乙酸 $C_2H_4O_2$	1.74	5.80	硝基苯 $C_6H_5NO_2$	4.22	14.1
甲酸甲酯 $C_2H_4O_2$	1.77	5.90	水 H_2O	1.85	6.17
乙醇 C_2H_6O	1.69	5.64	氨 NH_3	1.47	4.90
乙酸甲酯 $C_3H_6O_2$	1.72	5.74	二氧化硫 SO_2	1.6	5.34

*μ 单位 $D = 10^{-18}$ esu · cm。

**μ 单位 10^{-30} C · m;按 1 D = 3.33564 C · m 换算。

资料参见:[1]442,[2]E-58。

参考资料

[1] 印永嘉.物理化学简明手册.北京:高等教育出版社,1988.

[2] Weasst R C. CRC Handbook of Chemistry and Physics. Florida: CRC Press, Boca Raton, 1985-1986.